Lecture Notes in Computer Science　　9627

Commenced Publication in 1973
Founding and Former Series Editors:
Gerhard Goos, Juris Hartmanis, and Jan van Leeuwen

More information about this series at http://www.springer.com/series/7407

Mohammad Kaykobad · Rossella Petreschi (Eds.)

WALCOM: Algorithms and Computation

10th International Workshop, WALCOM 2016
Kathmandu, Nepal, March 29–31, 2016
Proceedings

Springer

Editors

Mohammad Kaykobad
Department of Computer Science
  and Engineering
Bangladesh University of Engineering
  and Technology
Dhaka
Bangladesh

Rossella Petreschi
Department of Computer Science
Sapienza University of Rome
Rome
Italy

ISSN 0302-9743 ISSN 1611-3349 (electronic)
Lecture Notes in Computer Science
ISBN 978-3-319-30138-9 ISBN 978-3-319-30139-6 (eBook)
DOI 10.1007/978-3-319-30139-6

Library of Congress Control Number: 2015955363

LNCS Sublibrary: SL1 – Theoretical Computer Science and General Issues

Printed on acid-free paper

This Springer imprint is published by SpringerNature
The registered company is Springer International Publishing AG Switzerland

Preface

This volume contains the proceedings of the WALCOM 2016 (International Workshop on Algorithms and Computation), held in Kathmandu, Nepal, March 29–31, 2016.

This was the tenth edition of an international workshop that was held for the first time in 2007 in Dhaka, Bangladesh. Four editions (2008, 2010, 2012, and 2015) were also held in Dhaka, and four other editions were held in India: 2009 (Kolkata), 2011 (New Delhi), 2013 (Kharagpur), and 2014 (Chennai).

WALCOM has established itself as a fully refereed conference for theoretical computer science research in Asia. It has also strengthened the ties between local and international scientific communities. We believe that this volume reflects the breadth and depth of this interaction. We received 68 submissions from 170 different authors in 28 countries. Each paper was assigned to three Program Committee members: The committee selected 27 papers based on approximately 190 reviewer reports. In addition to these contributed presentations, the conference included four invited talks.

We thank all the authors who submitted papers, the members of the Program Committee and the external reviewers. We are also grateful to the four invited speakers, Sajal Das, Missouri University of Science and Technology – Rolla (USA), Costas S. Iliopoulos, King's College London – London (UK), Giuseppe F. Italiano, University Tor Vergata – Rome (Italy), Giuseppe Persiano, University of Salerno – Fisciano (Italy), who kindly accepted our invitation to give plenary lectures at the workshop.

The assistance of many organizations and individuals was essential for the success of this meeting. We would like to thank all of our sponsors and supporting organizations and in particular Pramod Prodhan for taking the initiative of hosting it in scenic Nepal, Lena Sthapit of the Asian Institute of Technology and Management(AITM) and members of the Organizing Committee, AITM, KHInt, and IT Professional Forum of Nepal for their support for WALCOM 2016. Moreover, we thank the IEICE Technical Committee on Theoretical Foundations of Computing (COMP) and the Special Interest Group for Algorithms (SIGAL) of the Information Processing Society of Japan (IPSJ) for their cooperation.

We finally would like to thank EasyChair for providing us with a very friendly environment for handling the contributions and editing the proceedings of WALCOM 2016.

January 2016

Rossella Petreschi
Mohammad Kaykobad

Organization

Organizing Committee

Pramod Pradhan (Chair)	Asian Institute of Technology and Management (AITM), Nepal
Bhupa Das Rajbhandari	Asian Institute of Technology and Management (AITM), Nepal
Juddha Bahadur Gurung	National Information Technology Institute (NITI), Nepal
Mahesh Singh Kathayat	Convenor for Knowledge Management Seminar, IT Professional Forum (ITPF), Nepal
Udaya Lal Pradhan	National Information Technology Institute (NITI), Nepal
Prashant Lal Shrestha	IT Professional Forum (ITPF), Islington College and Knowledge Holding International (K-Hint), Nepal
Nayana Amatya	IT Professional Forum (ITPF), and Knowledge Holding International (K-Hint), Nepal
Jyoti Tandukar	Institute of Engineering, IT Professional Forum (ITPF), and Alternative Technology
Suresh K. Regmi	Knowledge Holding International (K-Hint) and Professional Computer System, Nepal
Bijendra Suwal	Nepal Investment Bank Limited and IT Professional Forum (ITPF), Nepal
Khusbu Sarkar Shrestha	Knowledge Holding International (K-Hint) Lochan Amatya – Nepal Telecom, IT Professional Forum (ITPF), Society of Electronics & Communication Engineers, Nepal
Shiv Bhusan Lal	IT Professional Forum (ITPF), and Knowledge Holding International (K-Hint), Nepal
Shashi Bhattarai	Development Dynamics, Knowledge Holding International (K-Hint) and IT Professional Forum (ITPF), Nepal
Madhav Narayan Shrestha	Asian Institute of Technology and Management (AITM), Nepal
Rabindra Raj Giri	Asian Institute of Technology and Management (AITM), Nepal
Basanta Prasad Joshi	Asian Institute of Technology and Management (AITM), Nepal
Hari Krishna Saiju	Asian Institute of Technology and Management (AITM), Nepal
Kushal Niroula	Asian Institute of Technology and Management (AITM), Nepal
Lena Sthapit	Asian Institute of Technology and Management (AITM), Nepal

Pravakar Pradhan	Asian Institute of Technology and Management (AITM), Nepal
Astha Bharijoo	Asian Institute of Technology and Management (AITM), Nepal
Sugandha K.C.	Asian Institute of Technology and Management (AITM), Nepal
Anil Byanjankar	Asian Institute of Technology and Management (AITM), Nepal
Surendra Joshi	Asian Institute of Technology and Management (AITM), Nepal

Steering Committee

Kyung-Yong Chwa	KAIST, Korea
Costas S. Iliopoulos	KCL, UK
M. Kaykobad	BUET, Bangladesh
Petra Mutzel	TU Dortmund, Germany
Shin-ichi Nakano	Gunma University, Japan
Subhas Chandra Nandy	Indian Statistical Institute, Kolkata, India
Takao Nishizeki	Tohoku University, Japan
C. Pandu Rangan	IIT, Madras, India
Md. Saidur Rahman	BUET, Bangladesh

Program Committee

Ljiljana Brankovic	University of Newcastle, Australia
Tiziana Calamoneri	Università La Sapienza Roma, Italy
Rezaul A. Chowdhury	State University of New York at Stony Brook, USA
Marek Chrobak	University of California, Riverside, USA
Gautam K. Das	Indian Institute of Technology Guwahati, India
Celina M.H. de Figueiredo	Universidade Federal do Rio de Janeiro, Brazil
Antoine Deza	McMaster University, Ontario, Canada
Raymond Greenlaw	United States Naval Academy, Annapolis, USA
Pinar Heggernes	University of Bergen, Norway
Seok-Hee Hong	University of Sydney, Australia
Kazuo Iwama	Kyoto University, Japan
Mohammad Kaykobad	Bangladesh University of Engineering and Technology, Dhaka
Dieter Kratsch	Université de Lorraine, Metz, France
Moshe Lewenstein	Bar Ilan University, Ramat Gan, Israel
Dániel Marx	Hungarian Academy of Sciences, Budapest, Hungary
Vangelis Paschos	LAMSADE, University of Paris-Dauphine, France
Rossella Petreschi	Università La Sapienza Roma, Italy
Nadia Pisanti	Università di Pisa, Italy; Erable Team, Inria, France
Sheung-Hung Poon	National Tsing Hua University, Hsin-Chu, Taiwan

Jakub Radoszewski   University of Warsaw, Poland
Saidur Md. Rahman   Bangladesh University of Engineering and Technology,
                                     Dhaka
Sohel Rahman   Bangladesh University of Engineering and Technology,
                                     Dhaka
Sasanka Roy   Chennai Mathematical Institute, India
Blerina Sinaimeri   Inria Grenoble, France
Etsuji Tomita   The University of Electro-Communications, Tokyo, Japan
Ryuhei Uehara   Japan Advanced Institute of Science and Technology,
                                     Nomi
Roger Wattenhofer   ETH Zurich, Switzerland
Gerhard J. Woeginger   TU Eindhoven, The Netherlands

Additional Reviewers

Bahreininejad, Ardeshir
Bandopadhyay, Sriparna
Bernasconi, Anna
Bigi, Giancarlo
Bläsius, Thomas
Boccardo, Davidson
Bonnet, Edouard
Brandstadt, Andreas
Bulteau, Laurent
Byrka, Jaroslaw
Da Fonseca, Guilherme D.
Dao, Minhson
De Agostino, Sergio
de Freitas, Rosiane
De, Minati
Dias, Zanoni
Eades, Peter
Epstein, Leah
Escoffier, Bruno
Felsner, Stefan
Fernau, Henning
Fertin, Guillaume
Finocchi, Irene
Franciosa, Paolo
Froese, Vincent
Gastaldello, Mattia
Giannakos, Aristotelis

Gourves, Laurent
Han, Xin
Inkulu, R.
Jaiswal, Ragesh
Jallu, Ramesh
Kammer, Frank
Karim, Md. Rezaul
Karmakar, Arindam
Kaufmann, Michael
Kaykobad, Mohammad
Kosowski, Adrian
Labarre, Anthony
Laura, Luigi
Lavor, Carlile
Le, Van Bang
Manlove, David
Mary, Arnaud
Mondal, Debajyoti
Monti, Angelo
Morgana, Aurora
Moscarini, Marina
Nakano, Shin-Ichi
Nandy, Subhas
Natarajan, Vijay
Niedermann, Benjamin
Nishat, Rahnuma Islam
Paixao, Joao

Peters, Daniel
Pirola, Yuri
Ray, Saurabh
Rutter, Ignaz
Sa, Vinicius
Sadakane, Kunihiko
Sarkar, Santanu
Scutella, Maria Grazia
Shafin, Md. Kishwar
Shende, Anil
Shparlinski, Igor
Simonetti, Luidi
Sinha Mahapatra,
      Priya Ranjan
Stamoulis, Georgios
Sterbini, Andrea
Subrahmanyam, Venkata
Suomela, Jukka
Telelis, Orestis
Tiedemann, Morten
Wandelt, Sebastian
Wrochna, Marcin
Yamashita, Masafumi
Zhang, Yong
Živný, Stanislav

Sponsors

Invited Talks (Abstracts)

Popping Superbubbles and Discovering Clumps: Recent Developments in Biological Sequence Analysis

Costas S. Iliopoulos, Ritu Kundu, Manal Mohamed,
and Fatima Vayani

Department of Informatics, King's College London, UK
{costas.iliopoulos, ritu.kundu, manal.mohamed,
fatima.vayani}@kcl.ac.uk

Abstract. The information that can be inferred or predicted from knowing the genomic sequence of an organism is astonishing. String algorithms are critical to this process. This paper provides an overview of two particular problems that arise during computational molecular biology research, and recent algorithmic developments in solving them.

2-Edge and 2-Vertex Connectivity
Problems in Directed Graphs

Giuseppe F. Italiano

Università di Roma "Tor Vergata", Italy
giuseppe.italiano@uniroma2.it

Abstract. We survey some recent results on 2-edge and 2-vertex connectivity problems in directed graphs. Despite being complete analogs of the corresponding notions on undirected graphs, in digraphs 2-vertex and 2-edge connectivity have a much richer and more complicated structure. It is thus not surprising that 2-connectivity problems on directed graphs appear to be more difficult than on undirected graphs. For undirected graphs it has been known for over 40 years how to compute all bridges, articulation points, 2-edge- and 2-vertex-connected components in linear time, by simply using depth first search. In the case of digraphs, however, the very same problems have been much more challenging and have been tackled only recently.

Social Pressure can Subvert Majority in Social Networks

Giuseppe Persiano

Università di Salerno, Italy
giuper@gmail.com

Abstract. It is often observed that agents tend to imitate the behavior of their neighbors in a social network. This imitating behavior might lead to the strategic decision of adopting a public behavior that differs from what the agent believes is the right one and this can subvert the behavior of the population as a whole.

In this paper, we consider the case in which agents express preferences over two alternatives and model social pressure with the majority dynamics: at each step an agent is selected and its preference is replaced by the majority of the preferences of her neighbors. In case of a tie, the agent does not change her current preference. A profile of the agents' preferences is stable if the preference of each agent coincides with the preference of at least half of the neighbors (thus, the system is in equilibrium).

We ask whether there are network topologies that are robust to social pressure. That is, we ask if there are graphs in which the majority of preferences in an initial profile always coincides with the majority of the preference in all stable profiles reachable from that profile. We completely characterize the graphs with this robustness property by showing that this is possible only if the graph has no edge or is a clique or very close to a clique. In other words, except for this handful of graphs, every graph admits at least one initial profile of preferences in which the majority dynamics can subvert the initial majority. We also show that deciding whether a graph admits a minority that becomes majority is NP-hard when the minority size is at most 1/4-th of the social network size.

The talk is based on joint work with: V. Auletta, I. Caragiannis, D. Ferraioli, and C. Galdi.

Beyond Cyber-Physical Era: What's Next?

Sajal K. Das

Daniel St. Clair Endowed Chair Professor
Chair, Department of Computer Science
Missouri University of Science and Technology
Rolla, MO, USA

Abstract. We live in an era of "Internet of Things" where our physical and personal environments are becoming increasingly smarter as they are immersed with sensing, networking, computing and communication capabilities. The availability of rich mobile devices like smartphones and wireless sensors have also empowered humans as an integral part of cyber-physical systems. This synergy has led to cyber-physical-social convergence exhibiting complex interactions, interdependencies and adaptations among devices, machines, systems/environments, users, human behavior, and social dynamics. In such a connected and mobile world, almost everything can act as information source, analyzer and decision maker. This talk will highlight some of the emerging research challenges and opportunities in cyber-physical-social convergence, and then present some novel solutions to tackle them. It will also reflect on a fundamental question: "What's Next?"

Contents

On-line Algorithms

Algorithms

Invited Talk

Popping Superbubbles and Discovering Clumps: Recent Developments in Biological Sequence Analysis

Costas S. Iliopoulos$^{(\boxtimes)}$, Ritu Kundu, Manal Mohamed, and Fatima Vayani

Department of Informatics, King's College London, London, UK
{costas.iliopoulos,ritu.kundu,manal.mohamed,fatima.vayani}@kcl.ac.uk

Abstract. The information that can be inferred or predicted from knowing the genomic sequence of an organism is astonishing. String algorithms are critical to this process. This paper provides an overview of two particular problems that arise during computational molecular biology research, and recent algorithmic developments in solving them.

1 Introduction

Since the publication of the first human genome in 2001 [15,24], research in the field of genomics has grown almost exponentially. Developments in next-generation sequencing technologies (see [1], for example) have made it possible to sequence new genomes at a fraction of the time and cost required only a few years ago.

The information that can be inferred or predicted from knowing the genomic sequence of an organism is astonishing. String algorithms are critical to the process of analysis and discovery, and molecular biologists are becoming increasingly reliant on accurate and efficient algorithms to process the vast amounts of data they produce.

Genomics research is important in several areas of study, including cell biology, medical and evolutionary genetics, synthetic biology, genomic medicine and many more.

The study of genomes requires a large and complex computational pipeline. From error correction during genome assembly, to motif discovery and gene prediction; each stage of the pipeline presents many problems. This paper provides an overview of two such cases, in the following format: Biological Motivation (Sects. 2.1 and 3.1), Definitions (Sects. 2.2 and 3.2), Algorithms (Sects. 2.4 and 3.3) and Discussions (Sects. 2.5 and 3.4) for the superbubble and clumps problems, respectively; concluded in Sect. 4.

2 Superbubbles

2.1 Biological Motivation

Copious data in the form of reads are generated by next-generation DNA sequencing. Reads are short sequences that overlap to form contiguous sequences,

© Springer International Publishing Switzerland 2016
M. Kaykobad and R. Petreschi (Eds.): WALCOM 2016, LNCS 9627, pp. 3–14, 2016.
DOI: 10.1007/978-3-319-30139-6_1

which are then concatenated to form the whole genome of an organism. This process is known as genome assembly.

Traditionally, assembly algorithms relied on the overlap-layout-consensus approach [3], where each node represents a read and each directed edge signifies an overlap between two reads. These methods have proved their use through numerous *de novo* genome assemblies [7].

More recently, a fundamentally different approach based on *de Bruijn graphs* was proposed [20]. Data elements in this approach are not reads, but strings of k nucleotides, called k-mers. In a *de Bruijn graph* [6], in contrast to an overlap graph, each node represents a $k - 1$ nucleotide long prefix and suffix of the k-mers, and each edge denotes a k-mer between its prefix and suffix nodes. In other words, the last nucleotide is the marginal information contained by a k-mer. Thus, the assembly problem in a de Bruijn graph is reduced to finding an Eulerian path, that is, a trail that visits each edge in the graph exactly once.

However, sequencing errors and genome repeats lead to adding false nodes and edges to the de Bruijn graph, and hence complicating it [25]. Simplification is achieved by filtering out motifs such as tips, bubbles, and cross links. In particular, a *bubble* is commonly caused by a small number of errors in the middle of reads, and it consists of multiple directed unipaths (where a unipath is a path in which all internal nodes are of degree 2) between two nodes. The simplicity of these motifs make them easy to identify and filter out efficiently, but the case of more complex motifs is quite different.

Recently, a complex generalisation of a bubble, the so-called *superbubble*, was proposed as an important subgraph class for analysing assembly graphs [19]. A *superbubble* is defined as a minimal subgraph H in the de Bruijn graph with exactly one start node s and one end node t such that: (1) H is a directed, acyclic, single-source (s), single-sink (t) graph; (2) there is no edge from a node not in H going to a node in $H \backslash \{s\}$ and (3) there is no edge from a node in $H \backslash \{t\}$ going to a node not in H. Efficient detection of superbubbles is essential for the application of genome assembly as many superbubbles are formed as a result of sequencing errors, inexact repeats, diploid/polyploid genomes, or frequent mutations [19].

2.2 Definitions

The concept of superbubbles was introduced and formally defined in [19] as follows.

Definition 1. *Let $G = (V, E)$ be a directed graph. For any ordered pair of distinct nodes s and t, $\langle s, t \rangle$ is called a superbubble if it satisfies the following:*

- **reachability:** *t is reachable from s;*
- **matching:** *the set of nodes reachable from s without passing through t is equal to the set of nodes from which t is reachable without passing through s;*
- **acyclicity:** *the subgraph induced by U is acyclic, where U is the set of nodes satisfying the matching criterion;*

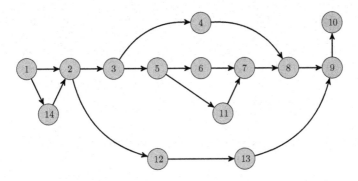

Fig. 1. A directed acyclic graph $G = (V, E)$ with set of nodes $V = \{1, 2, \cdots, 14\}$. Note that G has single source 1 and single sink 10.

– **minimality:** *no node in U other than t forms a pair with s that satisfies the conditions above;*

Nodes s and t, and $U\backslash\{s, t\}$ used in the above definition are the superbubbles entrance, exit and interior, respectively.

We note that a superbubble $\langle s, t \rangle$ in the above definition is equivalent to a single-source, single-sink, acyclic directed subgraph of G with source s and sink t, which does not have any cut nodes and preserves all in-degrees and out-degrees of nodes in $U\backslash\{s, t\}$, as well as the out-degree of s and in-degree of t.

Example 1. For the given directed graph in Fig. 1, $\langle 1, 2 \rangle$, $\langle 2, 9 \rangle$, $\langle 3, 8 \rangle$, $\langle 5, 7 \rangle$, $\langle 9, 10 \rangle$, and $\langle 12, 13 \rangle$ are the superbubbles.

We are now able to define the Problem 1 formally as follows:

PROBLEM 1: REPORT ALL SUPERBUBBLES
Input: a directed graph $G = (V, E)$.
Output: all ordered pairs of distinct nodes s and t, such that $\langle s, t \rangle$ is a superbubble.

2.3   Properties of Superbubbles

We next state a few important properties of superbubbles that form the basis of possible solutions.

Lemma 1 [19]. *Any node can be the entrance (respectively exit) of at most one superbubble.*

Note that Lemma 1 does not exclude the possibility that a node is an entrance of a superbubble and an exit of another superbubble.

Lemma 2 [23]. *Let G be a directed acyclic graph. We have the following two observations.*

(1) Suppose (p, c) is an edge in G, where p has one child and c has one parent, then $\langle p, c \rangle$ is a superbubble in G.

(2) For any superbubble $\langle s, t \rangle$ in G, there must exist some parent p of t such that p has exactly one child t.

Lemma 3 [5]. *For any superbubble $\langle s, t \rangle$ in a directed acyclic graph G, there must exist some child c of s such that c has exactly one parent s.*

2.4 Algorithms

The first solution to the superbubble-detection problem was proposed by Onodera *et al.* which runs in $\mathcal{O}(nm)$ time, where n is the number of nodes and m is the number of edges in the graph [19]. Subsequently, Sung *et al.* gave an improved $\mathcal{O}(m \log m)$-time algorithm to solve the problem [23]. Very recently, an even more efficient algorithm was designed by Brankovic *et al.* that has a linear time complexity [5]. We have named these solutions in accordance with the order of their proposal. The outline of each of the three algorithmic designs is provided as follows:

Solution 1: $\mathcal{O}(nm)$-time algorithm [19]. The basic idea in this approach is to visit nodes in an order that follows the standard topological sorting, starting from a given node s in the given directed graph $G = (V, E)$, to eventually report a node t such that $\langle s, t \rangle$ is a *superbubble* (if any). This procedure is iterated for all $s \in V$ to report all the superbubbles in G. The procedure aborts every time either a tip or a cycle is encountered. It works by arbitrarily visiting a node from a maintained dynamic set (initially containing s only) of the vertices. A visit consists of labeling the children of the picked node as 'seen'; pushing its children that have all the parents visited, into the dynamic set; testing whether or not the next node to be visited is an exit node; and reporting the node if the test is positive. The procedure takes $\mathcal{O}(n + m)$ time in the worst case, thus making the total time complexity of the algorithm to be $\mathcal{O}(n(n + m))$.

Solution 2: $\mathcal{O}(m \log m)$-time algorithm [23]. This algorithm begins by partitioning the given graph into a set of subgraphs. The partitioning method ensures that the set of superbubbles in all the subgraphs is the same as the set of superbubbles in the given graph. The subgraphs corresponding to each non-singleton strongly connected component constitute the elements of this set, along with an additional element consisting of the subgraph corresponding to the set of all the nodes involved in singleton strongly connected components. Superbubbles are then detected in each of the subgraphs of the set; if it is cyclic, it is first converted into a directed acyclic subgraph by means of depth-first search and by duplicating some nodes. Detection of the superbubbles in a directed acyclic subgraph is done by visiting each of its nodes in topological order. Each visit of a node t consists of iteratively merging t with its parents that have exactly one child, if t has at least two parents and some of the parents have exactly one child; and reporting $\langle s, t \rangle$ where s is the only parent of t and t is the only child of s.

Note that the cost of partitioning the graph and transforming it into directed acyclic subgraphs is linear with respect to the size of the graph. However, computing the superbubbles in each directed acyclic subgraph requires in total, $\mathcal{O}(m \log m)$ time, owing to the fact that for each parent of a node, the time required for merging is $\mathcal{O}(\log m)$.

Solution 3: $\mathcal{O}(n + m)$-time algorithm [5]. This approach proposes a linear-time algorithm to compute all superbubbles in a directed acyclic graph, thus improving the dominating factor of the time bound in the second solution. It is based on the topological ordering of vertices, obtained by the recursive form of the standard topological sort algorithm. The algorithm proceeds by identifying the potential entrance and exit candidates, in topological order; and checking the validity of a potential superbubble ending at each exit candidate. For a given pair of entrance and exit candidates, the possibility of a valid superbubble formed by the pair is checked using Range Minimum Queries (RMQ) on arrays of both the topologically furthest parent and child of each node. The total running time required to report all superbubbles in a directed acyclic graph is $\mathcal{O}(n+m)$. When it is combined with the two linear-time stages of the second solution, namely, partitioning of a given directed graph into a set of subgraphs and conversion of any cyclic subgraph into an acyclic subgraph, this solution reports all superbubbles in the given directed graph in linear time.

2.5 Discussion

The linear-time algorithm (Solution 3) provides an efficient and practical solution to identify superbubbles, a complex generalisation of bubble motifs, for analysing genome assembly graphs. An important research direction ahead will be to investigate other superbubble-like structures in assembly graphs, such as complex bulges [18]. It would also be interesting to look for ways of relaxing the restrictive definition of superbubbles, from a biological point of view.

3 Clumps

3.1 Biological Motivation

Pattern matching algorithms have been studied extensively since the inception of stringology. In molecular biology, there are several variations of the pattern matching problem that have unique applications. The following section will discuss two such variations, both of which report so-called *clumps* found within biological sequences.

DNA replication is arguably the most essential function of any organism. The success of cell proliferation relies on replication of DNA required by the daughter cells. The mechanism of DNA replication is initiated by the formation of the pre-replication complex at the origin of replication (Ori). A replicon is a

DNA molecule which has a single Ori. In most bacteria, the circular chromosome, which represents the entire genome, is the replicon, and its origin of replication is termed OriC. The bacterial pre-replication complex is a multi-protein complex of which a key protein, DnaA, is of significant interest in computational biology. This is because the activation of DNA replication is conditional upon the concentration of DnaA in the bacterial cell, and the formation of DnaA-OriC complexes are the first stage of DNA replication. Specifically, DnaA molecules bind to DnaA boxes, which are repeated sequences within OriC.

The most well-studied bacterial species is *E. coli*. The following is a summary of some review papers which cite such studies [16]. OriC in E. coli is 245 base pairs long. It contains three AT-rich 13-mers and within this region, three 6-mers (5'-AGATCT), which are DnaA boxes. Also within this region are seven out of the eleven 4-mer (GATC) deoxyadenosine methyltransferase recognition sites which occur in DnaA binding sites. The adjoining region contains more DnaA boxes, approximately five 9-mers (5'-TTWTNCACA). Note that in the preceding sequence, W represents T or A, and N represents any nucleic acid. More DnaA boxes, with slightly different sequences, are also present within these regions. Hereafter, we define this region as a *clump*.

As mentioned, the *E. coli* genome is highly annotated. However, many unannotated bacterial genomes exist, and therefore, an efficient way to find clumps is needed. In summary, in order to locate OriC in bacteria, a set of patterns must occur a certain number of times within a region in the genome. A formal definition of this problem is Problem 2.

Another variation of this problem, which has been studied recently [4,22], is to find clumps given a set of patterns, where the occurrences of the patterns may overlap. These will be referred to as clustered-clumps hereafter. This can be utilised, for example, for gene prediction. That is, to find genes within a genome based on the occurrence of specific DNA sequence motifs before or after them. Examples of such motifs include gene promoters; start and stop codons; and poly(A) tails. An example of overlapping motifs, specifically, recognition sites to which proteins bind, is presented in [14]. A formal definition of this problem is Problem 3.

3.2   Definitions

To provide an overview of our results we begin with a few definitions.

We think of a *string* X of *length* n as an array $X[1..n]$, where every $X[i]$, $1 \leq i \leq n$, is a *letter* drawn from some fixed *alphabet* Σ of size $|\Sigma| = \mathcal{O}(1)$. The *empty string* is denoted by ε. The set of all strings over an alphabet Σ (including empty string ε) is denoted by Σ^*. A string Y is a *factor* of a string X if there exist two strings U and V, such that $X = UYV$. Hence, We say that there is an *occurrence* of Y in X, or, simply, that Y *occurs in* X.

Consider the strings X, Y, U, and V, such that $X = UYV$. If $U = \varepsilon$, then Y is a *prefix* of X. If $V = \varepsilon$, then Y is a *suffix* of X. We denote by SA the *suffix array* of X of length n, that is, an integer array of size n storing the starting positions of all lexicographically sorted suffixes of X, that is, for all $1 < i \leq n$, we have $X[SA[i-1]..n] < X[SA[i]..n]$ [17]. Let $\mathsf{lcp}(i,j)$ denote the length of the longest common prefix between $X[SA[i]..n]$ and $X[SA[j]..n]$, for all positions i, j on X, and 0 otherwise. We denote by LCP the *longest common prefix* array of X defined by $\mathsf{LCP}[i] = \mathsf{lcp}(i-1,i)$, for all $1 < i \leq n$, and $\mathsf{LCP}[1] = 0$. SA and LCP of X can be computed in time and space $\mathcal{O}(n)$ [11].

A *degenerate symbol* $\tilde{\sigma}$ over an alphabet Σ is a non-empty subset of Σ, i.e., $\tilde{\sigma} \subseteq \Sigma$ and $\tilde{\sigma} \neq \emptyset$. $|\tilde{\sigma}|$ denotes the size of the set and we have $1 \leq |\tilde{\sigma}| \leq |\Sigma|$. A *degenerate string* is built over the potential $2^{|\Sigma|} - 1$ non-empty sets of letters belonging to Σ. In other words, a degenerate string $\tilde{X} = \tilde{X}[1..n]$, such that every $\tilde{X}[i]$ is a degenerate symbol, $1 \leq i \leq n$. For example, $\tilde{X} = \{a,b\}\{a\}\{c\}\{b,c\}\{a\}\{a,b,c\}$ is a degenerate string of length 6 over $\Sigma = \{a,b,c\}$. If $|\tilde{X}[i]| = 1$, that is, $\tilde{X}[i]$ represents a single symbol of Σ, we say that $\tilde{X}[i]$ is a *solid* symbol and i is a *solid position*. Otherwise $\tilde{X}[i]$ and i are said to be *non-solid symbol* and *non-solid position* respectively. For convenience we often write $\tilde{X}[i] = \sigma$ ($\sigma \in \Sigma$), instead of $\tilde{X}[i] = \{\sigma\}$, in case of solid symbols. Consequently, the degenerate string $\tilde{X}$ mentioned in the example previously will be written as $\{a,b\}ac\{b,c\}a\{a,b,c\}$. A string containing only solid symbols will be called a *solid string*. A *conservative degenerate string* is a degenerate string where its number of non-solid symbols is upper-bounded by a fixed positive constant k.

For degenerate strings, the notion of symbol equality is extended to single-symbol *match* between two degenerate symbols in the following way. Two degenerate symbols $\tilde{\sigma}_1$ and $\tilde{\sigma}_2$ are said to *match* (represented as $\tilde{\sigma}_1 \approx \tilde{\sigma}_2$) if $\tilde{\sigma}_1 \cap \tilde{\sigma}_2 \neq \emptyset$. Extending this notion to degenerate strings, we say that two degenerate strings $\tilde{X}$ and $\tilde{Y}$ *match* (denoted as $\tilde{X} \approx \tilde{Y}$) if $|\tilde{X}| = |\tilde{Y}|$ and $\tilde{X}[i] \approx \tilde{Y}[i]$, for $i = 1, \cdots, |\tilde{X}|$. Note that the relation $\approx$ is not transitive. A degenerate string $\tilde{Y}$ is said to *occur* at position i in another degenerate (resp. solid) string $\tilde{X}$ (resp. X) if $\tilde{Y} \approx \tilde{X}[i..i+|\tilde{Y}|-1]$ (resp. $\tilde{Y} \approx X[i..i+|\tilde{Y}|-1]$).

We are now able to define Problem 2 formally, which follows [8].

PROBLEM 2: FINDING CLUMPS
Input: A text T of length n, and integers $\ell < n$, $k < \ell/2$, $m < \ell$ and $d < m$.
Output: All (ℓ, k)-clumps in T, that is, all factors $T[i..i+\ell]$, which contain a conservative degenerate pattern $\tilde{P}$ such that $|\tilde{P}| = m$ and

$$Occ_{T[i..i+\ell]}(\tilde{P}) \geq k,$$

where $Occ_T(\tilde{P})$ is the number of occurrences of pattern $\tilde{P}$ in T, and the number of non-solid symbols in $\tilde{P} \leq d$.

In [2], a *clustered-clump* of a given set of patterns $\mathcal{P} = \{P_1, \cdots, P_{|\mathcal{P}|}\}$ is defined as follows: a *clustered-clump* is a maximal set of occurrences of patterns of $\mathcal{P}$ such that

- any consecutive letters of the clustered-clump is a *factor* of at least one occurrence of a pattern from $\mathcal{P}$.

– either the clustered-clump is composed of a *single* occurrence that overlaps no other occurrence, or each occurrence *overlaps* at least one other occurrence.

The clustered-clump in a text T is maximal in the sense that there exists no occurrence of the set $\mathcal{P}$ in T that overlaps this clustered-clump without being a factor of it.

Example 2. Consider the set $\mathcal{P} = \{\text{aba}, \text{bba}\}$ and the text $T = \text{bbbababababababb}$ bbabaababb, we have the following clumps underlined:

$$T = \text{bbbababababababbbbbabaababb}$$

Notice that the factor ababa at position 6 is not a clustered-clump since it is not maximal. Also, the factor bbabaaba at position 15, since its two-letter factor aa is neither a factor of an occurrence of aba, nor an occurrence of bba.

We are now able to define Problem 3 formally, which follows [2].

PROBLEM 3: FINDING CLUSTERED-CLUMPS
Input: A text T of length n, a set of conserved degenerate patterns $\tilde{\mathcal{P}} = \{\tilde{P}_1, \cdots, \tilde{P}_{|\tilde{\mathcal{P}}|}\}$, and integers d and m.
Output: All clustered-clumps in T, where the total number of non-solid symbols in $\tilde{\mathcal{P}} \leq d$ and $m = \sum_{1 \leq i \leq |\tilde{\mathcal{P}}|} |\tilde{P}_i|$.

3.3 Algorithms

Finding Clumps. The following algorithm follows [10,13], and to the best of our knowledge, is the first purely combinatorial solution to find OriC.

We first construct both the suffix array SA and longest common prefix array LCP of the string T. Then, a rank is assigned for each prefix, of length m, of each suffix, of length at least m, based on the order of the suffix array. In other words, each distinct factor of length m is associated with rank r such that two positions i and j in T are assigned rank r, if $T[i..i + m - 1] = T[j..j + m - 1]$ and $1 \leq i, j \leq n - m + 1$.

In the suffix array, the first i_0 suffixes, of length at least m, sharing the same prefix, of length m, will get rank 0; the next i_1 suffixes, of length at least m, sharing a unique prefix of length m, will get rank 1, and so on. In this way, each rank associates with a distinct factor of length m in T.

Next, based on the ranking, we construct a new string T' of length $n - m + 1$, such that $T'[i] = r$, if r is the rank given to suffix $T[i..n]$. Clearly, the ranks go up at most to the value $(n - m)$.

Example 3. Suppose $T = \text{AGCTTGCTAGCT}$ and $m = 3$. The following table shows the string T, its suffix array SA, longest common prefix array LCP and the newly-constructed string T'.

i	1	2	3	4	5	6	7	8	9	10	11	12
$T[i]$	A	G	C	T	T	G	C	T	A	G	C	T
$SA[i]$	9	1	11	7	3	10	6	2	12	8	5	4
$LCP[i]$	0	4	0	2	2	0	3	3	0	1	1	1
$T'[i]$	0	3	2	6	5	3	1	4	0	3		

Here, rank 0 represents AGC and occurs twice in T'; rank 1 represents CTA; rank 2 represents CTT; rank 3 represents GCT and occurs three times in T'; rank 4 represents TAG; rank 5 represents TGC; and rank 6 represents TTG.

Additionally, we construct and maintain a *Parikh vector* $\mathcal{V}(T')$, where the size of the vector is the highest rank value given to the m-length prefixes of suffixes in T. Each component of $\mathcal{V}(T')$ denotes the number of occurrences of the corresponding rank. In particular, for a given length ℓ, $\mathcal{V}_j(T')$ denotes $Occ_{T'[i..i+\ell-m]}(j)$. In this way, each component $\mathcal{V}_j(T')$ denotes the occurrences of distinct factor j of length m in $T[i..i+\ell-1]$.

The vector $\mathcal{V}(T')$ is initialised with the numbers of occurrences of all ranks in the prefix of T' of length $\ell' = \ell + m - 1$. After the initialisation, we proceed using a sliding window of length ℓ' to maintain the Parikh vector $\mathcal{V}(T')$.

Example 4. Following Example 3, and supposing $\ell = 7$, the tables below represents $\mathcal{V}(T')$ when it is initialised (Step 0) and after each of the first 5 steps of the computation.

Step 0		Step 1		Step 2		Step 3		Step 4		Step 5
0 1		0 0		0 0		0 0		0 1		0 1
1 0		1 0		1 1		1 1		1 1		1 1
2 1		2 1		2 1		2 0		2 0		2 0
3 1	$\Rightarrow$	3 2	$\Rightarrow$	3 1	$\Rightarrow$	3 1	$\Rightarrow$	3 1	$\Rightarrow$	3 2
4 0		4 0		4 0		4 1		4 1		4 1
5 1		5 1		5 1		5 1		5 1		5 0
6 1		6 1		6 1		6 1		6 0		6 0

At step i, when the window is shifted one position to the right, to position $i+1$, we have to update $\mathcal{V}(T')$ by possibly decrementing $\mathcal{V}_{T'[i]}(T')$ and incrementing $\mathcal{V}_{T'[i+\ell']}(T')$. If $\mathcal{V}_{T'[i+\ell']}(T')$ is at least k; then the factor $T[i..i+\ell-1]$ is a (ℓ, k)-clump with at least k occurrences of a some solid pattern P of length m.

We still need to compute the number of occurrences of every possible conservative degenerate pattern $\tilde{P}$ that matches the factor of rank $T'[i+\ell']$ and, at the same time, does not match the factor of rank $T'[i]$. Each such pattern $\tilde{P}$ is formed by merging components of $\mathcal{V}(T')$ which have at most d mismatches with the factor of rank $T'[i+\ell']$. For each such pattern $\tilde{P}$, we calculate the number of occurrences by adding the appropriate components of $\mathcal{V}(T')$.

Note that for an integer d, the number of possible $\tilde{P}$ that need to be checked at each position i is $\mathcal{O}(m^d)$. Careful preprocessing of all m-length factors allows efficient identification of all possible $\tilde{P}$ at each step; the details of which are excluded here. Thus, reporting all possible (ℓ, k)-clumps in a given text T can be achieved in $\mathcal{O}(nm^d)$ time.

Example 5. Following Examples 3 and 4, and supposing $d = 0$, two $(7, 2)$-clumps are reported at Step 1 & Step 5 both associated with the solid pattern GCT of rank 3. Instead suppose $d = 1$, then three more $(7, 2)$-clumps are reported at Step 0, 2 & 3 associated with $\{A,T\}GC$, $CT\{A,T\}$, and $T\{A,T\}G$.

Finding Clustered-Clumps. To the best of our knowledge, this problem has only been explored heretofore with a probabilistic approach [4, 22].

The solution we propose for this problem is based on the idea used in [21]. Let the number of patterns in the given set be r ($|\tilde{\mathcal{P}}| = r$), while m is the total lengths of such patterns. We begin by splitting each of the non-solid patterns, say $\tilde{P}_i$, into subpatterns $P_{i,j}$, $1 \le i \le r$ and $1 \le j \le sub(i)$, where each $P_{i,j}$ is a solid pattern over Σ and $sub(i)$ denotes the number of subpatterns obtained from a pattern $\tilde{P}_i$; we call the set of all new solid subpatterns $\mathcal{P}$. Effectively, we are breaking every pattern into subpatterns by chopping out the parts containing non-solid symbols so that each of the subpatterns is solid.

Example 6. Suppose $\tilde{\mathcal{P}} = \{AC\{TG\}AA\{CG\}TAA, AT\{C,G\}TT\{A,G\}C\}$. Then $\mathcal{P} = \{AC, AA, TAA, AT, TT, C\}$, and $sub(1) = sub(2) = 3$.

We next build the Aho-Corasick Automaton of the set $\mathcal{P}$; denoted $\mathcal{S}(\mathcal{P})$, where $\mathcal{S}(\mathcal{P})$ is the minimal deterministic finite automaton whose language is the set of suffixes of $\mathcal{P}$ (see [9, Sect. 6.6] for more description and for efficient construction). This data structure allows us to compute all the occurrences of the solid subpatterns in the text T in linear time.

In particular, we preprocess the occurrences of each of the subpatterns by constructing a matrix such that when the matrix is duly filled we can test in constant time whether or not a specific solid subpattern occurs in a given text position. If an occurrence of $P_{i,j}$ for $j > 1$ is found, then we need to check:

1. Whether the non-solid symbol in $\tilde{P}_i$ preceding $P_{i,j}$ matches the corresponding position in T.
2. Whether $P_{i,j-1}$ occurs in the corresponding position in T.

If both conditions are valid, then an occurrence of $P_{i,j}$ is reported correctly in the matrix. Notice that; in this way, an occurrence of $P_{i,sub(i)}$ corresponds to an occurrence of a non-solid pattern $\tilde{P}_i$ in T.

Using the information about the occurrences of the non-solid patterns in the text, we populate an array of size n that stores the length of the longest pattern occurring at each position of the text. It is easy to see that simple calculations in a single scan of this array can report the indices of all clustered-clumps in T.

Computing both $\mathcal{P}$ and $\mathcal{S}(\mathcal{P})$ takes $\mathcal{O}(m)$ time, while $\mathcal{O}(dn)$ time is required for finding the occurrences of all solid subpatterns (to fill the matrix) and the occurrences of non-solid patterns subsequently. Scanning of the array in the last step can be done in $\mathcal{O}(n)$ time. Thus, the solution finds all the clustered-clumps in the text in time equal to $\mathcal{O}(n + m)$ (for constant d).

3.4 Discussion

After publishing our algorithm design in detail, we intend to evaluate the efficiency and accuracy of our approach to predicting the locus of OriC, with that of OriFinder [12], a tool which analyses the distribution of bases in the genome and as well as the locations of motifs associated with OriC. We are currently extending our solutions for finding occurrences of clustered-clumps in texts with non-solid symbols. These kind of texts are common in biology and thus such an extension is likely to draw a lot of interest.

4 Conclusion

This paper has provided a review of recent developments in computational molecular biology, focusing on two unique problems in two different stages of discovery.

References

1. Balasubramanian, S., Klenerman, D., Barnes, C., Osborne, M.: Patent US20077232656 (2007)
2. Bassino, F., Clément, J., Fayolle, J., Nicodème, P.: Constructions for clumps statistics. CoRR abs/0804.3671 (2008). http://arxiv.org/abs/0804.3671
3. Batzoglou, S.: Algorithmic challenges in mammalian genome sequence assembly. In: Dunn, M., Jorde, L., Little, P., Subramaniam, S. (eds.) Encyclopedia of Genomics, Proteomics and Bioinformatics. Wiley, Hoboken (New Jersey) (2005)
4. Boeva, V., Clément, J., Régnier, M., Vandenbogaert, M.: Assessing the significance of sets of words. In: Apostolico, A., Crochemore, M., Park, K. (eds.) CPM 2005. LNCS, vol. 3537, pp. 358–370. Springer, Heidelberg (2005)
5. Brankovic, L., Iliopoulos, C.S., Kundu, R., Mohamed, M., Pissis, S.P., Vayani, F.: Linear-time superbubble identification algorithm for genome assembly. Theor. Comput. Sci. **609**(Part 2), 374–383 (2016). http://www.sciencedirect.com/science/article/pii/S0304397515009147
6. de Bruijn, N.G.: A combinatorial problem. Koninklijke Nederlandse Akademie v. Wetenschappen **49**, 758–764 (1946)
7. Butler, J., MacCallum, I., Kleber, M., Shlyakhter, I.A., Belmonte, M.K., Lander, E.S., Nusbaum, C., Jaffe, D.B.: ALLPATHS: de novo assembly of whole-genome shotgun microreads. Genome Res. **18**(5), 810–820 (2008)
8. Compeau, P.: Bioinformatics Algorithms: An Active Learning Approach. Active Learning Publishers, La Jolla (2014)
9. Crochemore, M., Hancart, C., Lecroq, T.: Algorithms on Strings, p. 392. Cambridge University Press, Cambridge (2007)
10. Ehlers, T., Manea, F., Mercaş, R., Nowotka, D.: k-abelian pattern matching. J. Discrete Algorithms **34**, 37–48 (2015)
11. Fischer, J.: Inducing the LCP-array. In: Dehne, F., Iacono, J., Sack, J.-R. (eds.) WADS 2011. LNCS, vol. 6844, pp. 374–385. Springer, Heidelberg (2011)
12. Gao, F., Zhang, C.T.: Ori-finder: a web-based system for finding orics in unannotated bacterial genomes. BMC Bioinform. **9**(1), 79 (2008)

13. Grossi, R., Iliopoulos, C.S., Mercaş, R., Pisanti, N., Pissis, S.P., Retha, A., Vayani, F.: Circular sequence comparison with q-grams. In: Pop, M., Touzet, H. (eds.) WABI 2015. LNCS, vol. 9289, pp. 203–216. Springer, Heidelberg (2015)

14. Kvietikova, I., Wenger, R.H., Marti, H.H., Gassmann, M.: The transcription factors ATF-1 and CREB-1 bind constitutively to the hypoxia-inducible factor-1 (HIF-1) DNA recognition site. Nucleic Acids Res. **23**(22), 4542–4550 (1995)

15. Lander, E.S., Linton, L.M., Birren, B., Nusbaum, C., Zody, M.C., Baldwin, J., Devon, K., Dewar, K., Doyle, M., FitzHugh, W., et al.: Initial sequencing and analysis of the human genome. Nature **409**(6822), 860–921 (2001)

16. Leonard, A.C., Grimwade, J.E.: Building a bacterial orisome: emergence of new regulatory features for replication origin unwinding. Mol. Microbiol. **55**(4), 978–985 (2005)

17. Manber, U., Myers, G.: Suffix arrays: a new method for on-line string searches. SIAM J. Comput. **22**(5), 935–948 (1993)

18. Nurk, S., Bankevich, A., Antipov, D., Gurevich, A.A., Korobeynikov, A., Lapidus, A., Prjibelski, A.D., Pyshkin, A., Sirotkin, A., Sirotkin, Y., Stepanauskas, R., Clingenpeel, S.R., Woyke, T., McLean, J.S., Lasken, R., Tesler, G., Alekseyev, M.A., Pevzner, P.A.: Assembling single-cell genomes and mini-metagenomes from chimeric MDA products. J. Comput. Biol. **20**(10), 714–737 (2013)

19. Onodera, T., Sadakane, K., Shibuya, T.: Detecting superbubbles in assembly graphs. In: Darling, A., Stoye, J. (eds.) WABI 2013. LNCS, vol. 8126, pp. 338–348. Springer, Heidelberg (2013)

20. Pevzner, P.A., Tang, H., Waterman, M.S.: An Eulerian path approach to DNA fragment assembly. Proc. Nat. Acad. Sci. U.S.A. **98**(17), 9748–9753 (2001)

21. Rahman, M.S., Iliopoulos, C.S.: Pattern matching algorithms with don't cares. In: van Leeuwen, J., Italiano, G.F., van der Hoek, W., Meinel, C., Sack, H., Plasil, F., Bielikova, M. (eds.) Proceedings of the 33rd International Conference on Current Trends in Theory and Practice of Computer Science (SOFSEM 2007), pp. 116–126. Institute of Computer Science AS CR, Prague (2007)

22. Régnier, M.: A unified approach to word statistics. In: Proceedings of the Second Annual International Conference on Computational Molecular Biology, RECOMB 1998, pp. 207–213. ACM, New York (1998). http://acm.org/10.1145/279069.279116

23. Sung, W., Sadakane, K., Shibuya, T., Belorkar, A., Pyrogova, I.: An $O(m \log m)$-time algorithm for detecting superbubbles. IEEE/ACM Trans. Comput. Biology Bioinform. **12**(4), 770–777 (2015)

24. Venter, J.C., Adams, M.D., Myers, E.W., Li, P.W., Mural, R.J., Sutton, G.G., Smith, H.O., Yandell, M., Evans, C.A., Holt, R.A., et al.: The sequence of the human genome. Science **291**(5507), 1304–1351 (2001)

25. Zerbino, D.R., Birney, E.: Velvet: algorithms for de novo short read assembly using de Bruijn graphs. Genome Res. **18**(5), 821–829 (2008)

Graphs Coloring

Tropical Dominating Sets
in Vertex-Coloured Graphs

Jean-Alexandre Anglès d'Auriac[1], Csilia Bujtás[2], Hakim El Maftouhi[1],
Marek Karpinski[3], Yannis Manoussakis[1(✉)], Leandro Montero[1],
Narayanan Narayanan[1], Laurent Rosaz[1], Johan Thapper[4], and Zsolt Tuza[2,5]

[1] Université Paris-Sud, L.R.I., Bât. 650, 91405 Orsay Cedex, France
{angles,yannis,lmontero,rosaz}@lri.fr, hakim.maftouhi@orange.fr,
narayana@gmail.com
[2] Department of Computer Science and Systems Technology,
University of Pannonia, Veszprém, Egyetem u. 10 8200, Hungary
{bujtas,tuza}@dcs.uni-pannon.hu
[3] Department of Computer Science, University of Bonn,
Friedrich-Ebert-Allee 144, 53113 Bonn, Germany
marek@cs.uni-bonn.de
[4] Université Paris-Est, Marne-la-Vallée, LIGM,
Bât. Copernic, 5 Bd Descartes, 77454 Marne-la-Vallée Cedex 2, France
thapper@u-pem.fr
[5] Alfréd Rényi Institute of Mathematics, Hungarian Academy of Sciences, Budapest,
Reáltanoda u. 13–15 1053, Hungary

Abstract. Given a vertex-coloured graph, a dominating set is said to be tropical if every colour of the graph appears at least once in the set. Here, we study minimum tropical dominating sets from structural and algorithmic points of view. First, we prove that the tropical dominating set problem is NP-complete even when restricted to a simple path. Last, we give approximability and inapproximability results for general and restricted classes of graphs, and establish a FPT algorithm for interval graphs.

Keywords: Dominating set · Vertex-coloured graph · Approximation

1 Introduction

Vertex-coloured graphs are useful in various situations. For instance, the Web graph may be considered as a vertex-coloured graph where the colour of a vertex represents the content of the corresponding page (red for mathematics, yellow for physics, etc.). Given a vertex-coloured graph G^c, a subgraph H^c (not necessarily induced) of G^c is said to be tropical if and only if each colour of G^c appears at least once in H^c. Potentially, any kind of usual structural problems (paths, cycles, independent and dominating sets, vertex covers, connected components, etc.) could be studied in their tropical version. This new tropical concept is

© Springer International Publishing Switzerland 2016
M. Kaykobad and R. Petreschi (Eds.): WALCOM 2016, LNCS 9627, pp. 17–27, 2016.
DOI: 10.1007/978-3-319-30139-6_2

close to, but quite different from, the colourful concept used for paths in vertex-coloured graphs [1,15,16]. It is also related to (but again different from) the concept of *colour patterns* used in bio-informatics [11]. Here, we study minimum tropical dominating sets in vertex-coloured graphs. A general overview on the classical dominating set problem can be found in [13].

Throughout the paper let $G = (V, E)$ denote a simple undirected non-coloured graph. Let $n = |V|$ and $m = |E|$. Given a set of colours $\mathcal{C} = \{1, ..., c\}$, $G^c = (V^c, E)$ denotes a vertex-coloured graph where each vertex has precisely one colour from $\mathcal{C}$ and each colour of $\mathcal{C}$ appears on at least one vertex. The colour of a vertex x is denoted by $c(x)$. A subset $S \subseteq V$ is a *dominating set* of G^c (or of G), if every vertex either belongs to S or has a neighbour in S. The *domination number* $\gamma(G^c)$ ($\gamma(G)$) is the size of a smallest dominating set of G^c (G). A dominating set S of G^c is said to be *tropical* if each of the c colours appears at least once among the vertices of S. The *tropical domination number* $\gamma^t(G^c)$ is the size of a smallest tropical dominating set of G^c. A *rainbow dominating set* of G^c is a tropical dominating set with exactly c vertices. More generally, a c-element set with precisely one vertex from each colour is said to be a *rainbow set*. We let $\delta(G^c)$ (respectively $\Delta(G^c)$) denote the minimum (maximum) degree of G^c. When no confusion arises, we write γ, γ^t, δ and Δ instead of $\gamma(G)$, $\gamma^t(G^c)$, $\delta(G^c)$ and $\Delta(G^c)$, respectively. We use the standard notation $N(v)$ for the (open) neighbourhood of vertex v, that is the set of vertices adjacent to v, and write $N[v] = N(v) \cup \{v\}$ for its closed neighbourhood. The set and the number of neighbours of v inside a subgraph H is denoted by $N_H(v)$ and by $d_H(v)$, independently of whether v is in H or in $V(G^c) - V(H)$. Although less standard, we shall also write sometimes $v \in G^c$ to abbreviate $v \in V(G^c)$.

Note that tropical domination in a vertex-coloured graph G^c can also be interpreted as "simultaneous domination" in two graphs which have a common vertex set. One of the two graphs is the non-coloured G itself, the other one is the union of c vertex-disjoint cliques each of which corresponds to a colour class in G^c. The notion of simultaneous dominating set[1] was introduced by Sampathkumar [17] and independently by Brigham and Dutton [5]. It was investigated recently by Caro and Henning [6] and also by further authors. Remark that $\delta \geq 1$ is regularly assumed for each factor graph in the results of these papers that is not the case in the present manuscript, as we do not forbid the presence of one-element colour classes.

The Tropical Dominating Set problem (TDS) is defined as follows.

Problem 1. TDS
Input: A vertex-coloured graph G^c and an integer $k \geq c$.
Question: Is there a tropical dominating set of size at most k?

---

[1] Also known under the names 'factor dominating set' and 'global dominating set' in the literature.

The Rainbow Dominating Set problem (RDS) is defined as follows.

Problem 2. RDS
Input: A vertex-coloured graph G^c.
Question: Is there a rainbow dominating set?

The paper is organized as follows. In Sect. 2 we give approximability and inapproximability results for TDS. We also show that the problem is FPT (fixed-parameter tractable) on interval graphs when parametrized by the number of colours.

2   Approximability and Fixed Parameter Tractability

We begin this section noting that the problem is intractable even for paths.

Theorem 1. *The RDS problem is NP-complete, even when the input is restricted to vertex-coloured paths.*

In the sequel, we assume familiarity with the complexity classes NPO and PO which are optimisation analogues of NP and P. A minimisation problem in NPO is said to be *approximable* within a constant $r \geq 1$ if there exists an algorithm A which, for every instance I, outputs a solution of measure $A(I)$ such that $A(I)/\mathsf{Opt}(I) \leq r$, where $\mathsf{Opt}(I)$ stands for the measure of an optimal solution. An NPO problem is in the class APX if it is approximable within *some* constant factor $r \geq 1$. An NPO problem is in the class PTAS if it is approximable within r for *every* constant factor $r > 1$. An APX-hard problem cannot be in PTAS unless P = NP. We use two types of reductions, L-reductions to prove APX-hardness, and PTAS-reductions to demonstrate inclusion in PTAS. In the Appendix we give a slightly more formal introduction and a description of reduction methods related to approximability. For more on these issues we refer to Ausiello et al. [3] and Crescenzi [8].

A problem is said to be *fixed parameter tractable* (FPT) with parameter $k \in \mathbb{N}$ if it has an algorithm that runs in time $f(k)\,|I|^{\mathcal{O}(1)}$ for any instance (I, k), where f is an arbitrary function that depends only on k.

In this section, we study the complexity of approximating and solving TDS conditioned on various restrictions on the input graphs and on the number of colours. First, we show that TDS is equivalent to MDS (Minimum Dominating Set) under L-reductions. In particular, this implies that the general problem lies outside APX. We then attempt to restrict the input graphs and observe that if MDS is in APX on some family of graphs, then so is TDS. However, there is also an immediate lower bound: TDS on any family of graphs that contains all paths is APX-hard. We proceed by adding an upper bound on the number of colours. We see that if MDS is in PTAS for some family of graphs with bounded degree, then so is TDS when restricted to $n^{1-\epsilon}$ colours for some $\epsilon > 0$. Finally, we show that TDS on interval graphs is FPT with the parameter being the number of colours and that the problem is in PO when the number of colours is logarithmic.

Proposition 1. *TDS is equivalent to MDS under L-reductions. It is approximable within* $\ln n + \Theta(1)$ *but NP-hard to approximate within* $(1 - \epsilon) \ln n$.

Proof. MDS is clearly a special case of TDS. For the opposite direction, we reduce an instance of TDS to an instance I of the Set Cover problem which is known to be equivalent to MDS under L-reductions [14]. In the Set Cover problem, we are given a ground set U and a collection of subsets $F_i \subseteq U$ such that $\bigcup_i F_i = U$. The goal is to cover U with the smallest possible number of sets F_i. Our reduction goes as follows. Given a vertex-coloured graph $G^c = (V^c, E)$, with the set of colours C, the ground set of I is $U = V^c \cup C$. Each vertex v of V gives rise to a set $F_v = N[v] \cup \{c(v)\}$, a subset of U. Every solution to I must cover every vertex $v \in V$ either by including a set that corresponds to v or by including a set that corresponds to a neighbour of v. Furthermore, every solution to I must include at least one vertex of every colour in C. It follows that every set cover can be translated back to a tropical dominating set of the same size. This shows that our reduction is an L-reduction.

The approximation guarantee follows from that of the standard greedy algorithm for Set Cover. The lower bound follows from the NP-hardness reduction to Set Cover in [9] in which the constructed Set Cover instances contain $o(N)$ sets, where N is the size of the ground set.

When the input graphs are restricted to some family of graphs, then membership in APX for MDS carries over to TDS.

Lemma 1. *Let $\mathcal{G}$ be a family of graphs. If MDS restricted to $\mathcal{G}$ is in APX, then TDS restricted to $\mathcal{G}$ is in APX.*

Proof. Assume that MDS restricted to $\mathcal{G}$ is approximable within r for some $r \geq 1$. Let G^c be an instance of TDS. We can find a dominating set of the uncoloured graph G of size at most $r\gamma(G)$ in polynomial time, and then add one vertex of each colour that is not yet present in the dominating set. This set is of size at most $r\gamma(G) + c - 1$. The size of an optimal solution of G^c is at least $\gamma(G)$ and at least c. Hence, the computed set will be at most $r + 1$ times the size of the optimal solution of G^c.

For $\Delta \geq 2$, let Δ-TDS denote the problem of minimising a tropical dominating set on graphs of degree bounded by Δ. The problem MDS is in APX for bounded-degree graphs, hence Δ-TDS is in APX by Lemma 1. The same lemma also implies that TDS restricted to paths is in APX. Next, we give explicit approximation ratios for these problems.

Proposition 2. *TDS restricted to paths can be approximated within* $5/3$.

Proof. Let $P^c = v_1, v_2, \ldots, v_n$ be a vertex-coloured path. For $i = 1, 2, 3$ let $\sigma_i = \{v_j \mid j \equiv i \pmod 3, \ 1 \leq j \leq n\}$. Select any subset σ_i' of V that contains precisely one vertex of each colour missing from σ_i. Let $S_i = \sigma_i \cup \sigma_i'$. By definition, S_i is a tropical set.

Taking into account that each colour must appear in a tropical dominating set, moreover any vertex can dominate at most two others, we see the following easy lower bounds:

$$n \leq 3\gamma^t(P^c),$$
$$2c \leq 2\gamma^t(P^c),$$
$$\frac{1}{5}(n + 2c) \leq \gamma^t(P^c).$$

Suppose for the moment that each of S_1, S_2, S_3 dominates G^c. Then, since each colour occurs in at most two of the σ'_i, we have $|S_1| + |S_2| + |S_3| \leq n + 2c$ and therefore

$$\gamma^t(P^c) \leq \min(|S_1|, |S_2|, |S_3|) \leq \frac{1}{3}(n + 2c).$$

Comparing the lower and upper bounds, we obtain that the smallest set S_i provides a 5/3-approximation. It is also clear that this solution can be constructed in linear time.

The little technical problem here is that the set S_i does not dominate vertex v_1 if $i = 3$, and it does not dominate v_n if $i \equiv n - 2 \pmod 3$. We can overcome this inconvenience as follows.

The set S_3 surely will dominate v_1 if we extend S_3 with either of v_1 and v_2. This means no extra element if we have the option to select e.g. v_1 into σ'_3. We cannot do this only if $c(v_1)$ is already present in σ_3. But then this colour is common in σ_1 and σ_3; that is, although we take an extra element for S_3, we can subtract 1 from the term $2c$ when estimating $|\sigma'_1| + |\sigma'_2| + |\sigma'_3|$. The same principle applies to the colour of v_n, too.

Even this improved computation fails by 1 when $n \equiv 2 \pmod 3$ and $c(v_1) = c(v_n)$, as we can then write just $2c - 1$ instead of $2c - 2$ for $|\sigma'_1| + |\sigma'_2| + |\sigma'_3|$. Now, instead of taking the vertex pair $\{v_1, v_n\}$ into S_3, we complete S_3 with v_2 and v_n. This yields the required improvement to $2c - 2$, unless $c(v_2)$, too, is present in σ_3. But then $c(v_2)$ is a common colour of σ_2 and σ_3, while $c(v_1)$ is a common colour of σ_1 and σ_3. Thus $|\sigma'_1| + |\sigma'_2| + |\sigma'_3| \leq 2c - 2$, and $|S_1| + |S_2| + |S_3| \leq n + 2c$ holds also in this case.

Remark 1. In an analogous way — which does not even need the particular discussion of unfavourable cases — one can prove that the square grid $P_n \square P_n$ admits an asymptotic 9/5-approximation. (This extends also to $P_n \square P_m$ where $m = m(n)$ tends to infinity as n gets large.) A more precise estimate on grids, however, may require a careful and tedious analysis.

Proposition 3. *Δ-TDS is approximable within* $\ln(\Delta + 2) + \frac{1}{2}$. *Moreover, there are absolute constants $C > 0$ and $\Delta_0 \geq 3$ such that for every $\Delta \geq \Delta_0$, it is NP-hard to approximate Δ-TDS within* $\ln \Delta - C \ln \ln \Delta$.

Proof. The second assertion follows from [7, Theorem 3]. For the first part, we apply reduction from Set Cover, similarly as in the proof of Proposition 1. So, for $G^c = (V^c, E)$ we define $U = V^c \cup C$ and consider the sets $F_v = N[v] \cup \{c(v)\}$ for the vertices $v \in V^c$. Every set cover in this set system corresponds to a tropical dominating set in G^c. Moreover, the Set Cover problem is approximable within $\sum_{i=1}^k \frac{1}{i} - \frac{1}{2} < \ln k + \frac{1}{2}$ [10], where k is an upper bound on the cardinality of any set of I. In our case, we have $k = \Delta + 2$ since $|N(v)| \leq \Delta$ for all v. Hence, TDS is approximable within $\ln(\Delta + 2) + \frac{1}{2}$.

We now show that TDS for paths is APX-complete.

Theorem 2. *TDS restricted to paths is* APX-*hard.*

Proof. We apply an L-reduction from the Vertex Cover problem (VC): Given a graph $G = (V, E)$, find a set of vertices $S \subseteq V$ of minimum cardinality such that, for every edge $uv \in E$, at least one of $u \in S$ and $v \in S$ holds. We write 3-VC for the vertex cover problem restricted to graphs of maximum degree three (subcubic graphs). The problem 3-VC is known to be APX-complete [2]. For a graph G, we write $\mathsf{Opt}_{VC}(G)$ for the minimum size of a vertex cover of G.

Let $G = (V, E)$ be a non-empty instance of 3-VC, with $V = \{v_1, \ldots, v_n\}$ and $E = \{e_1, \ldots, e_m\}$. Assume that G has no isolated vertices. The reduction sends G to an instance $\phi(G)$ of TDS which will have $m + n + 1$ colours: B (for black), E_i with $1 \le i \le m$ (for the ith edge), and S_j with $1 \le j \le n$ (for the jth vertex). The path has $9n + 3$ vertices altogether, starting with three black vertices of Fig. 1(a), we call this triplet V_0. Afterwards blocks of 6 and 3 vertices alternate, we call the latter $V_1, \ldots, V_n$, representing the vertices of G. Each V_j (other than V_0) is coloured as shown in Fig. 1(c). Assuming that v_j ($1 \le j \le n$) is incident to the edges e_{j_1}, e_{j_2}, and e_{j_3}, the two parts V_{j-1} and V_j are joined by a path representing these three incidences, and coloured as in Fig. 1(b). If v_j has degree less than 3, then the vertex in place of E_{j_3} is black; and if $d(v_j) = 1$, then also E_{j_2} is black.

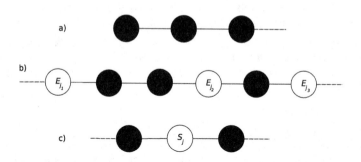

Fig. 1. Gadgets for the reduction of Theorem 2

Let $\sigma \subseteq V$ be an arbitrary solution to $\phi(G)$. First, we construct a solution σ' from σ with more structure, and with a measure at most that of σ. For every j, σ contains the vertex coloured S_j. Let σ' contain these as well. At least one of the first two vertices coloured B must also be in σ. Let σ' contain the second vertex coloured B. Now, if any V_j ($0 \le j \le n$) has a further (first or third) vertex which is an element of σ, then we can replace it with its predecessor or successor, achieving that they dominate more vertices in the path. This modification does not lose any colour because the first and third vertices of any V_j are black, and B is already represented in $\sigma \cap V_0$.

Now we turn to the 6-element blocks connecting a V_{j-1} with V_j. Since the third vertex of V_{j-1} and the first vertex of V_j are surely not in the modified σ, which still dominates the path, it has to contain at least two vertices of the 6-element block. And if it contains only two, then those necessarily are the second and fifth, both being black. Should this be the case, we keep them in σ'. Otherwise, if the modified σ contains more than two vertices of the 6-element block, then let σ' contain precisely E_{j_1}, E_{j_2}, and E_{j_3}. Since σ is a tropical dominating set, the same holds for σ'. It is also clear that $|\sigma'| \leq |\sigma|$.

Next, we create a solution $\psi(G, \sigma)$ to the vertex cover problem on G, using σ'. Let $v_j \in \psi(G, \sigma)$ if and only if $\{E_{j_1}, E_{j_2}, E_{j_3}\} \subseteq \sigma'$. Then, $|\psi(G, \sigma)| = |\sigma'| - 1 - 3n \leq |\sigma| - 1 - 3n$, and when σ is optimal, we have the equality $\mathsf{Opt}_{VC}(G) = \gamma^t(\phi(G)) - 1 - 3n$. Therefore,

$$|\psi(G, \sigma)| - \mathsf{Opt}_{VC}(G) \leq |\sigma| - \gamma^t(\phi(G)). \tag{1}$$

We may assume that G does not contain any isolated vertices. Under this assumption, we prove the lower bound $\mathsf{Opt}_{VC}(G) \geq n/4$ by induction, as follows: The bound clearly holds for an empty graph. Suppose that the bound holds for all graphs without isolated vertices with fewer than n vertices. Let σ^* be a minimal vertex cover of G and let $v \in V \setminus \sigma^*$. Then, all of v's neighbours are in σ^*. Let G' be the graph G with $N[v]$ removed as well as any isolated vertices resulting from this removal. Let n' be the number of vertices in G'. If v has $1 \leq n_v \leq 3$ neighbours, then $0 \leq n_i \leq 2n_v$ vertices become isolated when $N[v]$ is removed, so $\mathsf{Opt}_{VC}(G) = n_v + \mathsf{Opt}_{VC}(G') \geq n_v + n'/4 = n_v + (n - 1 - n_v - n_i)/4 \geq n_v + (n - 1 - 3n_v)/4 \geq n/4$.

This allows us to upper-bound the optimum of $\phi(G)$:

$$\begin{aligned}
\gamma^t(\phi(G)) &= \mathsf{Opt}_{VC}(G) + 1 + 3n \\
&\leq \mathsf{Opt}_{VC}(G) + 1 + 12 \cdot \mathsf{Opt}_{VC}(G) \leq 14 \cdot \mathsf{Opt}_{VC}(G). \tag{2}
\end{aligned}$$

It follows from (1) and (2) that ϕ and ψ constitute an L-reduction.

Corollary 1. *Fix $0 < \epsilon \leq 1$, and let $\mathcal{P}$ be the family of all vertex-coloured paths with at most n^ϵ colours, where n is the number of vertices. Then TDS restricted to $\mathcal{P}$ is NP-hard.*

Proof. We reduce from TDS on paths with an unrestricted number of colours which is NP-hard by Theorem 2. Let P^c be a vertex-coloured path on n vertices with $c \leq n$ colours. Let $Q^{c'}$ be the instance obtained by adding a path $v_1, v_2, \ldots, v_N$ with $N = \lceil (n+2)^{1/\epsilon} \rceil$ vertices to the end of P^c (this is a polynomial-time reduction for any fixed constant $\epsilon > 0$). Let A and B be two new colours. In the added path $v_1, v_2, \ldots, v_N$, let v_2 have colour A and all the other vertices have colour B. The instance $Q^{c'}$ has $n' = n + N$ vertices and $c' = c + 2 \leq n + 2 \leq N^\epsilon \leq (n')^\epsilon$ colours, so $Q^{c'} \in \mathcal{P}$.

Given a minimum tropical dominating set σ of $Q^{c'}$, we see that v_2 must be in σ to account for the colour A. We may further assume that v_1 is not in σ. If it were, then we could modify σ by removing v_1 and adding the last vertex of

P^c instead. It is now clear that taking σ restricted to $\{v_1, v_2, \ldots, v_N\}$ together with a tropical dominating set of P^c yields a tropical dominating set of $Q^{c'}$ and that σ restricted to P^c is a tropical dominating set of P^c. Hence, σ restricted to P^c is a minimum tropical dominating set of P^c.

We have seen that restricting the input to any graph family that contains at least the paths can take us into APX but not further. To find more tractable restrictions, we now introduce an additional restriction on the number of colours. The following lemma says that if the domination number grows asymptotically faster than the number of colours, then we can lift PTAS-inclusion of MDS to TDS.

Lemma 2. *Let $\mathcal{G}$ be a family of vertex-coloured graphs. Assume that there exists a computable function $f: \mathbb{Q} \cap (0, \infty) \to \mathbb{N}$ such that for every $r > 0$, $\gamma(G) > c/r$ whenever $G^c \in \mathcal{G}$ and $n(G^c) \geq f(r)$. Then, TDS restricted to $\mathcal{G}$ PTAS-reduces to MDS restricted to $\mathcal{G}$.*

Proof. To design a polynomial-time $(1+\varepsilon)$-approximation for any rational $\varepsilon > 0$, we pick $r = \varepsilon/2$; hence let $n_0 = f(\varepsilon/2)$. Let $G^c \in \mathcal{G}$ be a vertex-coloured graph. The reduction sends G^c to $\phi(G^c) = G$, the instance of MDS obtained from G^c by simply forgetting the colours. Let σ be any dominating set in G. Assuming that σ is a good approximation to $\gamma(G)$, we need to compute a good approximation $\psi(G^c, \sigma)$ to $\gamma^t(G^c)$. If $n(G^c) < n_0$, then we let $\psi(G^c, \sigma)$ be an optimal tropical dominating set of G^c. Otherwise, let $\psi(G^c, \sigma)$ be σ plus a vertex for each remaining non-covered colour. Since n_0 depends on ε but not on G^c or σ, it follows that ψ can be computed in time that is polynomial in $|V(G^c)|$ and $|\sigma|$.

We claim that ϕ and ψ provide a PTAS-reduction. This is clear if $n(G^c) < n_0$ since ψ then computes an optimal solution to G^c. Otherwise, assume that $n(G^c) \geq n_0$ and that $|\sigma| / \gamma(G) \leq 1 + \varepsilon/2$, i.e., σ is a good approximation. Then,

$$\frac{|\psi(G^c, \sigma)|}{\gamma^t(G^c)} \leq \frac{|\sigma| + c}{\gamma(G)} \leq \frac{2 + \varepsilon}{2} + \frac{c}{\gamma(G)} < 1 + \varepsilon,$$

where the last inequality follows from $n(G) \geq n_0$ and the definition of f.

Example 1. The problem MDS is in PTAS for planar graphs [4], but NP-hard even for planar subcubic graphs [12]. Let $\mathcal{G}$ be the family of planar graphs of maximum degree Δ, for any fixed $\Delta \geq 3$, and with a number of colours $c < n^{1-\epsilon}$ for some fixed $\epsilon > 0$. Let $f(r) = \lceil (\frac{\Delta+1}{r})^{1/\epsilon} \rceil$ and note that $\gamma(G) \geq n/(\Delta + 1) > cn^\epsilon/(\Delta + 1) \geq cf(r)^\epsilon/(\Delta + 1) \geq c/r$ whenever $n \geq f(r)$. It then follows from Lemma 2 that TDS is in PTAS when restricted to planar graphs of fixed maximum degree.

Example 2. As a second example, we observe how the complexity of TDS on a path varies when we restrict the number of colours. For an arbitrary number of colours, it is APX-complete by Lemma 1 and Theorem 2. If the number of colours is $\mathcal{O}(n^{1-\epsilon})$ for some $\epsilon > 0$, then it is in PTAS by Lemma 2, but NP-hard by Corollary 1. Finally, if the number of colours is $\mathcal{O}(\log n)$, then it can be shown to be in PO by a simple dynamic programming algorithm.

In the rest of this section, we look at the restriction where we consider the number of colours as a fixed parameter. We prove the following result.

Theorem 3. *There is an algorithm for TDS restricted to interval graphs that runs in time $\mathcal{O}(2^c n^2)$.*

This shows that TDS for interval graphs is FPT and, furthermore, that if $c = \mathcal{O}(\log n)$, then TDS is in PO.

Let G^c be a vertex-coloured interval graph with vertex set $V = \{1, \ldots, n\}$ and colour set C, and fix some interval representation $I_i = [l_i, r_i]$ for each vertex $1 \leq i \leq n$. Assume that the vertices are ordered non-decreasingly with respect to r_i. For $a, b \in V$, we use (closed) intervals $[a, b] = \{i \in V \mid a \leq i \leq b\}$ to denote subsets of vertices with respect to this order.

Define an *i-prefix dominating set* as a subset $U \subseteq V$ of vertices that contains i and dominates $[1, i]$ in G^c. We say that U is *proper* if, for every $i, j \in U$, we have neither $I_i \subseteq I_j$ nor $I_j \subseteq I_i$.

Let $f \colon \mathcal{P}(C) \times [0, n] \to \mathbb{N} \cup \{\infty\}$ be the function defined so that, given a subset $S \subseteq C$ of colours and a vertex $i \in V$, $f(S, i)$ is the least number of vertices in a proper i-prefix dominating set that covers precisely the colours in S, or ∞ if there is no such set. The value of $f(S, 0)$ is defined to be 0 when $S = \emptyset$ and ∞ otherwise. Our proof is based on a recursive definition of f (Lemma 5) and the fact that f determines γ^t (Lemma 4). First, we need a technical lemma.

Lemma 3. *Let $U \subseteq V$ and let i be the largest element in U. If U is i-prefix dominating, then it dominates precisely the same vertices as $[1, i]$. In particular, U dominates G if and only if $[1, i]$ does.*

Proof. Assume to the contrary that there is a $j \in [1, i] - U$ that dominates some $k > i$, and that k is not dominated by U. This means that j is connected to k in G, so $l_k \leq r_j$. But then we have $l_k \leq r_j \leq r_i \leq r_k$, so $[l_i, r_i] \cap [l_k, r_k] \neq \emptyset$, hence $i \in U$ dominates k, a contradiction.

Lemma 4. *For every interval graph G^c, we have*

$$\gamma^t(G^c) = \min\{f(S, i) + |C - S| \mid S \subseteq C, i \in V, [1, i] \text{ dominates } G^c\}.$$

Proof. $f(S, i)$ is the size of some set $U \subseteq V$ that covers the colours S and that, by Lemma 3, dominates G^c. We obtain a tropical dominating set by adding a vertex of each missing colour in $C - S$. Therefore, each expression $f(S, i) + |C - S|$ on the right-hand side corresponds to the size of a tropical dominating set, so $\gamma^t(G^c)$ is at most the minimum of these.

For the opposite inequality, let U be a minimum tropical dominating set of G^c. Remove from U all vertices i for which there is some $j \in U$ with $I_i \subseteq I_j$, and call the resulting set U'. By construction U' still dominates G^c. Let S be the set of colours covered by U'. Then U' is a minimum set with these properties, so by the definition of f, $|U'| = f(S,i)$, where i is the greatest element in U'. Since $U' \subseteq [1,i]$, it follows that $[1,i]$ dominates G^c. Therefore, the right-hand side is at most $f(S,i) + |C - S| = |U'| + |C - S| \leq |U| = \gamma^t(G^c)$.

The following lemma gives a recursive definition of the function f that permits us to compute it efficiently when the number of colours in C grows at most logarithmically.

Lemma 5. *For every interval graph G^c, the function f satisfies the following recursion:*

$$f(S,0) = \begin{cases} 0 & \text{if } S = \emptyset, \\ \infty & \text{otherwise;} \end{cases}$$

$$f(S,i) = 1 + \min\{f(S',j) \mid S' \cup \{c(i)\} = S, j \in P_i\}, \qquad \text{for } i \in V,$$

where $j \in P_i$ if and only if either $j = 0$ and $\{i\}$ is i-prefix dominating, or $j \in V$, $j < i$, $[1,j] \cup \{i\}$ is i-prefix dominating, and $I_i \not\subseteq I_j$, $I_j \not\subseteq I_i$.

Proof. The proof is by induction on i. The base case $i = 0$ holds by definition. Assume that the lemma holds for all $0 \leq i \leq k - 1$ and all $S \subseteq C$.

Let U be a minimum proper k-prefix dominating set that covers precisely the colours in S. We want to show that $|U| = f(S,k)$. If $U = \{k\}$, then $S = \{c(k)\}$, and it follows immediately that $f(S,k) = 1$. Otherwise, $U - \{k\}$ is non-empty. Let $j < k$ be the greatest vertex in $U - \{k\}$. Assume that $U - \{k\}$ is not j-prefix dominating. Then, there is some $i < j$ that is not dominated by j but that is dominated by k, hence $l(k) \leq r(i) < l(j)$. Therefore $I_j \subseteq I_k$, so U is not proper, a contradiction. Hence, $U - \{k\}$ is a proper j-prefix dominating set. By induction, $|U - \{k\}| \geq \min\{f(S',j) \mid S' \cup \{c(k)\} = S\}$. This shows the inequality $|U| \geq f(S,k)$.

For the opposite inequality, it suffices to show that if $[1,j] \cup \{k\}$ is k-prefix dominating, U' is any proper j-prefix dominating set, and $I_k \not\subseteq I_j, I_j \not\subseteq I_k$, then $U' \cup \{k\}$ is a proper k-prefix dominating set. It follows from Lemma 3 that $U' \cup \{k\}$ is k-prefix dominating. Since $I_k \not\subseteq I_j$, we must have $r_i \leq r_j < r_k$ for all $i < j$, hence $I_k \not\subseteq I_i$. Assume that $I_i \subseteq I_k$ for some $i < j$. Then, since $I_j \not\subseteq I_k$, we have $l_j < l_k \leq l_i \leq r_i \leq r_j$, which contradicts U' being proper. It follows that $U' \cup \{k\}$ is proper.

Proof of Theorem 3. The sets P_i for $i \in V$ in Lemma 5 can be computed in time $\mathcal{O}(n^2)$ as follows. Let $a_i \in V$ be the least vertex such that i dominates $[a_i, i]$, and let $b_j \in V$ be the least vertex such that $[1,j]$ does not dominate b_j, or ∞ if $[1,j]$ dominates G. Note that i does not dominate any vertex strictly smaller than a_i since the vertices are ordered non-decreasingly with respect to the right endpoints of their intervals. Therefore, $P_i = \{j < i \mid a_i \leq b_j, I_i \not\subseteq I_j, I_j \not\subseteq I_i\}$. The vectors a_i and b_j are straightforward to compute in time $\mathcal{O}(n^2)$, hence P_i can be computed in time $\mathcal{O}(n^2)$ using this alternative definition.

When P_i is computed for all $i \in V$, the recursive definition of f in Lemma 5 can be used to compute all values of f in time $\mathcal{O}(2^c n^2)$, and it can easily be modified to compute, for each S and i, some specific i-prefix dominating set of size $f(S, i)$, also in time $\mathcal{O}(2^c n^2)$. Therefore, by Lemma 4, one can find a minimum tropical dominating set in time $\mathcal{O}(2^c n^2)$. □

References

1. Akbari, S., Liaghat, V., Nikzad, A.: Colorful paths in vertex-colorings of graphs. Electron. J. Comb. **18**, P17 (2011)
2. Alimonti, P., Kann, V.: Some APX-completeness results for cubic graphs. Theor. Comput. Sci. **237**(1–2), 123–134 (2000)
3. Ausiello, G., Crescenzi, P., Gambosi, G., Kann, V.: Complexity and Approximation. Springer, Heidelberg (1999)
4. Baker, B.: Approximation algorithms for NP-complete problems on planar graphs. J. ACM **41**(1), 153–180 (1994)
5. Brigham, R.C., Dutton, R.D.: Factor domination in graphs. Discrete Math. **86**(1–3), 127–136 (1990)
6. Caro, Y., Henning, M.A.: Simultaneous domination in graphs. Graphs Combin. **30**(6), 1399–1416 (2014)
7. Chlebík, M., Chlebíková, J.: Approximation hardness of dominating set problems in bounded degree graphs. Inf. Comput. **206**(11), 1264–1275 (2008)
8. Crescenzi, P.: A short guide to approximation preserving reductions. In: Proceedings of the IEEE Conference on Computational Complexity, pp. 262–273 (1997)
9. Dinur, I., Steurer, D.: Analytical approach to parallel repetition. In: Proceedings of the 46th Annual ACM Symposium on Theory of Computing (STOC 2014), pp. 624–633 (2014)
10. Duh, R., Fürer, M.: Approximation of k-set cover by semi-local optimization. In: Proceedings of the 29th Annual ACM Symposium on the Theory of Computing (STOC 1997), pp. 256–264 (1997)
11. Fellows, M., Fertin, G., Hermelin, D., Vialette, S.: Upper and lower bounds for finding connected motifs in vertex-colored graphs. J. Comput. Syst. Sci. **77**(4), 799–811 (2011)
12. Garey, M., Johnson, D.: Computers and Intractability: A Guide to the Theory of NP-Completeness. W. H. Freeman, New York (1979)
13. Haynes, T., Hedetniemi, S., Slater, P.: Fundamentals of Domination in Graphs. Marcel Dekker, New York (1998)
14. Kann, V.: On the Approximability of NP-complete Optimization Problems. Ph.D. thesis, Department of Numerical Analysis and Computing Science, Royal Institute of Technology, Stockholm (1992)
15. Li, A.: A generalization of the gallai-roy theorem. Graphs Comb. **17**, 681–685 (2001)
16. Lin, C.: Simple proofs of results on paths representing all colors in proper vertex-colorings. Graphs Comb. **23**, 201–203 (2007)
17. Sampathkumar, E.: The global domination number of a graph. J. Math. Phys. Sci. **23**(5), 377–385 (1989)

On Hamiltonian Colorings of Block Graphs

Devsi Bantva[(⊠)]

Lukhdhirji Engineering College, Morvi 363 642, Gujarat, India
devsi.bantva@gmail.com

Abstract. A hamiltonian coloring c of a graph G of order p is an assignment of colors to the vertices of G such that $D(u,v) + |c(u) - c(v)| \geq p - 1$ for every two distinct vertices u and v of G, where $D(u,v)$ denotes the detour distance between u and v. The value $hc(c)$ of a hamiltonian coloring c is the maximum color assigned to a vertex of G. The hamiltonian chromatic number, denoted by $hc(G)$, is the min$\{hc(c)\}$ taken over all hamiltonian coloring c of G. In this paper, we present a lower bound for the hamiltonian chromatic number of block graphs and give a sufficient condition to achieve the lower bound. We characterize symmetric block graphs achieving this lower bound. We present two algorithms for optimal hamiltonian coloring of symmetric block graphs.

Keywords: Hamiltonian coloring · Hamiltonian chromatic number · Block graph · Symmetric block graph

1 Introduction

A *hamiltonian coloring* c of a graph G of order p is an assignment of colors (nonnegative integers) to the vertices of G such that $D(u,v) + |c(u) - c(v)| \geq p - 1$ for every two distinct vertices u and v of G, where $D(u,v)$ denotes the detour distance which is the length of the longest path between u and v. The value of $hc(c)$ of a hamiltonian coloring c is the maximum color assigned to a vertex of G. The *hamiltonian chromatic number* $hc(G)$ of G is min$\{hc(c)\}$ taken over all hamiltonian coloring c of G. It is clear from definition that two vertices u and v can be assigned the same color only if G contains a hamiltonian $u - v$ path. Moreover, if G is a hamiltonian-connected graph then all the vertices can be assigned the same color. Thus the hamiltonian chromatic number of a connected graph G measures how close G is to being hamiltonian-connected, minimum the hamiltonian chromatic number of a connected graph G is, the closer G is to being hamiltonian-connected. The concept of hamiltonian coloring was introduced by Chartrand *et al.* [2] as a variation of *radio k-coloring* of graphs.

At present, the hamiltonian chromatic number is known only for handful of graph families. Chartrand *et al.* investigated the exact hamiltonian chromatic numbers for complete graph K_n, cycle C_n, star $K_{1,n}$ and complete bipartite graph $K_{r,s}$ in [2,3]. Also an upper bound for $hc(P_n)$ was established by Chartrand *et al.* in [2] but the exact value of $hc(P_n)$ which is equal to the radio antipodal number $ac(P_n)$ given by Khennoufa and Togni in [5]. In [6],

© Springer International Publishing Switzerland 2016
M. Kaykobad and R. Petreschi (Eds.): WALCOM 2016, LNCS 9627, pp. 28–39, 2016.
DOI: 10.1007/978-3-319-30139-6_3

Shen *et al.* have discussed the hamiltonian chromatic number for graphs G with $\max\{D(u,v) : u, v \in V(G), u \neq v\} \leq \frac{p}{2}$, where p is the order of graph G and they gave the hamiltonian chromatic number for a special class of caterpillars and double stars. The researchers emphasize that determining the hamiltonian chromatic number is interesting but a challenging task even for some basic graph families.

Without loss of generality, we initiate with label 0, then the *span* of any hamiltonian coloring c which is defined as $\max\{|c(u) - c(v)| : u, v \in V(G)\}$, is the maximum integer used for coloring. However, in [2,3,6] only positive integers are used as colors. Therefore, the hamiltonian chromatic number defined in this article is one less than that defined in [2,3,6] and hence we will make necessary adjustment when we present the results of [2,3,6] in this article. Moreover, for standard graph theoretic terminology and notation we follow [8].

In this paper, we present a lower bound for the hamiltonian chromatic number of block graphs and give a sufficient condition to achieve the lower bound. As an illustration, we present symmetric block graphs (those block graphs whose all blocks are cliques of size n, each cut vertex is exactly in k blocks and the eccentricity of end vertices is same) achieving this lower bound. We present two algorithms for optimal hamiltonian coloring of symmetric block graphs.

2   A Lower Bound for Hamiltonian Chromatic Number of Block Graphs

A *block graph* is a connected graph all of whose blocks are cliques. The detour distance between u and v, denoted by $D(u,v)$, is the longest distance between u and v in G. The *detour eccentricity* $\epsilon_D(v)$ of a vertex v is the detour distance from v to a vertex farthest from v. The *detour center* $C_D(G)$ of G is the subgraph of G induced by the vertex/vertices of G whose detour eccentricity is minimum. In [4], Chartrand *et al.* shown that the detour center $C_D(G)$ of every connected graph G lies in a single block of G. The vertex/vertices of detour center $C_D(G)$ are called *detour central vertex/vertices* for graph G. In a block graph G, if u is on the $w - v$ path, where w is the nearest detour central vertex for v, then u is an *ancestor* of v, and v is a *descendent* of u. Let u_i, $i = 1, 2, ..., n$ are adjacent vertices of a block attached to a central vertex. Then the subgraph induced by u_i, $i = 1, 2, ..., n$ and all its descendent is called a *branch* at w. Two branches are called *different* if they are induced by vertices of two different blocks attached to the same central vertex, and *opposite* if they are induced by vertices of two different blocks attached to different central vertices. For a block graph G, define *detour level function* $\mathcal{L}$ on $V(G)$ by

$$\mathcal{L}(u) := \min\{D(w, u) : w \in V(C_D(G))\}, \text{ for any } u \in V(G).$$

The *total detour level of a graph* G, denoted by $\mathcal{L}(G)$, is defined as

$$\mathcal{L}(G) := \sum_{u \in V(G)} \mathcal{L}(u). \tag{1}$$

Note that if $|C_D(G)| = \omega$ then the detour distance between any two vertices u and v in a block graph G satisfies

$$D(u,v) \leq \mathcal{L}(u) + \mathcal{L}(v) + \omega - 1. \qquad (2)$$

Moreover, equality holds in (2) if u and v are in different branches when $\omega = 1$ and in opposite branches when $\omega \geq 2$.

Define $\xi = \min\{|V(B_i)| - 1 : B_i$ is a block attached to detour central vertex$\}$ when $\omega = 1$; otherwise $\xi = 0$.

We first give a lower bound for the hamiltonian chromatic number of block graphs. A hamiltonian coloring c on $V(G)$, induces an ordering of $V(G)$, which is a line up of the vertices with equal or increasing images. We denote this ordering by $V(G) = \{u_0, u_1, u_2, ..., u_{p-1}\}$ with

$0 = c(u_0) \leq c(u_1) \leq c(u_2) \leq ... \leq c(u_{p-1})$.

Notice that, c is a hamiltonian coloring, then the span of c is $c(u_{p-1})$.

Theorem 1. *Let G be a block graph of order p and ω, ξ and $\mathcal{L}(G)$ are defined as earlier then*

$$hc(G) \geq (p-1)(p-\omega) - 2\mathcal{L}(G) + \xi. \qquad (3)$$

Proof. It suffices to prove that any hamiltonian coloring c of block graph G has no span less than the right hand side of (3). Let c be an arbitrary hamiltonian coloring for G, where $0 = c(u_0) \leq c(u_1) \leq c(u_2) \leq ... \leq c(u_{p-1})$. Then $c(u_{i+1}) - c(u_i) \geq p - 1 - D(u_i, u_{i+1})$, for all $0 \leq i \leq p - 2$. Summing up these $p - 1$ inequalities, we get

$$c(u_{p-1}) - c(u_0) \geq (p-1)^2 - \sum_{i=0}^{p-1} D(u_i, u_{i+1}) \qquad (4)$$

We consider following two cases.

Case-1: $\omega = 1$. In this case, note that $\mathcal{L}(u_0) + \mathcal{L}(u_{p-1}) \geq \xi$ and by substituting (2) into (4) we obtain,

$$c(u_{p-1}) - c(u_0) \geq (p-1)^2 - \sum_{i=0}^{p-1} D(u_i, u_{i+1})$$

$$\geq (p-1)^2 - \sum_{i=0}^{p-1} (\mathcal{L}(u_i) + \mathcal{L}(u_{i+1}) + \omega - 1)$$

$$= (p-1)^2 - 2 \sum_{u \in V(G)} \mathcal{L}(u) + \mathcal{L}(u_0) + \mathcal{L}(u_{p-1}) - (p-1)(\omega - 1)$$

$$\geq (p-1)(p-\omega) - 2\mathcal{L}(G) + \xi$$

Case-2: $\omega \geq 2$. In this case, note that $\mathcal{L}(u_0) + \mathcal{L}(u_{p-1}) \geq 0$ and by substituting (2) into (4) we obtain,

$$
\begin{aligned}
c(u_{p-1}) - c(u_0) &\geq (p-1)^2 - \sum_{i=0}^{p-1} D(u_i, u_{i+1}) \\
&\geq (p-1)^2 - \sum_{i=0}^{p-1} (\mathcal{L}(u_i) + \mathcal{L}(u_{i+1}) + \omega - 1) \\
&= (p-1)^2 - 2 \sum_{u \in V(G)} \mathcal{L}(u) + \mathcal{L}(u_0) + \mathcal{L}(u_{p-1}) - (p-1)(\omega - 1) \\
&\geq (p-1)(p-\omega) - 2\mathcal{L}(G) \\
&= (p-1)(p-\omega) - 2\mathcal{L}(G) + \xi
\end{aligned}
$$

Thus, any hamiltonian coloring has span not less than the right hand side of (3) and hence we obtain $hc(G) \geq (p-1)(p-\omega) - 2\mathcal{L}(G) + \xi$.

The next result gives sufficient condition with optimal hamiltonian coloring for the equality in (3).

Theorem 2. *Let G be a block graph of order p and, ω, ξ and $\mathcal{L}(G)$ are defined as earlier then*

$$
hc(G) = (p-1)(p-\omega) - 2\mathcal{L}(G) + \xi, \tag{5}
$$

if there exists an ordering $\{u_0, u_1,...,u_{p-1}\}$ with $0 = c(u_0) \leq c(u_1) \leq ... \leq c(u_{p-1})$ of vertices of block graph G such that

1. $\mathcal{L}(u_0) = 0$, $\mathcal{L}(u_{p-1}) = \xi$ *when $\omega = 1$ and $\mathcal{L}(u_0) = \mathcal{L}(u_{p-1}) = 0$ when $\omega \geq 2$,*
2. u_i *and u_{i+1} are in different branches when $\omega = 1$ and opposite branches when $\omega \geq 2$,*
3. $D(u_i, u_{i+1}) \leq \frac{p}{2}$, *for $0 \leq i \leq p - 2$.*

Moreover, under these conditions the mapping c defined by

$$
c(u_0) = 0 \tag{6}
$$

$$
c(u_{i+1}) = c(u_i) + p - 1 - \mathcal{L}(u_i) - \mathcal{L}(u_{i+1}) - \omega + 1, 0 \leq i \leq p - 2 \tag{7}
$$

is an optimal hamiltonian coloring of G.

Proof. Suppose (1), (2) and (3) hold for an ordering $\{u_0, u_1,...,u_{p-1}\}$ of the vertices of G and c is defined by (6) and (7). By Theorem 1, it is enough to prove that c is a hamiltonian coloring whose span is $c(u_{p-1}) = (p-1)(p-\omega) - 2\mathcal{L}(G) + \xi$.

Without loss of generality assume that $j - i \geq 2$ then

$$
\begin{aligned}
c(u_j) - c(u_i) &= \sum_{t=i}^{j-1} [c(u_{t+1}) - c(u_t)] \\
&\geq \sum_{t=i}^{j-1} [p - 1 - \mathcal{L}(u_t) - \mathcal{L}(u_{t+1}) - w + 1] \\
&\geq \sum_{t=i}^{j-1} [p - 1 - D(u_t, u_{t+1})] \\
&= (j - i)(p - 1) - \sum_{t=i}^{j-1} D(u_t, u_{t+1}) \\
&\geq (j - i)(p - 1) - (j - i)\left(\tfrac{p}{2}\right) \\
&= (j - i)\left(\tfrac{p-1}{2}\right) \\
&= p - 2
\end{aligned}
$$

Note that $D(u_i, u_{i+1}) \geq 1$; it follows that $|c(u_j) - c(u_i)| + D(u_i, u_{i+1}) \geq p - 1$. Hence, c is a hamiltonian coloring for G. The span of c is given by

$$
\begin{aligned}
\mathrm{span}(c) &= \sum_{t=0}^{p-2} [c(u_{t+1}) - c(u_t)] \\
&= \sum_{t=0}^{p-2} [p - 1 - \mathcal{L}(u_t) - \mathcal{L}(u_{t+1}) - w + 1] \\
&= (p - 1)^2 - \sum_{t=0}^{p-2} [\mathcal{L}(u_t) + \mathcal{L}(u_{t+1})] - (p - 1)(w - 1) \\
&= (p - 1)(p - w) - 2 \sum_{u \in V(G)} \mathcal{L}(u) + \mathcal{L}(u_0) + \mathcal{L}(u_{p-1}) \\
&= (p - 1)(p - w) - 2\mathcal{L}(G) + \xi
\end{aligned}
$$

Therefore, $hc(G) \leq (p - 1)(p - w) - 2\mathcal{L}(G) + \xi$. This together with (3) implies (5) and that c is an optimal hamiltonian coloring.

3   Hamiltonian Chromatic Number of Symmetric Block Graphs

In this section, we continue to use the terminology and notation defined in previous section. We use Theorems 1 and 2 to determine the hamiltonian chromatic number of symmetric block graphs.

A *symmetric block graph*, denoted by $B_{n,k}$(or $B_{n,k}(d)$ if diameter is d), is a block graph with at least two blocks such that all blocks are cliques of size n, each cut vertex is exactly in k blocks and the eccentricity of end vertices is same (see Fig. 1). It is straight forward to verify that the detour center of symmetric block graph of diameter d is a vertex when d is even and a block of size n when d is odd. Consequently, the number of detour central vertex/vertices for a symmetric block graph $B_{n,k}$ of diameter d is either 1 or n depending upon d is even or odd. We observe that $B_{2,k}(2)$ are stars $K_{1,k}$, $B_{n,k}(2)$ are one point union of k complete graphs (a one point union of k complete graphs, also denoted by K_n^k, is a graph obtained by taking v as a common vertex such that any two copies of K_n are edge disjoint and do not have any vertex common except v), $B_{2,2}(d)$ are paths P_{d+1} and $B_{2,k}(d)$ are symmetric trees (see [7]). The hamiltonian chromatic number of stars $K_{1,k}$ is reported by Chartrand $et\ al.$ in [2]. The hamiltonian chromatic number of paths which is equal to the antipodal radio number of paths given by Khennoufa and Togni in [5] and the hamiltonian chromatic number of symmetric trees is investigated by Bantva in [1]. Hence we consider $k \geq 2$ and $d, n \geq 3$. However, for completeness we first give the hamiltonian chromatic number for $B_{n,k}(2)$ in Theorem 6 and next we consider general case.

Theorem 3. $[2]$ For $n \geq 3$, $hc(K_{1,n}) = (n-1)^2$.

Theorem 4. $[5]$ For any $n \geq 5$,

$$hc(P_n) = ac(P_n) = \begin{cases} 2p^2 - 2p + 2, & if\ n = 2p + 1, \\ 2p^2 - 4p + 4, & if\ n = 2p. \end{cases}$$

Theorem 5. $[1]$ Let T be a symmetric tree of order $p \geq 4$ and $\Delta(T) \geq 3$. Then

$$hc(T) = (p-1)(p-1-\epsilon(T)) + \epsilon'(T) - 2\mathcal{L}(T),$$

where $\epsilon(T) = 0$ when $C(T) = \{w\}$ and $\epsilon(T) = 1$ when $C(T) = \{w, w'\}$; and $\epsilon'(T) = 1 - \epsilon(T)$

The next result gives the hamiltonian chromatic number for one point union of k copies of complete graph K_n.

Theorem 6. For $n, k \geq 2$,

$$hc(K_n^k) = \begin{cases} (n-1)^2, & if\ k = 2, \\ k(k-2)(n-1)^2 + n - 1, & if\ k \geq 3. \end{cases}$$

Proof. Let K_n^k be one point union of k complete graph. To prove the result we consider following two cases.

Case - 1: $k = 2$. Let $G = K_n^2$ with vertex set $\{x_1, x_2, ..., x_{n-1}, y_1, y_2, ..., y_{n-1}, z\}$, where x_i and y_i, $1 \leq i \leq n-1$ be the vertices of block on each side and z is the common vertex of two blocks in G. Let c be a minimum hamiltonian coloring of G with $0 \in c(V(G))$. Since G contains hamiltonian path between x_i and y_i for $1 \leq i \leq n-1$, we can color x_i and y_i with same color. Since $D(z, x_i) = D(z, y_i) = D(x_i, x_j) = D(y_i, y_j) = n-1$ and $D(x_i, y_j) = 2n-2 = p-1$, for $1 \leq i, j \leq n-1$ and $i \neq j$. It follows that $|c(z) - c(x_j)| \geq n-1$ and $|c(x_i) - c(x_j)| \geq n-1$. This implies that $hc(G) = hc(c) \geq 0 + (n-1)(n-1) = (n-1)^2$.

Next we show that $hc(G) \leq (n-1)^2$. To prove this, it is enough to give hamiltonian coloring with span equal to $(n-1)^2$. Define a coloring c of G by

$$c(z) = 0$$
$$c(x_i) = c(y_i) = i(n-1), \ 1 \le i \le n-1$$

Since c is a hamiltonian coloring, $hc(G) \le hc(c) = c(x_{n-1}) = c(y_{n-1}) = (n-1)(n-1) = (n-1)^2$ and hence $hc(G) = hc(K_n^{(2)}) = (n-1)^2$.

Case - 2: $k \ge 3$. Let $G = K_n^k$ with vertex set $\{v_i^j, w : 1 \le i \le n-1, 1 \le j \le k\}$ such that for each $j = 1, 2, ..., k$, v_i^j where $1 \le i \le n-1$ are in same block and w is the common vertex of G. Define a coloring c of G by

$$c(w) = 0$$
$$c(v_1^1) = (k-1)(n-1)$$
$$c(v_{i+1}^1) = c(v_i^1) + k(n-1), \ 2 \le i \le n-1$$
For $j = 1, 2, ..., k-1$
$$c(v_i^{j+1}) = c(v_i^j) + (k-2)(n-1), \ 1 \le i \le n-1.$$

Since c is a hamiltonian coloring, $hc(G) \le hc(c) = (k-1)(n-1)+(k-1)(k-2)(n-1) + k(k-2)(n-1)(n-2) = k(k-2)(n-1)^2 + n - 1$.

Now we show that $hc(G) \ge k(k-2)(n-1)^2 + n - 1$. Let c be a minimum hamiltonian coloring of G. Since G contains no hamiltonian path no two vertices can be colored the same. A hamiltonian coloring induces an ordering on $V(G)$ with increasing images. We may assume that $0 = c(u_0) < c(u_1) < ... < c(u_{p-1})$. We consider three subcases.

Subcase - 1: $c(w) = c(u_0) = 0$. Since $D(u_0, u_1) = D(w, u_1) = n - 1$, $c(u_1) \ge (k-1)(n-1)$. Also $D(v_x^j, v_y^j) = n - 1$ for $1 \le j \le k$, $x \ne y$ and $D(v_x^t, v_y^l)$ for $1 \le x, y \le n-1$, $1 \le t, l(t \ne l) \le k$. It follows that $c(u_3) \ge (k-2)(n-1)$ and $c(u_{i+1}) \ge c(u_i) + (k-2)(n-1)$ for all $3 \le i \le k(n-1) - 1$. This implies that $c(u_{p-1}) \ge (k-1)(n-1)+(k(n-1)-1)(k-2)(n-1) = k(k-2)(n-1)^2 + n - 1$. Therefore $hc(c) \ge k(k-2)(n-1)^2 + n - 1$.

Subcase - 2: $c(w) = c(u_{p-1}) = hc(c)$. Since $D(v_x^j, v_y^j) = n - 1$ for $1 \le j \le k$, $x \ne y$ and $D(v_x^t, v_y^l)$ for $1 \le x, y \le n-1$, $1 \le t, l(t \ne l) \le k$. For each i with $1 \le i \le k(n-1) - 1$, $c(v_1) = 0$ and $c(v_{i+1}) = (k-2)(n-1)$ and $c(w) = c(u_{p-1}) = c(u_{k(n-1)-1}) + (k-1)(n-1)$. This implies that $c(u_{p-1}) = c(w) \ge (k(n-1)-1)(k-2)(n-1) + (k-1)(n-1) = k(k-2)(n-1)^2 + n - 1$. Therefore $hc(c) \ge k(k-2)(n-1)^2 + n - 1$.

Subcase - 3: $c(u_i) \le c(w) \le c(u_{i+1})$ for some i with $1 \le i \le k(n-1)-1$. Since $D(v_x^j, v_y^j) = n-1$ for $1 \le j \le k$, $x \ne y$ and $D(v_x^t, v_y^l) = n-2$ for $1 \le x, y \le n-1$, $1 \le t, l(t \ne l) \le k$. Define $c(u_0) = 0$ and $c(u_{i+1}) = c(u_i) + (k-2)(n-1)$ for $1 \le i \le m$ and $m \le p-3$. Then $c(u_{m+1}) = c(w) = c(u_m) + (k-1)(n-1)$ and $c(u_{m+2}) = c(u_{m+1}) + (k-1)(n-1)$, $c(u_{i+1}) = c(u_i) + (k-2)(n-1)$ for $m+2 \le i \le p-1$. Therefore $hc(c) \ge k(k-2)(n-1)^2 + 2(n-1)$.

Hence from Subcase - 1, 2 and 3, $hc(G) = k(k-2)(n-1)^2 + n - 1$.

Thus, from Case - 1 and 2, we have

$$hc(K_n^k) = \begin{cases} (n-1)^2, & \text{if } k = 2, \\ k(k-2)(n-1)^2 + n - 1, & \text{if } k \ge 3. \end{cases}$$

We now determine the hamiltonian chromatics number for $B_{n,k}(d)$ for $k \geq 2$, $n, d \geq 3$ using Theorem 2. Note that $B_{3,2}(3)$, $B_{3,3}(3)$ and $B_{3,2}(4)$ block graphs does not satisfies condition (c) of Theorem 2 but it is easy to verify that the hamiltonian chromatic numbers for these three graphs are coincide with the numbers produce by the formula stated in Theorem 7. Moreover, labels assigned by Algorithms given in proof of Theorem 7 is the optimal hamiltonian coloring for these graphs.

Theorem 7. *Let $k \geq 1$, $n \geq 2$, $d \geq 3$ be integers, $r = \lfloor \frac{d}{2} \rfloor$ and $\Phi_r(x) = 1 + x + x^2 + ... + x^{r-1}$. Then $hc\,(B_{n+1,k+1}(d))$*

$$
= \begin{cases} n^2(k+1)\left[\Phi_r(kn)\left((k+1)\Phi_r(kn) - 2r\right) + \frac{2(\Phi_r(kn)-r)}{kn-1}\right] + n, & \text{if } d \text{ is even,} \\ kn^2(n+1)\left[\Phi_r(kn)\left(k(n+1)\Phi_r(kn) - 2r + 1\right) + \frac{2(\Phi_r(kn)-r)}{kn-1}\right], & \text{if } d \text{ is odd.} \end{cases} \tag{8}
$$

Proof. The order p and total detour level of $B_{n+1,k+1}(d)$ is given by

$$
p = \begin{cases} 1 + \displaystyle\sum_{i=1}^{r}(k+1)k^{i-1}n^i, & \text{if } d \text{ is even,} \\ 1 + n + \displaystyle\sum_{i=1}^{r}k^i n^{i+1}, & \text{if } d \text{ is odd.} \end{cases} \tag{9}
$$

$$
\mathcal{L}(G) = \begin{cases} n^2(k+1)\left(r\Phi_r(kn) + \frac{r-\Phi_r(kn)}{kn-1}\right), & \text{if } d \text{ is even,} \\ kn^2(n+1)\left(r\Phi_r(kn) + \frac{r-\Phi_r(kn)}{kn-1}\right), & \text{if } d \text{ is odd.} \end{cases} \tag{10}
$$

Substituting (9) and (10) into (3) gives the right hand side of (8).

We now prove that the right hand side of (8) is the actual value for the hamiltonian chromatics number of symmetric block graph. For this purpose we give a systematic hamiltonian coloring whose span is the right hand side of (8). We consider following two cases.

Case - 1: d **is even.** In this case, symmetric block graphs has only one detour central vertex say w. We apply the following algorithm to find a hamiltonian coloring of symmetric block graph of even diameter whose span is right-hand side of (8).

Algorithm 1: An optimal hamiltonian coloring of symmetric block graphs $B_{n+1,k+1}(d)$, where d is even.

Input: A symmetric block graph $B_{n+1,k+1}$ of even diameter.

Idea: Find an ordering of vertices of block graphs $B_{n+1,k+1}$ of even diameter which satisfies Theorem 2 and labeling defined by (6)–(7) is a hamiltonian coloring whose span is right-hand side of (8).

Initialization: Start with a central vertex w.

Iteration: Define $c : V(B_{n+1,k+1}) \to \{0, 1, 2,..\}$ as follows:

Step-1: Let v^1, v^2, ..., $v^{(k+1)n}$ be the vertices adjacent to w such that any $k+1$ consecutive vertices in the list are in different blocks.

Step-2: Now kn descendent vertices of each v^t, $t = 1, 2, ..., (k+1)n$ by v_0^t, $v_1^t,...,v_{kn-1}^t$ such that any k consecutive vertices in the list are in different blocks. Next the kn descendent vertices of each v_l^t, $0 \le l \le kn - 1, 1 \le t \le (k+1)n$ by v_{l0}^t, $v_{l1}^t,...,v_{l(kn-1)}^t$ such that any k consecutive vertices lies in different blocks; inductively kn descendent vertices of $v_{i_1,i_2,...,i_l}^t$ $(0 \le i_1, i_2, ..., i_l \le kn - 1, 1 \le t \le (k+1)n)$ are indexed by $v_{i_1,i_2,...,i_l,i_{l+1}}^t$ where $i_{l+1} = 0, 1, ..., kn - 1$ such that any k consecutive vertices in the list are in different blocks.

Step-3: Rename $v_{i_1,i_2,...,i_l,i_{l+1}}^t$ by v_j^t, $(1 \le t \le (k+1)n)$, where

$$j = 1 + i_1 + i_2(kn) + ... + i_l(kn)^{l-1} + \sum_{l+1 \le t \le \frac{d}{2}} (kn)^t.$$

Step-4: Give ordering $\{u_0, u_1,...,u_{p-1}\}$ of vertices of symmetric block graphs as follows.

For $1 \le j \le p - (k+1)n - 1$, let

$$u_j := \begin{cases} v_s^t, \text{ where } s = \lceil \frac{j}{(k+1)n} \rceil, \text{ if } j \equiv t(\text{mod } (k+1)n), 1 \le t \le (k+1)n - 1, \\ v_s^{(k+1)n}, \text{ where } s = \lceil \frac{j}{(k+1)n} \rceil, \text{ if } j \equiv 0(\text{mod } (k+1)n) \end{cases}$$

For $p - (k+1)n \le j \le p - 1$, let

$$u_j := v^{j-p+(k+1)n+1}.$$

Then above defined ordering $\{u_0, u_1,...,u_{p-1}\}$ of vertices satisfies Theorem 2.

Step-5: Define $c : V(B_{n+1,k+1}) \to \{0, 1, 2,...\}$ by $c(u_0) = 0$ and $c(u_{i+1}) = c(u_i) + p - 1 - \mathcal{L}(u_i) - \mathcal{L}(u_{i+1}) - w + 1, 0 \le i \le p - 2$.

Output: The span of c is $\text{span}(c) = c(u_{p-1}) = c(u_0) + (p-1)^2 - 2 \left(\sum_{u \in V(G)} \mathcal{L}(u) \right) + n - 1 = (p-1)^2 - 2\mathcal{L}(B_{n+1,k+1}) + n - 1$ which is exactly the right-hand side of (8) by using (9) and (10) in the case of symmetric block graphs.

Case - 2: d is odd. In this case, symmetric block graphs has $n+1$ central vertices say v^1, v^2, ..., v^{n+1}. We apply the following algorithm to find a hamiltonian coloring of symmetric block graph of odd diameter whose span is right-hand side (8).

Algorithm 2: An optimal hamiltonian coloring of symmetric block graphs $B_{n+1,k+1}(d)$, where d is odd.

Input: A symmetric block graph $B_{n+1,k+1}(d)$ of odd diameter d.

Idea: Find an ordering of vertices of block graphs $B_{n+1,k+1}$ of odd diameter which satisfies Theorem 2 and labeling defined by (6)–(7) is a hamiltonian coloring whose span is right-hand side of (8).

Initialization: Starts with central vertices $v^1, v^2, ..., v^{n+1}$.

Iteration: Define $c : V(B_{n+1,k+1}) \to \{0, 1, 2,...\}$ as follows:

Step-1: Now kn descendent vertices of each v^t, $t = 1, 2, ..., (k+1)n$ by v_0^t, $v_1^t,...,v_{kn-1}^t$ such that any k consecutive vertices in the list are in different blocks. Next kn descendent vertices of each v_l^t, $0 \le l \le kn-1, 1 \le t \le n+1$ by $v_{l0}^t, v_{l1}^t,...,v_{l(kn-1)}^t$ such that any k consecutive vertices are in different blocks; inductively kn descendent vertices of $v_{i_1,i_2,...,i_l}^t$ $(0 \le i_1, i_2, ..., i_l \le kn-1, 1 \le t \le n+1)$ are indexed by $v_{i_1,i_2,...,i_l,i_{l+1}}^t$ where $i_{l+1} = 0, 1, ..., kn-1$ such that any k consecutive vertices in the list are in different blocks.

Step-2: We rename $v_{i_1,i_2,...,i_l,i_{l+1}}^t$ by v_j^t, $(1 \le t \le n+1)$, where

$$j = 1 + i_1 + i_2(kn) + ... + i_l(kn)^{l-1} + \sum_{l+1 \le t \le \frac{d-1}{2}} (kn)^t.$$

Step-3: Define an ordering $\{u_0, u_1,...,u_{p-1}\}$ as follows:
For $1 \le j \le p-n-1$, let

$$u_j := \begin{cases} v_s^t, \text{ where } s = \lceil j/(n+1) \rceil, \text{ if } j \equiv t(\bmod\ (n+1)) \text{ with } 1 \le t \le n, \\ v_s^{n+1}, \text{ where } s = \lceil j/(n+1) \rceil, \text{ if } j \equiv 0(\bmod\ (n+1)) \end{cases}$$

For $p - n \le j \le p - 1$,

$$u_j := v^{j-p+n+1}.$$

Then above defined ordering $\{u_0, u_1,...,u_{p-1}\}$ of vertices satisfies Theorem 2.
Step-4: Define $c : V(B_{n+1,k+1}) \to \{0, 1, 2,...\}$ by $c(u_0) = 0$ and $c(u_{i+1}) = c(u_i) + p - 1 - L(u_i) - L(u_{i+1}) - w + 1, 0 \le i \le p - 1$.

Output: The span of c is $c(u_{p-1}) = hc(G) = c(u_0) + (p-1)^2 - 2\left(\sum_{u \in V(G)} L(u)\right) -$
$(p-1)(n-1) = (p-1)^2 - 2\,\mathcal{L}(G) - (p-1)(n-1)$ which is exactly the right-hand side of (8) by using (9) and (10) in the case of symmetric block graphs. Thus, from Case - 1 and Case - 2, we obtain $hc(B_{n+1,k+1}(d))$

$$= \begin{cases} n^2(k+1)\left[\Phi_r(kn)\left((k+1)\Phi_r(kn) - 2r\right) + \frac{2(\Phi_r(kn)-r)}{kn-1}\right] + n, & \text{if } d \text{ is even,} \\ kn^2(n+1)\left[\Phi_r(kn)\left(k(n+1)\Phi_r(kn) - 2r + 1\right) + \frac{2(\Phi_r(kn)-r)}{kn-1}\right], & \text{if } d \text{ is odd.} \end{cases}$$

Example 1. An optimal hamiltonian coloring of $B_{4,2}(4)$ using the procedure of Theorem 7 is shown in Fig. 1(a).

For $B_{4,2}(4)$, $k = 1$, $n = 3$, $d = 4$, $r = \lfloor \frac{d}{2} \rfloor = 2$ and $\Phi_{\lfloor \frac{d}{2} \rfloor}(kn) = \Phi_2(3) = 1 + 3 = 4$. By Theorem 7, $hc(B_{4,2}(4))$

$$= n^2(k+1) \left[\Phi_{\lfloor \frac{d}{2} \rfloor}(kn) \left((k+1)\Phi_{\lfloor \frac{d}{2} \rfloor}(kn) - 2\lfloor \tfrac{d}{2} \rfloor \right) + \frac{2\left(\Phi_{\lfloor \frac{d}{2} \rfloor}(kn) - \lfloor \frac{d}{2} \rfloor \right)}{kn-1} \right] + n$$

$$= 3^2 \cdot (1+1) \left[4((1+1) \cdot 4 - 2 \cdot 2) + \frac{2(4-2)}{3-1} \right] + 3 = 327.$$

Example 2. An optimal hamiltonian coloring of $B_{4,2}(5)$ using the procedure of Theorem 7 is shown Fig. 1(b).

For $B_{4,2}(5)$, $k = 1$, $n = 3$, $d = 5$, $r = \lfloor \frac{d}{2} \rfloor = 2$ and $\Phi_{\lfloor \frac{d}{2} \rfloor}(kn) = \Phi_2(3) = 1 + 3 = 4$. By Theorem 7, $hc(B_{4,2}(4))$

$$= kn^2(n+1) \left[\Phi_{\lfloor \frac{d}{2} \rfloor}(kn) \left(k(n+1)\Phi_{\lfloor \frac{d}{2} \rfloor}(kn) - 2\lfloor \tfrac{d}{2} \rfloor + 1 \right) + \frac{2\left(\Phi_{\lfloor \frac{d}{2} \rfloor}(kn) - \lfloor \frac{d}{2} \rfloor \right)}{kn-1} \right]$$

$$= 1 \cdot 3^2 \cdot (3+1) \left[4(1 \cdot (3+1) \cdot 4 - 2 \cdot 2 + 1) + \frac{2(4-2)}{3-1} \right] = 1944.$$

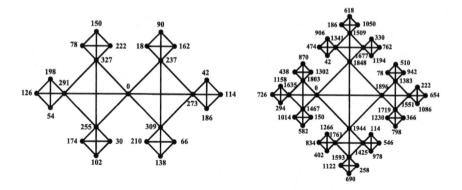

Fig. 1. Optimal hamiltonian coloring of $B_{4,2}(4)$ and $B_{4,2}(5)$.

References

1. Bantva, D.: On hamiltonian colorings of trees, communicated
2. Chartrand, G., Nebeský, L., Zhang, P.: Hamiltonian coloring of graphs. Discrete Appl. Math. **146**, 257–272 (2005)
3. Chartrand, G., Nebeský, L., Zhang, P.: On hamiltonian colorings of graphs. Discrete Math. **290**, 133–143 (2005)
4. Chartrand, G., Escuadro, H., Zhang, P.: Detour distance in graphs. J. Combin. Math. Combin. Comput. **53**, 75–94 (2005)

5. Khennoufa, R., Togni, O.: A note on radio antipodal colourings of paths. Math. Bohemica **130**(3), 277–282 (2005)
6. Shen, Y., He, W., Li, X., He, D., Yang, X.: On hamiltonian colorings for some graphs. Discrete Appl. Math. **156**, 3028–3034 (2008)
7. Vaidya, S.K., Bantva, D.D.: Symmetric regular cacti - properties and enumeration. Proyecciones J. Math. **31**(3), 261–275 (2012)
8. West, D.B.: Introduction to Graph theory. Prentice-Hall of India, New Delhi (2001)

Vertex-Coloring with Star-Defects

Patrizio Angelini[1], Michael A. Bekos[1], Michael Kaufmann[1($\boxtimes$)],
and Vincenzo Roselli[2]

[1] Institut Für Informatik, Universität Tübingen, Tübingen, Germany
{angelini,bekos,mk}@informatik.uni-tuebingen.de
[2] Dipartimento di Ingegneria, Università Roma Tre, Rome, Italy
roselli@dia.uniroma3.it

Abstract. *Defective coloring* is a variant of traditional vertex-coloring, according to which adjacent vertices are allowed to have the same color, as long as the monochromatic components induced by the corresponding edges have a certain structure. Due to its important applications, as for example in the bipartisation of graphs, this type of coloring has been extensively studied, mainly with respect to the size, degree, and acyclicity of the monochromatic components.

In this paper we focus on defective colorings in which the monochromatic components are acyclic and have small diameter, namely, they form stars. For outerplanar graphs, we give a linear-time algorithm to decide if such a defective coloring exists with two colors and, in the positive case, to construct one. Also, we prove that an outerpath (i.e., an outerplanar graph whose weak-dual is a path) always admits such a two-coloring. Finally, we present NP-completeness results for non-planar and planar graphs of bounded degree for the cases of two and three colors.

1 Introduction

Graph coloring is a fundamental problem in graph theory, which has been extensively studied over the years (see, e.g., [4] for an overview). Most of the research in this area has been devoted to the *vertex-coloring problem* (or *coloring problem*, for short), which dates back to 1852 [18]. In its general form, the problem asks to label the vertices of a graph with a given number of colors, so that no two adjacent vertices share the same color. In other words, a coloring of a graph partitions its vertices into independent sets, usually referred to as *color classes*, as all their vertices have the same color. A central result in this area is the so-called *four color theorem*, according to which every planar graph admits a coloring with at most four colors; see e.g. [11]. Note that the problem of deciding whether a planar graph is 3-colorable is NP-complete [10], even for graphs of maximum degree 4 [7].

Several variants of the coloring problem have been proposed. One of the most studied is the so-called *defective coloring*, which was independently introduced

This work has been supported by DFG grant Ka812/17-1a and by the MIUR project AMANDA, prot. 2012C4E3KT_001.

M. Kaykobad and R. Petreschi (Eds.): WALCOM 2016, LNCS 9627, pp. 40–51, 2016.
DOI: 10.1007/978-3-319-30139-6_4

by Andrews and Jacobson [1], Harary and Jones [13], and Cowen et al. [6]. In the defective coloring problem edges between vertices of the same color class are allowed, as long as the monochromatic components induced by vertices of the same color maintain some special structure. In this respect, one can regard the classical vertex-coloring as a defective one in which every monochromatic component is an isolated vertex. In this work we focus on defective colorings in which each component is acyclic and has small diameter. In particular, we say that a graph G is tree-diameter-λ κ-colorable if the vertices of G can be colored with κ colors, so that all monochromatic components are acyclic and of diameter at most λ, where $\kappa \geq 1$ and $\lambda \geq 0$. Clearly, a classical κ-coloring corresponds to a tree-diameter-0 κ-coloring.

We present algorithmic and complexity results for tree-diameter-λ κ-colorings for small values of κ and $\lambda = 2$. For simplicity, we refer to this problem as *(star, κ)-coloring*, as each monochromatic component is a *star* (i.e., a tree with diameter two; see Fig. 1d). Similarly, we refer to the tree-diameter-λ κ-coloring problem when $\lambda = 1$ as *(edge, κ)-coloring* problem. By definition, a (edge, κ)-coloring is also a (star, κ)-coloring. Fig. 1a–c show a trade-off between number of colors and structure of the monochromatic components.

Our work can be seen as a variant of the *bipartisation* of graphs, namely the problem of making a graph bipartite by removing a small number of elements (e.g., vertices or edges), which is a central graph theory problem with many applications [12,14]. The bipartisation by removal of a (not-necessarily minimal) number of *non-adjacent* edges corresponds to the (edge, 2)-coloring problem. In the (star, 2)-coloring problem, we also solve some kind of bipartisation by removing independent stars. Note that we do not ask for the minimum number of removed stars but for the existence of a solution.

(a)   (b)   (c)   (d)

Fig. 1. (a-c) Different colorings of the same graph: (a) a traditional 4-coloring, (b) an (edge, 3)-coloring (c) a (star, 2)-coloring; (d) a star with three leaves; its *center* has degree 3.

To the best of our knowledge, this is the first time that the defective coloring problem is studied under the requirement of having color classes of small diameter. Previous research was focused either on their size or their degree [1,6,13,16,17]. As byproducts of these previous works, one can obtain results for the (edge, κ)-coloring problem. More precisely, from a result of Lovász [16], it follows that all graphs of maximum degree 4 or 5 are (edge, 3)-colorable. However, determining whether a graph of maximum degree 7 is (edge, 3)-colorable is NP-complete [6]. In the same work, Cowen et al. [6] prove that there exist planar graphs that are not (edge, 3)-colorable and that the corresponding decision problem is NP-complete, even in the case of planar graphs of maximum degree 10. Results for graphs embedded on general surfaces are also known [3,5,6]. Closely related is also the so-called *tree-partition-width* problem, a variant of the defective coloring problem in which the graphs induced by each color class must be acyclic [8,9,21], i.e., there is no restriction on their diameter. Our contributions are:

- In Sect. 2, we present a linear-time algorithm to determine whether an outerplanar graph is (star, 2)-colorable. Note that outerplanar graphs are 3-colorable [20], and hence (star, 3)-colorable, but not necessarily (star, 2)-colorable. On the other hand, we prove that it is always possible to construct (star, 2)-colorings for outerpaths, which form a special subclass of outerplanar graphs whose weak-dual[1] is a path.
- In Sect. 3, we prove that the (star, 2)-coloring problem is NP-complete, even for graphs of maximum degree 5 (note that the corresponding (edge, 2)-coloring problem is NP-complete, even for graphs of maximum degree 4 [6]). Since all graphs of maximum degree 3 are (edge, 2)-colorable [16], this result leaves open only the case for graphs of maximum degree 4. We also prove that the (star, 3)-coloring problem is NP-complete, even for graphs of maximum degree 9 (recall that the corresponding (edge, 3)-coloring problem is NP-complete, even for graphs of maximum degree 7 [6]). Since all graphs of maximum degree 4 or 5 are (edge, 3)-colorable [16], our result implies that the computational complexity of the (star, 3)-coloring problem remains unknown only for graphs of maximum degree 6, 7, and 8. For planar graphs, we prove that the (star, 2)-coloring problem remains NP-complete even for triangle-free planar graphs (recall that triangle-free planar graphs are always 3-colorable [15], while the test of 2-colorability can be done in linear time).

2   Coloring Outerplanar Graphs and Subclasses

In this section we consider (star, 2)-colorings of outerplanar graphs. To demonstrate the difficulty of the problem, we first give an example (see Fig. 1) of a small outerplanar graph not admitting any (star, 2)-coloring. Therefore, in Theorem 1 we study the complexity of deciding whether a given outerplanar graph admits such a coloring and present a linear-time algorithm for this problem; note that outerplanar graphs always admit 3-colorings [20]. Finally, we show that a notable subclass of outerplanar graphs, namely outerpaths, always admit (star, 2)-colorings by providing a constructive linear-time algorithm (see Theorem 2).

Lemma 1. *There exist outerplanar graphs that are not (star, 2)-colorable.*

Proof. We prove that the outerplanar graph of Fig. 2a is not (star, 2)-colorable. In particular, we show that in any 2-coloring of this graph there exists a monochromatic path of four vertices. Assume w.l.o.g. that vertex u has color gray.

---

[1] Recall that the *weak-dual* of a plane graph is the subgraph of its dual induced by neglecting the face-vertex corresponding to its unbounded face.

Then, at least two vertices out of $u_1, \ldots, u_8$ are gray, as otherwise there would be a path of four white vertices. Hence, u is the center of a gray star.

This implies that both u_2 and u_3 are white, as otherwise paths $u_{21}, \ldots, u_{24}$ and $u_{31}, \ldots, u_{34}$ would have to consist of only white vertices. In this case, however, either one of paths $u_{21}, \ldots, u_{24}$ and $u_{31}, \ldots, u_{34}$ consists only of gray vertices, or there exists a path from one of $u_{21}, \ldots, u_{24}$ via u_2 and u_3 to one of $u_{31}, \ldots, u_{34}$, that consists only of white vertices. Hence, all the aforementioned cases lead to a monochromatic path of four vertices.   □

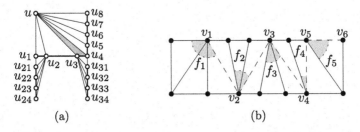

Fig. 2. (a) An outerplanar graph that is not (star, 2)-colorable. (b) An outerpath, whose spine edges are drawn as dashed segments. Dotted arcs highlighted in gray correspond to edges belonging to the fan of each spine vertex. Note that $|f_6| = 0$.

Lemma 1 implies that not all outerplanar graphs are (star, 2)-colorable. In the following we give a linear-time algorithm to decide whether an outerplanar graph is (star, 2)-colorable and in case of an affirmative answer to compute the actual coloring.

Theorem 1. *Given an outerplanar graph G, there exists a linear-time algorithm to test whether G admits a (star, 2)-coloring and to construct a (star, 2)-coloring, if one exists.*

Proof. We assume that G is embedded according to its outerplanar embedding. We can also assume that G is biconnected. This is not a loss of generality, as we can always reduce the number of cut-vertices by connecting two neighbors a and b of a cut-vertex c belonging to two different biconnected components with a path having two internal vertices. Clearly, if the augmented graph is (star, 2)-colorable, then the original one is (star, 2)-colorable. For the other direction, given a (star, 2)-coloring of the original graph, we can obtain a corresponding coloring of the augmented graph by coloring the neighbors of a and b with different color than the ones of a and b, respectively.

Denote by T the weak dual of G and root it at a leaf ρ of T. For a node μ of T, we denote by $G(\mu)$ the subgraph of G corresponding to the subtree of

T rooted at μ. We also denote by $f(\mu)$ the face of G corresponding to μ in T. If $\mu \neq \rho$, consider the parent ν of μ in T and their corresponding faces $f(\nu)$ and $f(\mu)$ of G, and let (u,v) be the edge of G shared by $f(\nu)$ and $f(\mu)$. We say that (u,v) is the *attachment edge* of $G(\mu)$ to $G(\nu)$. The attachment edge of the root ρ is any edge of face $f(\rho)$ that is incident to the outer face (since G is biconnected and ρ is a leaf, this edge always exists). Consider a (star, 2)-coloring of $G(\mu)$. In this coloring, each of the endpoints u and v of the attachment edge of $G(\mu)$ plays exactly one of the following roles: (*i*) *center* or (*ii*) *leaf* of a colored star; (*iii*) *isolated vertex*, that is, it has no neighbor with the same color; or (*iv*) *undefined*, that is, the only neighbor of u (resp. v) which has its same color is v (resp. u). Note that if the only neighbor of u (resp. v) which has its same color is different from v (resp. from u), we consider u (resp. v) as a center. Two (star, 2)-colorings of $G(\mu)$ are *equivalent* w.r.t. the attachment edge (u,v) of $G(\mu)$ if in the two (star, 2)-colorings each of u and v has the same color and plays the same role. This definition of equivalence determines a partition of the colorings of $G(\mu)$ into a set of equivalence classes. Since both the number of colors and the number of possible roles of each vertex u and v are constant, the number of different equivalence classes is also constant (note that, when the role is undefined, u and v must have the same color).

In order to test whether G admits a (star, 2)-coloring, we perform a bottom-up traversal of T. When visiting a node μ of T we compute the maximal set $C(\mu)$ of equivalence classes such that, for each class $C \in C(\mu)$, graph $G(\mu)$ admits at least a coloring belonging to C. Note that $|C(\mu)| \leq 38$. In order to compute $C(\mu)$, we consider the possible equivalence classes one at a time, and check whether $G(\mu)$ admits a (star, 2)-coloring in this class, based on the sets $C(\mu_1), \ldots, C(\mu_h)$ of the children $\mu_1, \ldots, \mu_h$ of μ in T, which have been previously computed. In particular, for an equivalence class C we test the existence of a (star, 2)-coloring of $G(\mu)$ belonging to C by selecting an equivalence class $C_i \in C(\mu_i)$ for each $i = 1, \ldots, h$ in such a way that:

1. the color and the role of u in C_1 are the same as the ones u has in C;
2. the color and the role of v in C_h are the same as the ones v has in C;
3. for any two consecutive children μ_i and μ_{i+1}, let x be the vertex shared by $G(\mu_i)$ and $G(\mu_{i+1})$. Then, x has the same color in C_i and C_{i+1} and, if x is a leaf in C_i, then x is isolated in C_{i+1} (or vice-versa); and
4. for any three consecutive children μ_{i-1}, μ_i, and μ_{i+1}, let x (resp. y) be the vertex shared by $G(\mu_{i-1})$ and $G(\mu_i)$ (resp. by $G(\mu_i)$ and $G(\mu_{i+1})$). Then, x (resp. y) has the same color in C_i and C_{i-1} (resp. C_{i+1}); also, if x and y are both undefined in C_i, then in C_{i-1} and C_{i+1} none of x and y is a leaf, and at least one of them is isolated.

Note that the first two conditions ensure that the coloring belongs to C, while the other two ensure that it is a (star, 2)-coloring. Since the cardinality of $C(\mu_i)$ is bounded by a constant, the test can be done in linear time. If the test succeeds, add C to $C(\mu)$.

Once all 38 equivalence classes are tested, if $C(\mu)$ is empty, then we conclude that G is not (star, 2)-colorable. Otherwise we proceed with the traversal of T.

At the end of the traversal, if $C(\rho)$ is not empty, we conclude that G is (star, 2)-colorable. A (star, 2)-coloring of G can be easily constructed by traversing T top-down, by following the choices performed during the bottom-up visit. □

In the following, we consider a subclass of outerplanar graphs, namely outerpaths, and we prove that they always admit (star, 2)-colorings. Note that the example that we presented in Lemma 1 is "almost" an outerpath, meaning that the weak-dual of this graph contains only degree-1 and degree-2 vertices, except for one specific vertex that has degree 3 (see the face of Fig. 2a highlighted in gray). Recall that the weak-dual of an outerpath is a path (hence, it consists of only degree-1 and degree-2 vertices).

Let G be an outerpath (see Fig. 2b). We assume that G is inner-triangulated. This is not a loss of generality, as any (star, 2)-coloring of a triangulated outerpath induces a (star, 2)-coloring of any of its subgraphs. We first give some definitions. We call *spine vertices* the vertices $v_1, v_2, \ldots, v_m$ that have degree at least four in G. We consider an additional spine vertex v_{m+1}, which is the (unique) neighbor of v_m along the cycle delimiting the outer face that is not adjacent to v_{m-1}. Note that the spine vertices of G induce a path, that we call *spine* of G^2. The *fan* f_i of a spine vertex v_i consists of the set of neighbors of v_i in G, except for v_{i-1} and for those following and preceding v_i along the cycle delimiting the outer face[3]; note that $|f_i| \geq 1$ for each $i = 1, \ldots, m$, while $|f_{m+1}| = 0$. For each $i = 1, \ldots, m+1$, we denote by G_i the subgraph of G induced by the spine vertices $v_1, \ldots, v_i$ and by the fans $f_1, \ldots, f_{i-1}$. Note that $G_{m+1} = G$. We denote by c_i the color assigned to spine vertex v_i, and by $c(G_i)$ a coloring of graph G_i. Finally, we say that an edge of G is *colored* if its two endpoints have the same color.

Theorem 2. *Every outerpath admits a (star, 2)-coloring, which can be computed in linear time.*

Proof. Let G be an outerpath with spine $v_1, \ldots, v_k$. We describe an algorithm to compute a (star, 2)-coloring of G. At each step $i = 1, \ldots, k$ of the algorithm we consider the spine edge (v_{i-1}, v_i), assuming that a (star, 2)-coloring of G_i has already been computed satisfying one of the following conditions (see Fig. 3):

Q_0: The only colored vertex is v_1;

Q_1: $c_i \neq c_{i-1}$, vertex v_{i-1} is the center of a star with color c_{i-1}, and no colored edge is incident to v_i;

Q_2: $c_i = c_{i-1}$, and no colored edge other than (v_{i-1}, v_i) is incident to v_{i-1} or v_i;

Q_3: $c_i \neq c_{i-1}$, vertex v_{i-1} is a leaf of a star with color c_{i-1}, and no colored edge is incident to v_i;

---

[2] Note that the spine of G coincides with the spine of the caterpillar obtained from the outerpath G by removing all the edges incident to its outer face, neglecting the additional spine vertex v_{m+1}.

[3] Fan f_i contains all the leaves of the caterpillar incident to v_i, plus the following spine vertex v_{i+1}.

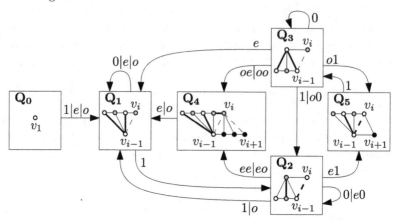

Fig. 3. Schematization of the algorithm. Each node represents the (unique) condition satisfied by G_i at some step $0 \le i \le k$. An edge label $0, 1, e, o$ represents the fact that the cardinality of a fan f_i is 0, 1, even $\ne 0$, or odd $\ne 1$. If the label contains two characters, the second one describes the cardinality of f_{i+1}. An edge between Q_j and Q_h with label $x \in \{1, e, o\}$ (with label xy, where $y \in \{0, 1, e, o\}$) represents the fact that, if G_i satisfies condition Q_j and $|f_i| = x$ (resp. $|f_i| = x$ and $|f_{i+1}| = y$), then f_i is colored so that G_{i+1} satisfies Q_h.

Q_4: $c_i \ne c_{i-1}$, vertex v_{i-1} is the center of a star with color c_{i-1}, and vertex v_i is the center of a star with color c_i; further, $i < k$ and $|f_i| > 1$;

Q_5: $c_i = c_{i-1}$, vertex v_{i-1} is the center of a star with color c_{i-1}, and no colored edge other than (v_{i-1}, v_i) is incident to v_i; further, $i < k$ and $|f_i| = 1$.

In the first step of the algorithm, we assign an arbitrary color to v_1, and hence $c(G_1)$ satisfies Q_0. For $i = 1, \dots, k$ we color f_i depending on the condition satisfied by $c(G_i)$. In particular, we color the vertices of f_i in such a way that $c(G_{i+1})$ is a (star, 2)-coloring satisfying one of the aforementioned conditions. However, due to space constraints the detailed case analysis is given in [2]; refer to Fig. 3 for a schematization.   □

3   NP-completeness for (Planar) Graphs of Bounded Degree

In this section, we study the computational complexity of the (star, 2)-coloring and (star, 3)-coloring problems for (planar) graphs of bounded degree.

Theorem 3. *It is NP-complete to determine whether a graph admits a (star, 2)-coloring, even in the case where its maximum degree is no more than 5.*

Proof. The problem clearly belongs to NP; a non-deterministic algorithm only needs to guess a color for each vertex of the graph and then in linear time can trivially check whether the graphs induced by each color-set are forests of stars. To prove that the problem is NP-hard, we employ a reduction from the well-known Not-All-Equal 3-SAT problem or NAESAT for short [19, p. 187]. An instance of NAESAT consists of a 3-CNF formula ϕ with variables $x_1, \ldots, x_n$ and clauses $C_1, \ldots, C_m$. The task is to find a truth assignment of ϕ so that no clause has all three literals *equal* in truth value (that is, not all are true). We show how to construct a graph G_ϕ of maximum vertex-degree 5 admitting a (star, 2)-coloring if and only if ϕ is satisfiable. Intuitively, graph G_ϕ reflecting formula ϕ consists of a set of subgraphs serving as variable gadgets that are connected to simple 3-cycles that serve as clause gadgets in an appropriate way; see Fig. 4c for an example.

Consider the graph of Fig. 4a, which contains two adjacent vertices, denoted by u_1 and u_2, and four vertices, denoted by v_1, v_2, v_3 and v_4, that form a path, so that each of u_1 and u_2 is connected to each of v_1, v_2, v_3 and v_4. We claim that in any (star, 2)-coloring of this graph u_1 and u_2 have different colors. For a proof by contradiction assume that u_1 and u_2 have the same color, say white. Since u_1 and u_2 are adjacent, none of v_1, v_2, v_3 and v_4 is white. So, $v_1, \ldots, v_4$ form a monochromatic component in gray which is of diameter 3; a contradiction. Hence, u_1 and u_2 have different colors, say gray and white, respectively. In addition, the colors of v_1, v_2, v_3 and v_4 alternate along the path $v_1 \to v_2 \to v_3 \to v_4$, as otherwise there would exist two consecutive vertices v_i and v_{i+1}, with $i = 1, 2, 3$, of the same color, which would create a monochromatic triangle with either u_1 or u_2.

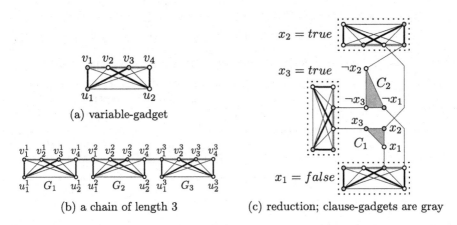

(a) variable-gadget

(b) a chain of length 3

(c) reduction; clause-gadgets are gray

Fig. 4. Illustration of: (a) a graph with 6 vertices, (b) a chain of length 3, (c) the reduction from NAESAT to (star, 2)-coloring: $\phi = (x_1 \vee x_2 \vee x_3) \wedge (\neg x_1 \vee \neg x_2 \vee \neg x_3)$. The solution corresponds to the assignment $x_1 = false$ and $x_2 = x_3 = false$. Sets O_{x_1}, E_{x_2} and E_{x_3} (E_{x_1}, O_{x_2} and O_{x_3}, resp.) are colored gray (white, resp.).

For $k \geq 1$, we form a *chain of length k* that contains k copies $G_1, G_2, \ldots, G_k$ of the graph of Fig. 4a, connected to each other as follows (see Fig. 4b). For $i = 1, 2, \ldots, k$, let u_1^i, u_2^i, v_1^i, v_2^i, v_3^i and v_4^i be the vertices of G_i. Then, for $i = 1, 2, \ldots, k - 1$ we introduce between G_i and G_{i+1} an edge connecting vertices v_4^i and v_1^{i+1} (dotted in Fig. 4b). This edge ensures that v_4^i and v_1^{i+1} are not of the same color, since otherwise we would have a monochromatic path of length four. Hence, the colors of the vertices of the so-called *spine-path* $v_1^1 \rightarrow v_2^1 \rightarrow v_3^1 \rightarrow v_4^1 \rightarrow \ldots \rightarrow v_1^k \rightarrow v_2^k \rightarrow v_3^k \rightarrow v_4^k$ alternate along this path. In other words, if the odd-positioned vertices of the spine-path are white, then the even-positioned ones will be gray, and vice versa. In addition, all vertices of the spine-path have degree 4 (except for v_1^1 and v_4^k, which have degree 3).

For each variable x_i of ϕ, graph G_ϕ contains a so-called *variable-chain* $\mathcal{C}_{x_i}$ of length $\lceil \frac{n_i-2}{2} \rceil$, where n_i is the number of occurrences of x_i in ϕ, $1 \leq i \leq n$; see Fig. 4c. Let $O[\mathcal{C}_{x_i}]$ and $E[\mathcal{C}_{x_i}]$ be the sets of odd- and even-positioned vertices along the spine-path of $\mathcal{C}_{x_i}$, respectively. For each clause $C_i = (\lambda_j \vee \lambda_k \vee \lambda_\ell)$ of ϕ, $1 \leq i \leq m$, where $\lambda_j \in \{x_j, \neg x_j\}$, $\lambda_k \in \{x_k, \neg x_k\}$, $\lambda_\ell \in \{x_\ell, \neg x_\ell\}$ and $j, k, \ell \in \{1, \ldots, n\}$, graph G_ϕ contains a 3-cycle of corresponding *clause-vertices* which, of course, cannot have the same color (*clause-gadget*; highlighted in gray in Fig. 4c). If λ_j is positive (negative), then we connect the clause-vertex corresponding to λ_j in G_ϕ to a vertex of degree less than 5 that belongs to set $E[\mathcal{C}_{x_j}]$ ($O[\mathcal{C}_{x_j}]$) of chain $\mathcal{C}_{x_j}$. Similarly, we create connections for literals λ_k and λ_ℓ; see the edges leaving the triplets for clauses C_1 and C_2 in Fig. 4c.

The length of $\mathcal{C}_{xi}$, $1 \leq i \leq n$, guarantees that all connections are accomplished so that no vertex of $\mathcal{C}_{x_i}$ has degree larger than 5. Thus, G_ϕ is of maximum degree 5. Since G_ϕ is linear in the size of ϕ, the construction can be done in $O(n + m)$ total time.

We show that G_ϕ is (star, 2)-colorable if and only if ϕ is satisfiable. Assume first that ϕ is satisfiable. If x_i is true (false), then we color $E[\mathcal{C}_{x_i}]$ white (gray) and $O[\mathcal{C}_{x_i}]$ gray (white). Hence, $E[\mathcal{C}_{x_i}]$ and $O[\mathcal{C}_{x_i}]$ admit different colors, as desired. Further, if x_i is true (false), then we color gray (white) all the clause-vertices of G_ϕ that correspond to positive literals of x_i in ϕ and we color white (gray) those corresponding to negative literals. Thus, a clause-vertex of G_ϕ cannot have the same color as its neighbor at the variable-gadget. Since in the truth assignment of ϕ no clause has all three literals true, no three clause-vertices belonging to the same clause have the same color.

Suppose that G_ϕ is (star, 2)-colorable. By construction, each of $E[\mathcal{C}_{x_i}]$ and $O[\mathcal{C}_{x_i}]$ is either white or gray, $i = 1, \ldots, n$. If $P[\mathcal{C}_{x_i}]$ is white, then we set $x_i = true$; otherwise, we set $x_i = false$. Assume, to the contrary, that there is a clause of ϕ whose literals are all true or all false. By construction, the corresponding clause-vertices of G_ϕ, which form a 3-cycle in G_ϕ, have the same color, which is a contradiction. □

The above construction highly depends on the presence of triangles (refer, e.g., to the clause gadgets). In the following theorem, we prove that the (star, 2)-coloring problem remains NP-complete, even in the case of triangle-free planar graphs.

Theorem 4. *It is NP-complete to determine whether a triangle-free planar graph admits a (star, 2)-coloring.*

Proof. We follow the same construction as for Theorem 3 but to ensure planarity we replace the crossings with appropriate crossing-gadgets and to avoid trian-

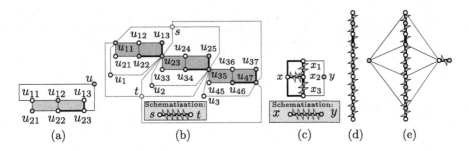

Fig. 5. (a) clause-gadget, (b) transmitter-gadget, (c) variable-gadget, (d) a chain of length 11, (e) crossing-gadget.

gular faces we use slightly more complicated variable- and clause-gadgets (see Fig. 5). A detailed description of our proof is given in [2].   □

Next, we prove that the (star, 2)-coloring problem is NP-complete even if one allows one more color and the input graph is either of maximum degree 9 or planar of maximum degree 16. Recall that all planar graphs are 4-colorable.

Theorem 5. *It is NP-complete to determine whether a graph G admits a (star, 3)-coloring, even in the case where the maximum degree of G is no more than 9 or in the case where G is a planar graph of maximum degree 16.*

Proof. Membership in NP can be proved as in Theorem 3. To prove that the problem is NP-hard, we employ a reduction from the well-known 3-COLORING problem, which is NP-complete even for planar graphs of maximum vertex-degree 4 [7]. So, let G be an instance of the 3-COLORING problem. To prove the first part of the theorem, we will construct a graph H of maximum vertex-degree 9 admitting a (star, 3)-coloring if and only if G is 3-colorable.

Central in our construction is the complete graph on six vertices K_6, which is (star, 3)-colorable; see Fig. 6a. We claim that in any (star, 3)-coloring of K_6 each vertex is adjacent to exactly one vertex of the same color. For a proof by contradiction, assume that there is a (star, 3)-coloring of K_6 in which three vertices, say u, v and w, have the same color. From the completeness of K_6, it follows that u, v and w form a monochromatic components of diameter 3, which is a contradiction.

Graph H is obtained from G by attaching a copy of K_6 at each vertex u of G, and by identifying u with a vertex of K_6, which we call *attachment-vertex*. Hence, H has maximum degree 9. As H is linear in the size of G, it can be constructed in linear time.

If G admits a 3-coloring, then H admits a (star, 3)-coloring in which each attachment-vertex has the same color as the corresponding vertex of G, and the colors of the other vertices are determined based on the color of the attachment-vertices. To prove that a (star, 3)-coloring of H determines a 3-COLORING of G, it is enough to prove that any two adjacent attachment-vertices v and w in H

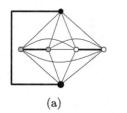 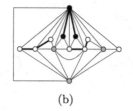 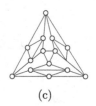

(a)   (b)   (c)

Fig. 6. (a) The complete graph on six vertices K_6. (b) The attachment-graph for the planar case. (c) A planar graph of max-degree 4 that is not (star, 2)-colorable.

have different colors, which holds since both v and w are incident to a vertex of the same color in the corresponding copies of K_6 associated with them.

For the second part of the theorem, we attach at each vertex of G the planar graph of Fig. 6b using as attachment its topmost vertex, which is of degree 12 (instead of K_6 which is not planar). Hence, the constructed graph H is planar and has degree 16 as desired. Furthermore, it is not difficult to be proved that in any (star, 3)-coloring of the graph of Fig. 6b its attachment-vertex is always incident to (at least one) another vertex of the same color, that is, it has exactly the same property with any vertex of K_6. Hence, the rest of the proof is analogous to the one of the first part of the theorem.   □

4   Conclusions

In this work, we presented algorithmic and complexity results for the (star, 2)-coloring and the (star, 3)-coloring problems. There exist several open questions raised by our work.

- We proved that it is NP-complete to determine whether a graph of maximum degree 5 is (star, 2)-colorable. An obvious extension is the question about the complexity of the (star, 2)-colorability problem for graphs of maximum degree 4. The question is of relevance even for planar graphs of maximum degree 4. Note that not all planar graphs of maximum degree 4 are (star, 2)-colorable (Fig. 6c shows such a counterexample), while all graphs of maximum degree 3 are even (edge, 2)-colorable [16].
- Are there other meaningful classes of graphs, besides the outerpaths, that are always (star, 2)-colorable?
- We proved NP-completeness for the question whether a graph of maximum degree 9 is (star, 3)-colorable. The corresponding complexity question is still open for the classes of graphs of maximum degree 6, 7 and 8. Recall that graphs of maximum degree 4 or 5 are always (star, 3)-colorable [16].
- One possible way to expand the class of graphs that admit defective colorings is to allow larger diameters for the graphs induced by the same color class.

Acknowledgments. We thank the participants of the special session GNV of IISA'15 inspiring this work.

References

1. Andrews, J., Jacobson, M.S.: On a generalization of chromatic number. Congressus Numerantium **47**, 33–48 (1985)
2. Angelini, P., Bekos, M.A., Kaufmann, M., Roselli, V.: Vertex-coloring with star-defects. Arxiv report arxiv.org/abs/1512.02505 (2015)
3. Archdeacon, D.: A note on defective coloring of graphs in surfaces. J. Graph Theor. **11**(4), 517–519 (1987)
4. Chartrand, G., Lesniak, L.M.: Graphs and Digraphs. Wadsworth, Monterey (1986)
5. Cowen, L.J., Cowen, R.H., Woodall, D.R.: Defective colorings of graphs in surfaces: partitions into subgraphs of bounded valency. J. Graph Theor. **10**(2), 187–195 (1986)
6. Cowen, L.J., Goddard, W., Jesurum, C.E.: Defective coloring revisited. J. Graph Theor. **24**(3), 205–219 (1997)
7. Dailey, D.P.: Uniqueness of colorability and colorability of planar 4-regular graphs are NP-complete. Discrete Math. **30**(3), 289–293 (1980)
8. Ding, G., Oporowski, B.: On tree-partitions of graphs. Discrete Math. **149**(13), 45–58 (1996)
9. Edelman, A., Hassidim, A., Nguyen, H.N., Onak, K.: An efficient partitioning oracle for bounded-treewidth graphs. In: Goldberg, L.A., Jansen, K., Ravi, R., Rolim, J.D.P. (eds.) RANDOM 2011 and APPROX 2011. LNCS, vol. 6845, pp. 530–541. Springer, Heidelberg (2011)
10. Garey, M.R., Johnson, D.S.: Computers and Intractability: A Guide to the Theory of NP-Completeness. W.H. Freeman, San Francisco (1979)
11. Gonthier, G.: Formal proof–the four-color theorem. Not. Am. Math. Soc. **55**(11), 1382–1393 (2008)
12. Hadlock, F.: Finding a maximum cut of a planar graph in polynomial time. SIAM J. Comput. **4**(3), 221–225 (1975)
13. Harary, F., Jones, K.: Conditional colorability II: bipartite variations. Congressus Numerantium **50**, 205–218 (1985)
14. Karp, R.M.: Reducibility among combinatorial problems. In: Miller, R.E., Thatcher, J.W., Bohlinger, J.D. (eds.) Complexity of Computer Computations, pp. 85–103. Springer, Heidelberg (1972)
15. Kowalik, L.: Fast 3-coloring triangle-free planar graphs. Algorithmica **58**(3), 770–789 (2010)
16. Lovász, L.: On decomposition of graphs. Studia Scientiarum Mathematicarum Hungarica **1**, 237–238 (1966)
17. Lovász, L.: Three short proofs in graph theory. J. Comb. Theor. Ser. B **19**(3), 269–271 (1975)
18. Maritz, P., Mouton, S.: Francis guthrie: a colourful life. Math. Intell. **34**(3), 67–75 (2012)
19. Papadimitriou, C.H.: Computational Complexity. Academic Internet Publ., London (2007)
20. Proskurowski, A., Syso, M.M.: Efficient vertex- and edge-coloring of outerplanar graphs. SIAM J. Algebraic Discrete Methods **7**(1), 131–136 (1986)
21. Wood, D.R.: On tree-partition-width. Eur. J. Comb. **30**(5), 1245–1253 (2009). Part Special Issue on Metric Graph Theory

Graphs Exploration

Lower Bounds for Graph Exploration Using Local Policies

Aditya Kumar Akash[1], Sándor P. Fekete[2], Seoung Kyou Lee[3],
Alejandro López-Ortiz[4], Daniela Maftuleac[4(✉)], and James McLurkin[3]

[1] IIT Bombay, Mumbai, India
adityakumarakash@gmail.com
[2] TU Braunschweig, Braunschweig, Germany
s.fekete@tu-bs.de
[3] Rice University, Houston, TX, USA
{sl28,jmclurkin}@rice.edu
[4] University of Waterloo, Waterloo, ON, Canada
{alopez-o,dmaftule}@uwaterloo.ca

Abstract. We give lower bounds for various natural node- and edge-based local strategies for exploring a graph. We consider this problem both in the setting of an arbitrary graph as well as the abstraction of a geometric exploration of a space by a robot, both of which have been extensively studied. We consider local exploration policies that use time-of-last-visit or alternatively least-frequently-visited local greedy strategies to select the next step in the exploration path. Both of these strategies were previously considered by Cooper et al. (2011) for a scenario in which counters for the last visit or visit frequency are attached to the edges. In this work we consider the case in which the counters are associated with the nodes, which for the case of dual graphs of geometric spaces could be argued to be intuitively more natural and likely more efficient. Surprisingly, these alternate strategies give worst-case superpolynomial/exponential time for exploration, whereas the least-frequently-visited strategy for edges has a polynomially bounded exploration time, as shown by Cooper et al. (2011).

1 Introduction

We consider the problem of a mobile agent or robot exploring an arbitrary graph. This is a well-studied problem in the literature, both in geometric and combinatorial settings. The robot or agent may wish to explore an arbitrary graph, e.g. a social network or the graph derived from the exploration of a geometric space. In the latter case, this is often modeled as an exploration task in the dual graph, where nodes correspond to rooms or regions, and edges corresponds to paths from one region to another [1,4,6]. In either setting, the goal is to explore every node in the graph (i.e., a corresponding region in space) in the smallest possible worst-case time. More formally, the question is this:
Given an unknown graph G and a local exploration policy, what is the time when the last node is visited as a function of the size of the graph G?

© Springer International Publishing Switzerland 2016
M. Kaykobad and R. Petreschi (Eds.): WALCOM 2016, LNCS 9627, pp. 55–67, 2016.
DOI: 10.1007/978-3-319-30139-6_5

There are several natural local strategy candidates for exploring a graph. We consider only strategies that use a local policy at each node for selecting the immediate neighbor that is visited next. The selection of neighbor can be done using one of the following policies: (1) *Least Recently Visited vertex* (LRV-v), (2) *Least Recently Visited edge* (LRV-e), (3) *Least Frequently Visited vertex* (LFV-v), and (4) *Least Frequently Visited edge* (LFV-e).

In the strategies above, we assume that each vertex or node holds an associated value, reflecting the last time it was visited (for the case of least recently visited strategies) or a counter of the total times it has been explored (for least frequently visited policies). Then the robot selects the neighboring vertex or adjacent edge with lowest value, i.e., oldest time stamp or least frequently visited.

Because we are hoping to minimize the time to visit every vertex (the dual of a region in the geometric space), it would seem more natural to consider first the LRV-v strategy or failing that, the LFV-v strategy. However, up until now, the only strategies with known theoretical worst-case bounds are LRV-e and LFV-e.

However, it has been an open problem whether these natural node-based policies are efficient. In an experimental study [8], we consider the task of patrolling (i.e. repeatedly visiting) a polygonal space that has been triangulated in a pre-established fashion. This problem can be modelled as exploration of the dual graph of the triangulation. This was the original motivation to study the problem of exploring graphs in general, and the dual graphs of triangulation in particular. In that paper we sketch an exponential lower bound for LRV-v. In this work, we give a superpolynomial lower bound for LFV-v for exploring a graph. This is in sharp contrast to the edge case, for which LFV-e has polynomially bounded exploration time. In particular, we show that there exist a graph on n vertices and $n-1$ edges corresponding to the dual of a convex decomposition of a polygon where the convex polygons are fat and of limited area and such that the exploration time is $\Theta(n^{\sqrt{n}/2})$ in the worst case. In the process we show full proofs for lower bounds for the LRV-v, sketched in the experimental study [8], and give lower bounds for the LRV-e and worst-case behavior for LFV-e in graphs of degree 3, thus extending the results by Yanovski et al. and Cooper et al. [3,10] which are shown only for graphs of higher degree in the so-called ANT model. This model has also been studied by Bonato et al. [2] with expected coverage time for random graphs Table 1.

Related Work. We study policies that require only local information, which can be maintained by simple devices. The policy *Least Recently Visited* is known to have worst-case exploration times that are exponential in the size of the graph, as shown by Cooper et al. [3]. More recently, the present authors (inspired by empirical considerations) studied LRV-v, LFV-v and LFV-e in the context of robot swarms and studied the observed average case using simulations.

The exponential lower bound for LFV-v was shown by Koenig et al. [5] while Malpani et al. [9] give exponential lower bounds for LRV-e on general graphs.

Table 1. Summary of results

Local policy	Graph class	Lower bound	Upper bound
LRV-v	General graphs	$\exp(\Theta(n))$ follows from Theorem 1	$\exp(\Theta(n))$
	Duals of triangulations	$\exp(\Theta(n))$ Theorem 1	$\exp(\Theta(n))$
LRV-e	General graphs	$\exp(\Theta(n))$ [3]	$\exp(\Theta(n))$
	Duals of triangulations	$\Omega(n^2)$ Theorem 2	$\exp(O(n))$
LFV-v	General graphs	$\Omega(n^{\sqrt{n}/2})$ [5,9]	$O(n\,\delta^d)$ Theorem 4
	Duals of triangulations	$\Omega(n^2)$ Theorem 5	$O(n^{\sqrt{n}/2})$ Theorem 9
LFV-e	General graphs	$\Omega(n^2)$ follows from Theorem 6	$O(n \cdot d)$ [3]
	Duals of triangulations	$\Omega(n^2)$ Theorem 6	$O(n \cdot d)$ Corollary 1

Summary of Results. In this work we suggest that LFV-e should be the preferred choice and complement this result by giving (1) an exponential lower bound for the worst case for LRV-v of triangulations, (2) a quadratic lower bound for the worst case for LRV-e of triangulations, (3) an exact bound on the maximum frequency difference of two neighboring nodes in LFV-v, (4) a quadratic lower bound for LFV-v in graphs of degree 3, (5) a quadratic lower bound for LFV-e in graphs of degree 3 and, most importantly, (6) a superpolynomial lower bound for the worst-case of LFV-v when the graph corresponds to a small convex decomposition of a polygon.

2   Worst-Case Behavior of LRV-e and LRV-v

The worst-case behavior of LRV-e in arbitrary graphs can be exponential in the number of nodes in the graph, provided we allow a maximum degree of at least 4. That is, for every n, there exists a graph with n vertices in which the largest exploration time for an edge is $\exp(\Theta(n))$ [3]. Figure 1 illustrates one such graph (with vertices of degree 4). The starting edge is leftmost in the graph and the last edge to be visited is the rightmost one. The diamond-like subgraph is such that when reached by a left-to-right path results in the path not passing through to the edges on the right on every two-out-of-three occasions. In this sense the gadget reflects back 2/3rds of all paths.

If we connect $\Theta(n)$ such gadgets in series, we will require a total of $(3/2)^{\Theta(n)}$ paths, starting from the left for at least one of them to reach the rightmost edge in the series. Given that our scenario is based on visiting (dual) vertices, it is natural to consider the worst-case behavior of LRV-v for the special class of planar graphs of maximum degree 3 that can arise as duals of triangulations. Until now, this has been an open problem. Moreover, it also makes sense to consider the worst-case behavior of LRV-e for the same special graph class, which is not covered by the work of Cooper et al. [3].

Theorem 1. *There are dual graphs of triangulations (in particular, planar graphs with n vertices of maximum degree 3), in which LRV-v leads to a largest exploration time for a node that is exponential in n.*

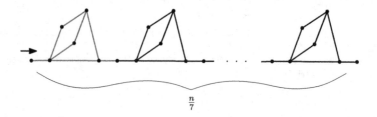

Fig. 1. Graph with n vertices with a chain of $n/7$ gadgets. A single gadget is colored in red for illustration purposes. Exploring takes exponential time in the worst case [3] (Color figure online).

Proof. Consider the graph G_D with n vertices in Fig. 3, which contains $(n-1)/9$ identical components (each containing 9 vertices) connected in a chain. This graph is the dual of the triangulation of the polygon in black lines shown in Fig. 2. Observe that every vertex has degree at most 3. We prove the claimed exponential time bound by recursively calculating the time taken to complete one cycle in the transition diagram shown in Fig. 4.

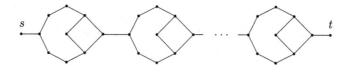

Fig. 2. This figure depicts (1) a hexagonal polygonal region with holes in black lines (2) its triangulation G_P in red lines and (3) the dual graph of the triangulation G_D shown in blue lines (Color figure online).

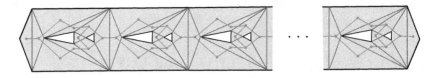

Fig. 3. The dual graph G_D illustrated in blue in Fig. 2 (Color figure online).

We monitor the movement of a robot from this situation onwards. Let T_n denote the time taken to complete one cycle of G_D, i.e., the time taken by a robot to start from and return to the first vertex of the first component of G_D. Similarly, let T_i denote the cycle which starts on the leftmost vertex s reaching the i gadget on the left-to-right path and back to s. Hence the graph will be fully explored when we first reach the last component. This requires three consecutive visits to the next-to-last (penultimate) component in the path, the first two visits are reflected back to the starting node s, and the last goes through.

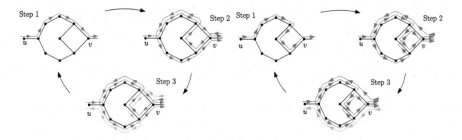

Fig. 4. Two possible LRV-v alternating paths on each component of the graph G_D.

From the possible paths illustrated in Fig. 4, we can observe that the vertex u is visited only during the beginning and end of the cycle, while the vertex v is visited twice in this cycle. It is not hard to check that the summation of visits to all edges in one component during one cycle is 22. Using this we can see a simple recursion as follows:

$$T_i \geq 22 + 2 \cdot T_{i-1}, \quad T_0 = 0$$

Solving this equation, we get $T_n \geq 22 \cdot (2^n - 1)$, and hence the last vertex t in the path is visited after at least $2 \cdot T_{n-1} \geq 22 \cdot (2^n - 2)$ steps, which is exponential in the number n of nodes of graph G_D, as claimed.   □

As it turns out, the lower bound for LRV-e in [3], i.e., for time stamping vertices instead of edges, also holds for graphs of max degree 3 as follows.

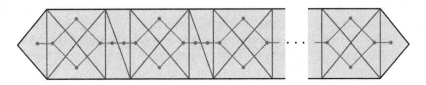

Fig. 5. This figure depicts (1) a hexagonal polygonal region in black lines (2) its triangulation G_P in red lines and (3) the dual graph of the triangulation G_D shown in blue lines (Color figure online).

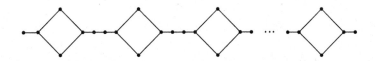

Fig. 6. The dual graph G_D consisting of a chain of $n/6$ cycles from Fig. 5.

Theorem 2. *There are dual graphs of triangulations (in particular, planar graphs with n vertices of maximum degree 3), in which LRV-e leads to a largest exploration time for a node that is quadratic in n.*

Proof. Consider the graph G_D of Fig. 6 which consists of a chain of $n/6$ cycles of length 4 connected in series. As illustrated in Fig. 7, each component is traversed initially following the colored oriented paths from step 1 and further alternating the paths from step $2k$ and $2k + 1$, for k positive integer. When all nodes have time stamp zero we can choose to visit the nodes in any arbitrary order.[1]

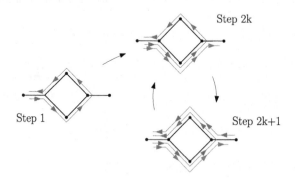

Fig. 7. LRV-e strategy on each component of the graph G_D.

In other words, the first time a component is traversed, the path changes direction and goes back to the start. The rest of the times when the component is traversed, the direction does not change. Thus, in order to traverse the ith component in the chain, we need to traverse the first $i - 1$ components in the chain. The total time, i.e. the number of steps, to reach the rightmost vertex in the chain comes to $\sum_{i=1}^{n/6}(i - 1) = \frac{n(n-6)}{12} = \Theta(n^2)$.   □

3   Worst-Case Behavior of LFV-v and LFV-e

First, we provide evidence that a polynomial upper bound on the worst-case latency (i.e. time between consecutive visits) is unlikely for LFV-v. We start by showing some interesting properties of graphs explored under LFV-v. It would seem at first that the nodes in a path followed by the robot form a non-increasing sequence of frequency values. This is so as we seemingly always select a node of lowest frequency. However, if all neighbors of a node have the same or higher frequency, then the destination node will have strictly larger frequency than the present node (see Figs. 8 and 9).

---

[1] In general, this property holds whenever there are several neighboring nodes with the lowest time stamp. For example, in a star starting from the center we visit all neighboring nodes in arbitrary order until all of them have time stamp 1. At this point we can once again choose an arbitrary order to visit the neighbors anew.

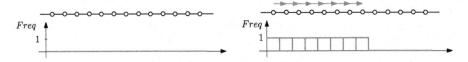

Fig. 8. A path being traversed from left to right with its frequency histogram below. Initially all nodes have frequency zero. Then half way through the path traversal nodes to left have frequency 1 and nodes to the right are still at zero.

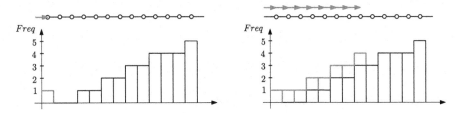

Fig. 9. A path with a corresponding staircase pattern in the histogram.

We also observe that it is possible to create dams or barriers by having a flower configuration in the path (see Fig. 10). We reach the center of the flower and then take the loops or petals, thus increasing the count of the center (see histogram on Fig. 10). Then the robot moves past the center node of the flower, which forms a barrier that impedes the robot from traversing from right to left past the center of the flower, until the count of the nodes to the right of the path has risen to match that of the barrier.

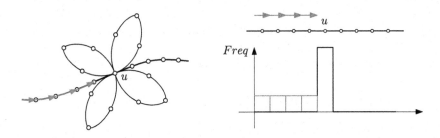

Fig. 10. A path with a "flower" configuration which creates a barrier.

With these three basic configurations (path, staircase, flower) in hand, we can combine them to create a graph in which the starting node s has $\delta(s)$ neighbors as shown in Fig. 11 where we can see that of the $\delta(s)$ neighbors $\delta(s) - 1$ have simple paths leading back to s. These paths go via a distinguished neighbor called u which is shared by all the paths from which they connect by a single shared edge to s. Each of the paths is a staircase with barriers (see Figs. 9 and 10). That is for each time we go from s to one of the first $\delta - 1$ neighbors we then climb a staircase up to u. Then from u we enter the other staircases from the "high" side until stopped by a barrier, which makes us return to u and eventually revisiting s from this last neighbor. This shows the following theorem.

Theorem 3. *There exists a configuration for LFV-v in which some neighbors of the starting vertex have a frequency count of k, while the starting point has a frequency count of $k\delta$. Moreover, the value of k can be as high as $\Theta(n/\delta)$.*

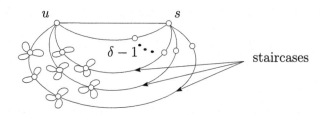

Fig. 11. A configuration in which the frequency of the starting point s is much larger than the majority of its neighbors.

This result provides some indication that the worst-case ratio between smallest and largest frequency labels of vertices may be exponential, which would arise if we could construct an example in which the ratio of the respective frequencies of two neighbors is the degree δ. Observe that δ can be $\Theta(n)$ in the worst case. From this it can be shown that at most δ^d steps are required to explore the graph, where d and δ are the diameter and the maximum degree of the graph.

Lemma 1. *Consider a graph G_D explored using the strategy LFV-v. Let g denote the frequency of the starting node s at time t and let $\delta(s)$ be the number of neighbors of s. Then there are at least $g \bmod \delta$ neighbors with frequency at least $\lfloor g/\delta \rfloor + 1$ and the remaining neighboring nodes have frequency at least $\lfloor g/\delta \rfloor$.*

Proof. By induction on g. Denote as $g' = g - 1$, $f = \lfloor g/\delta \rfloor$, $f' = \lfloor g'/\delta \rfloor$.
Basis of induction. $g = 0$. In this case $f = \lfloor 0/\delta \rfloor = 0$, so we have trivially at least $(0 \bmod \delta) = 0$ nodes with frequency at least 1 and the rest of the nodes have frequency at least 0. For good measure the reader may wish to prove the case $g = 1$.
Induction step. When g increases by one, we have either (1) $f = f'$ or (2) $f = f' + 1$.
In case (1) the robot explores a neighbor with min frequency at least $f' = f$ whose frequency increases to $f + 1$, thus increasing the number of neighbors with that frequency by 1 (if no such neighbor exists this means all neighbors already have frequency at least $f + 1$ and hence it trivially holds that at least $(g \bmod \delta)$ neighbors have frequency at least $f + 1$).
In case (2) when $f = f' + 1$ we have $\lfloor (g - 1)/\delta \rfloor + 1 = \lfloor g/\delta \rfloor$ which implies $(g \bmod \delta) = 0$ and $(g - 1 \bmod \delta) = \delta - 1$. Hence all but one of the neighbors

are guaranteed to have frequency at least $f' + 1$ (which is equal to f) and there is at most one neighbor with frequency f' which is the min and gets visited thus increasing its frequency to $f' + 1 = f$. This means that now all neighbors have frequency at least f and trivially at least $(\delta(s) \bmod \delta) = 0$ neighbors have frequency at least $f + 1$, as claimed.   □

Theorem 4. *The highest frequency node in a graph with unvisited nodes, using LFV-v, has frequency bounded by δ^d, where δ is the degree of the node and d the diameter of the graph.*

Proof. Consider any shortest path from an unvisited node to the node with highest frequency. The path is of length at most the diameter d of the graph. In each step the increase in frequency is at most a factor δ over the unvisited node hence the frequency of the most visited node is bounded by δ^d.   □

However, there is no known example of a dual of a triangulation graph displaying this worst-case behavior.

Theorem 5. *There exist graphs with n vertices of maximum degree 3, in which the largest exploration frequency for a node, using LFV-v, is $\Theta(n^2)$.*

Proof. This proof follows the outline of the proof of Theorem 2 for the same graph G_D represented in Fig. 5. As illustrated in Fig. 12, each of G_D's components is traversed initially following the colored oriented paths from step 1 and further alternating the paths from step $2k$ and $2k + 1$.

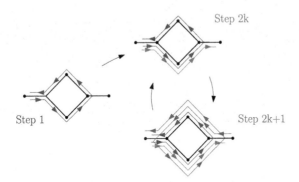

Fig. 12. LFV-v strategy on each component of the graph G_D.

In other words, the first time a component is traversed, the path changes direction and goes back to the start. The rest of the times when the component is traversed, the direction does not change. Thus, in order to traverse the ith component in the chain, we need to revisit the first $i - 1$ components in the chain, which is $\Theta(n^2)$.   □

Note that using LFV-e on the graph shown in Fig. 6, each component of the graph is traversed using the exact same strategy as shown in Fig. 12 for LFV-v.

Theorem 6. *There exist graphs with n vertices of maximum degree 3, in which the largest exploration frequency for an edge, using LFV-e, is $\Theta(n^2)$.*

Theorem 7. *[3] In a graph G with at most m edges and diameter d, the latency of each edge when carrying out LFV-e is at most $O(m \cdot d)$.*

This allows us to establish a good upper bound on LFV-e in our setting.

Corollary 1. *Let $G_D = (V_D, E_D)$ be the dual graph of a triangulation, with $|V_D| = n$ vertices and diameter d. Then the latency of each vertex when carrying out LFV-e is at most $O(n \cdot d)$.*

Proof. Since G_D is planar, it follows that $n \in \Theta(m)$, where $m = |E_D|$ is the number of edges of G_D. Because patrolling an edge requires visiting both of its vertices, the claim follows from the upper bound of Theorem 7.   □

We note that this bound can be tightened for regions with small aspect ratio, for which the diameter is bounded by the square root of the area.

Corollary 2. *For regions with diameter $d \in O(\sqrt{n})$, the latency of each dual vertex when carrying out LFV-e is at most $O(n^{1.5})$.*

4   A Graph with Superpolynomial Exploration Time

Koenig et al. gave a graph requiring superpolynomial exploration time, thus proving the theorem:

Theorem 8. *[5] LFV-v has worst-case exploration time $\Omega(n^{\sqrt{n}/2})$ on an n vertex graph. This holds even if the graph is planar and has sublinear maximum degree.*

We illustrate a similar construction for completeness. The graph is a caterpillar tree where the central path has $\ell + 2 = \lfloor \sqrt{n} \rfloor$ vertices, and without loss of generality we assume ℓ to be odd (see Fig. 13). The root which we term node 0, has $b + c + 1$ leaves (for some constant $c > 10$ and value b to be determined later) as children plus one edge connecting to the path, for a total degree of $b + c + 2$. The ith node in the path has $b - i + 1$ leaves as children and is connected in a path; hence node i has degree $b - i + 3$, for $1 \leq i \leq \ell$. The last node in the path has $b + 1$ leaves as children and degree $b + 2$.

We start from the root and as all nodes have visit frequency 0 we can choose to visit the nodes in any arbitrary order. Recall that this property holds whenever there are several neighboring nodes with the lowest frequency.

From the root we visit all leaves save one, thus increasing the frequency of the root to $b + c + 1$ and then proceed down the path. Then for a node i, $1 \leq i \leq \ell$ in the path, if i is odd we arbitrarily follow the path leaving the leaves with

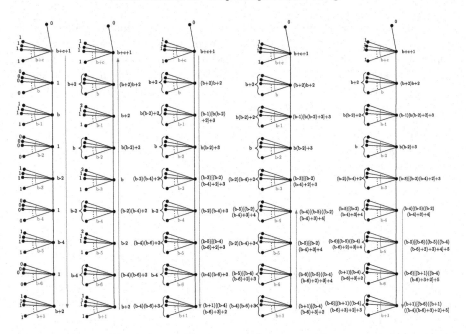

Fig. 13. A caterpillar tree requiring superpolynomial exploration time. The exploration starts with the red node and the red arrows indicate the direction of the exploration. The upper node is visited only after $\Theta(n^{\sqrt{n}/2})$ steps (Color figure online).

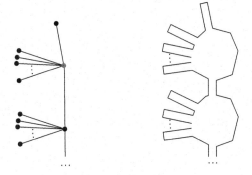

Fig. 14. A dual graph of a convex polygonal decomposition of a simple polygon. Observe that each subpolygon can be made as fat as required and of arbitrarily limited area.

frequency 0, while if i is even we visit all leaves first and then proceed down the path, thus increasing the frequency of node i to $b - i + 2$. In the last node in the path we visit all of its $b + 1$ children and then return to it with a final frequency of $b + 2$.

It is not hard to verify that subsequent traversals lead to an $\Omega(n^{\sqrt{d}/2})$ lower bound. See ARXIV paper for details [7].

We note that this graph does not correspond to the dual of triangulation since it is not of max degree 3. However, in contrast to the graph in Fig. 1 by Cooper et al., this graph corresponds to the dual of the simple polygon without holes under a small convex decomposition. We illustrate this in Fig. 14 where each edge is fattened into a rectangle-like portion and each node of degree $d \geq 3$ becomes a d-sided convex polygonal area.

Theorem 9. *LFV-v has worst-case exploration time $\Omega(n^{\sqrt{n}/2})$ when exploring the dual graph of a convex decomposition of a polygon, even under the restriction that the convex polygonal areas be fat and of limited area.*

5   Conclusions

In this paper we give (1) an exponential lower bound for the worst case for LRV-v of triangulations, (2) a quadratic lower bound for the worst case for LRV-e of triangulations, (3) an exact bound on the maximum degree difference between two neighboring nodes in LFV-v, (4) a quadratic lower bound for LFV-v in graphs of degree 3, (5) a quadratic lower bound for LFV-e in graphs of degree 3 and, most importantly, (6) a superpolynomial lower bound for the worst-case of LFV-v when the graph corresponds to a small convex decomposition of a polygon.

We conjecture that for graphs of maximum degree 3, the performance of LFV-v is quadratic and its average coverage time is linear.

References

1. Becker, A., Fekete, S.P., Kröller, A., Lee, S.K., McLurkin, J., Schmidt, C.: Triangulating unknown environments using robot swarms. In: Proceedings 29th Annual ACM Symposium Computational Geometry, pp. 345–346 (2013)
2. Bonato, A., del Río-Chanona, R.M., MacRury, C., Nicolaidis, J., Pérez-Giménez, X., Prałat, P., Ternovsky, K.: Workshop on algorithms and models for the web graph. In: Gleich, D.F., Komjáthy, J., Litvak, N. (eds.) WAW 2006. LNCS, vol. 4936, pp. 18–23. Springer, Heidelberg (2008)
3. Cooper, C., Ilcinkas, D., Klasing, R., Kosowski, A.: Derandomizing random walks in undirected graphs using locally fair exploration strategies. Distrib. Comput. **24**(2), 91–99 (2011)
4. Fekete, S.P., Lee, S.K., López-Ortiz, A., Maftuleac, D., McLurkin, J.: Patrolling a region with a structured swarm of robots with limited individual capabilities. In: International Workshop on Robotic Sensor Networks (WRSN) (2014)
5. Koenig, S., Szymanski, B., Liu, Y.: Efficient and inefficient ant coverage methods. Annals of Math. Artif. Intell. Spec. Issue Ant Rob. **31**(1), 41–76 (2001)
6. Lee, S.K., Becker, A., Fekete, S.P., Kröller, A., McLurkin, J.: Exploration via structured triangulation by a multi-robot system with bearing-only low-resolution sensors. In: IEEE International Conference on Robotics and Automation, ICRA 2014, Hong Kong, China, pp. 2150–2157 (2014)

7. Maftuleac, D., Lee, S., Fekete, S.P., Akash, A. K., López-Ortiz, A., McLurkin, J.: Local policies for efficiently patrolling a triangulated region by a robot swarm. CoRR, abs/1410.2295 (2014)
8. Maftuleac, D., Lee, S., Fekete, S.P., Akash, A.K., López-Ortiz, A., McLurkin, J.: Local policies for efficiently patrolling a triangulated region by a robot swarm. In: International Conference on Robotics and Automation (ICRA), pp. 1809–1815 (2015)
9. Malpani, N., Chen, Y., Vaidya, N.H., Welch, J.L.: Distributed token circulation in mobile ad hoc networks. IEEE Trans. Mobile Comput. 4(2), 154–165 (2005)
10. Yanovski, V., Wagner, I.A., Bruckstein, A.M.: A distributed ant algorithm for efficiently patrolling a network. Algorithmica 3(37), 165–186 (2003)

Optimal Distributed Searching in the Plane with and Without Uncertainty

Alejandro López-Ortiz and Daniela Maftuleac$^{(\boxtimes)}$

Cheriton School of Computer Science, University of Waterloo,
Waterloo, ON N2L 3G1, Canada
{alopez-o,dmaftule}@uwaterloo.ca

Abstract. We consider the problem of multiple agents or robots searching in a coordinated fashion for a target in the plane. This is motivated by Search and Rescue operations (SAR) in the high seas which in the past were often performed with several vessels, and more recently by swarms of aerial drones and/or unmanned surface vessels. Coordinating such a search in an effective manner is a non trivial task. In this paper, we develop first an optimal strategy for searching with k robots starting from a common origin and moving at unit speed. We then apply the results from this model to more realistic scenarios such as differential search speeds, late arrival times to the search effort and low probability of detection under poor visibility conditions. We show that, surprisingly, the theoretical idealized model still governs the search with certain suitable minor adaptations.

1 Introduction

Searching for an object on the plane with limited visibility is often modelled by a search on a lattice. In this case it is assumed that the search agent identifies the target upon contact. An axis parallel lattice induces the Manhattan or L_1 metric on the plane. One can measure the distances traversed by the search agent or robot using this metric. Traditionally, search strategies are analysed using the competitive ratio used in the analysis of on-line algorithms. For a single robot the competitive ratio is defined as the ratio between the distance traversed by the robot in its search for the target and the length of the shortest path between the starting position of the robot and the target. In other words, the competitive ratio measures the detour of the search strategy as compared to the optimal shortest route.

In 1989, Baeza-Yates et al. [1–3] proposed an optimal strategy for searching on a lattice with a single searcher with a competitive ratio of $2n + 5 + \Theta(1/n)$ to find a point at an unknown distance n from the origin. The strategy follows a spiral pattern exploring n-balls in increasing order, for all integer n. This model has been historically used for search and rescue operations in the high seas where a grid pattern is established and search vessels are dispatched in predetermined patterns to search for the target [4,14].

© Springer International Publishing Switzerland 2016
M. Kaykobad and R. Petreschi (Eds.): WALCOM 2016, LNCS 9627, pp. 68–79, 2016.
DOI: 10.1007/978-3-319-30139-6_6

Historically, searches were conducted using a limited number (at most a handful) of vessels and aircrafts. This placed heavy constraints in the type of solutions that could be considered, and this is duly reflected in the modern search and rescue literature [6,8,11].

However, the comparably low cost of surface or underwater unmanned vessels allows for searches using hundreds, if not thousands of vessels.[1] Motivated by this consideration, we study strategies for searching optimally in the plane with a given, arbitrarily large number of robots.

Additionally, the search pattern reflects probabilities of detection and discovery according to some known distribution that reflects the specifics of the search at hand. For example, the search of the *SS Central America* reflected the probabilities of location using known survivor accounts and ocean currents. These probabilities were included in the design stage of the search pattern, with the ship and its gold cargo being successfully recovered in 1989 after more than 130 years of previous unsuccessful search efforts [13]. The case of parallel searchers without communication was studied by Feinerman et al. [5] and Koenig et al. [7]. In this paper they study robots without communication or a unique ID. In contrast we assume both a unique ID from the outset and centralized communication for the case when new agents join the search.

In this paper, we address the problem of searching in the plane with multiple centrally-coordinated agents under probability of detection and discovery. We begin with the theoretical model for two and four robots of López-Ortiz and Sweet [9] that abstracts out issues of visibility and differing speeds of searchers. Searching for an object on the plane with limited visibility is commonly modelled by a search on a lattice. Under this setting, visual contact on the plane corresponds to identifying the target upon contact on the grid.

Models. There are several parameters to model the cost of a SAR search. First is the total cost of the search effort as measured in vessel and personnel hours times the number of hours in the search, both for the worst case and average case as compared to the shortest path to the target. The second is the effectiveness of the search in terms of the probability of finding the target. Lastly, the time to discovery or speed-to-destination as time is of the essence in most search rescue scenarios. That is to say, a multiple robot search is preferable to a single agent search with the same overall cost as the time to discovery is lower. In this paper we aim to minimize the time to destination under a fixed number of robots as compared to the shortest path.

Summary of Results and Structure of the Paper. We construct a theoretical model and give an optimal strategy for searching with k robots with unit speed,

---

[1] For example, the cost of an unmanned search vehicle is in the order of tens of thousands of dollars which can be amortized over hundreds of searches, while the cost of conventional search efforts range from the low hundred thousands of dollars up to sixty million dollars for high profile searches such as Malaysia Airlines MH370 and Air France 447. This suggests that somewhere in the order of a few hundred to a few tens of thousands of robots can be brought to bear in such a search.

starting simultaneously from a common origin. We then progressively enrich this model with practical parameters, specifically different search speeds, different arrival times to the search effort and poor visibility conditions. We show that the principles from the theoretical solution also govern the more realistic search scenario under these conditions subject to a few minor adaptations. Lastly, we deal with cases with a varying probability of location as well as probability of detection (POD).

We first consider the case where all searchers start from a common point which we term the origin, and second, when they start from arbitrary points on the lattice. The robots proceed in a coordinated fashion determined at start time. Once the search begins, there is no need for further communication or interaction. Each robot has a unique serial identifier known to each robot and used in determining the search path to follow.

Initially we consider the case where all k searchers move at the same speed and give an optimal strategy for finding a target with $k = 4r$ searchers, for some r positive integer.

This is then generalized to any number of robots (not just multiples of four) and using the same ideas, we show that the techniques developed also generalize to searchers with various speeds. Lastly, we show that the proposed theoretical strategy also governs a search under actual weather conditions, in which there is a non-negligible probability of the target being missed in a search. We use tables from the extensive literature on SAR (Search-and-Rescue) operations to conduct simulations and give scenarios in which the proposed strategy can greatly aid in the quest for a missing person or object in a SAR setting [6].

2   Parallel Searching

López-Ortiz and Sweet [9] consider the case of searches using two and four robots (see Fig. 1(a)). In this case, the robots move in symmetric paths around the origin and prove the following theorem.

Theorem 1. *[9] Searching in parallel with $k = 2, 4$ robots for a point at an unknown distance n in the lattice is $(2n + 4 + 4/(3n))/k + o(1/n^2)$ competitive.*

This is in fact optimal for the two and four robots case, as the next theorem shows. Let the *n-ball* consist of those points of distance n from the origin.

Theorem 2. *Searching in parallel with k robots for a point at an unknown distance n in the lattice requires at least $(2n^2 + 4n + 4/3)/k + \Omega(1/n)$ steps, which implies a competitive ratio of at least $(2n + 4 + 4/(3n))/k + \Omega(1/n^2)$.*

Proof. Following the notation of [9], let $A(n)$ be the combined total distance traversed by all robots up and until the last point at distance n is visited. We claim that in the worst case $A(n) \geq 2n^2 + 5n + 3/2$, for some $n > 1$. Define $g(n)$ as the number of points visited on the $(n + 1)$-ball before the last visit to a point on the n-ball. First, note that there are $2n^2 + 2n + 1$ points in the interior of the closed ball of radius n and that visiting any m points requires at least $m-1$ steps. Hence, $A(n) = 2n^2 + 2n + g(n)$. If $g(n)$ points have already been visited, this means that after the last point at distance n is visited, there remain $4(n+1) - g(n)$ points to visit in the n-ball. Now, visiting m points in a ball requires at least $2m-1$ steps with one robot, and $2m-k$ with k robots. Thus, visiting the remaining points requires at least $2(4(n+1) - g(n)) - k$ steps. Hence, $A(n+1) = A(n)^2 + 2(4(n+1) - g(n)) - k$ as claimed. Now we consider the competitive ratio at distance n and $n+1$ for each of the robots as they visit the last point at such distance in their described path. We denote by $A_i(n)$ the portion of the points $A(n)$ visited by the ith robot. Hence, the competitive ratio for robot i at distances n and $n+1$ is given by $A_i(n)/n$ and $A_i(n+1)/(n+1)$. Observe that $\sum_{i=1}^{k} A_i(n) = A(n)$, for any n and hence, there exist i and j such that $A_i(n) \geq A(n)/k$ and $A_j(n+1) \geq A(n+1)/k$. Lastly, the competitive ratio, as a worst case measure is minimized when $A_i(n)/n = A_j(n+1)/(n+1)$, or equivalently, when $A(n)/n = A(n+1)/(n+1)$ with solution $g(n) = 2n + (4-k)n/(3n+1)$. Substituting in the expression for $A(n)$, we obtain $A(n) = 2n^2 + 4n + (4-k)n/(3n+1) = 2n^2 + 4n + 4/3 + \Theta(1/n)$ with a robot searching, in the worst case at least $A(n)/k$ steps for a competitive ratio of $\frac{2n+4+4/(3n)}{k} + \Omega(1/n^2)$. $\qquad\square$

3   Search Strategy

3.1   Even-Work Strategy for Parallel Search with $k = 4r$ Robots

A natural generalization of the $k = 2$ and $k = 4$ robot cases suggest a spiral strategy consisting of k nested spirals searching in an outward fashion. However, because the pattern must replicate or echo the shape of inner paths, all attempts lead to an unbalanced distribution of the last search levels and thus a suboptimal strategy. A better competitive ratio gives us the strategy described in this section that we call *even-work strategy*. Each of the r robots covers an equal region of a quadrant using the pattern in Fig. 4. The entire strategy consists of four rotations of this pattern, one for each quadrant in the plane.

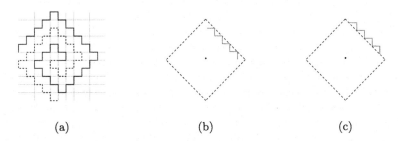

(a)   (b)   (c)

Fig. 1. (a) Search with two robots. (b), (c) Covering the n-ball: best case scenario and worst case scenario.

Fig. 2. Allocation of additional search tasks as radius increases (x-axis). The y-axis indicates which robot is activated in that ball.

Theorem 3. *Searching in parallel with $k = 4r$ robots for a point at an unknown distance n in the lattice has asymptotic competitive ratio of at most $(2n+7.42)/k$.*

Proof. We know the lower bound for asymptotic competitive ratio is $2n/k+5/k$. We want to describe the upper bound of even-work strategy of $2n/k + 7.428/k$. From the lower bound, we can deduce that for each ball the number of extra points (i.e., points outside the ball) covered by the robots is 5 in the best case (Fig. 1(b)). In the worst case, the robots perform 8 units of extra amount of work (Fig. 1(c)). So in order to cover all the points on a ball, the robots traverse a total of 13 units of extra distance. Thus, $13/5 = 2.6$ is an upper bound on the amount of work per point. When robots move from the ball of radius n to $n + 1$, a single robot must pick up the extra point to be explored. We balance the distribution of the new work as shown in Fig. 2. In this figure, the x-axis marks the distance from the origin of the current ball being explored, while the y-axis indicates the robot that is tasked with the exploration of the extra points in ball $n + 1$ over the n ball. There are four such points in total, or one per side. After covering the ball n, we have $2n^2 + 2n$ points covered inside it. The lower bound gives us $2n^2 + 5n$ amount of work to cover all the points at distance n. When we look at the last $4n$ points (on the n ball), for each of the $4n$ points, we have $3n$ work. Thus, $7/4$ amount of work per point (lower bound). From where we get the relation: $\frac{\lceil n/k \rceil (1+8/5)}{\lceil n/k \rceil (1+3/4)} = \frac{13/5}{7/4} = 1.486$, and $5 \cdot 1.486 = 7.428$. □

3.2 Parallel Search with Any Number of Robots

This case illustrates how the abstract search strategy for a number of robots multiple of four can readily be adapted to an arbitrary number of robots. Let k be the number of robots, where k is not necessarily divisible by 4.

We first design the strategy for $4k$ robots obtaining 4 times as many regions as robots. We then assign to every robot 4 consecutive regions as shown in Fig. 3 for the case $k = 7$. Observe that now some of the regions span more than one quadrant and how the search path for each robot transitions from region to region while exploring the same ball of radius n in all four regions assigned to it. Observe that from Theorems 2 and 3, it follows that this strategy searches the plane optimally as well.

Theorem 4. *Searching in parallel with k robots for a point at an unknown distance in the lattice has an asymptotic competitive ratio of at most $(2n+7.42)/k$.*

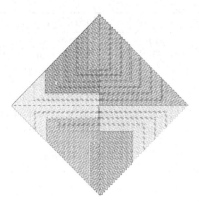

Fig. 3. Parallel search with $k = 7$ robots on the ball of radius 74.

The precise description of the search paths is shown in pseudocode of Sect. 4.1. Each robot only needs to know its unique ID and the total number of robots involved in the search. The code was implemented in Maple and used for drawing the figures in this paper.

4 From Theory to Practice

4.1 The Search Strategy

In Fig. 4, we show the search strategy with $k = 4r$ robots. Since the robots traverse at unit speed, the total area explored by each robot is t while the combination of all robots is kt. While we envision the swarm of robots being usually deployed from a single vessel and as such all of them starting from the same original position, for certain searches additional resources are brought to bear as more searchers join the search-and-rescue effort. In this setting we must consider an agent or agents joining a search effort already under way.

Fig. 4. Parallel search with $k = 4r$ robots, where $r = 7$, at time $t = 40, 80, 160, 240$. This figure illustrates the search only in one of the four quadrants.

Algorithm 1. Strategy(r, n)

Input: Let $k = 4r$ the number of robots, and let n be the covered distance.

Output: parallel search strategy of r robots in a quadrant.

Initialization$(r, 0)$;

Robot-1(n).

for $i = 2$ **to** $r - 1$ **do**

  Middle-robots(i, n).

end for

Robot-r(n).

Algorithm 2. Initialization(x, y)

Input: Starting point (x, y).

Output: Constructs the initial pattern for a robot

2 up; 1 right; 2 down; 3 right.

Algorithm 3. Robot-1(n)

Input: $k = 4r$ robots, n the covered distance.

Output: The parallel search strategy of the first robot in a quadrant.

for $v = 1$ **to** n **do**

  for $j = 1$ **to** $2(r - 1)$ **do**

    Stairs$(8(v - 1) + 3, horiz, NW)$.

    Stairs$(8(v - 1) + 5, horiz, SE)$.

  end for

  for $j = 1$ **to** 2 **do**

    Stairs$(4j + 8(v - 1), horiz, NW)$.

    Stairs$(4j + 2 + 8(v - 1), vert, SE)$.

  end for

end for

Algorithm 4. Stairs$(n, d, direction)$

Input: Let n be the number of steps in the stair, d - the initial horizontal or vertical step and $direction$ either NW for North-West or SE for South-East.

Output: The stairs in direction $direction$ starting with the first step d.

if $direction = NW$ **then**

  1 up.

  Init-Stair(n, d, NW).

  1 up.

else

  1 right.

  Init-Stair(n, d, SE).

  1 right.

end if

Algorithm 5. Init-Stair$(n, d, direction)$

Input: Let n be the number of steps in the stair, d - the initial horizontal or vertical step and $direction$ either NW for North-West or SE for South-East.

Output: The n stairs in direction $direction$ starting with the first step d.

if $n > 1$ **then**

  if $d = horiz$ **then**

    if $direction = NW$ **then**

      2 left.

    else

      2 right.

    end if

    Init-Stair$(n - 1, vert, direction)$.

  else

    if $direction = NW$ **then**

      2 up.

    else

      2 down.

    end if

    Init-Stair$(n - 1, horiz, direction)$.

  end if

end if

Algorithm 6. Middle-robots(i, n)

Input: $k = 4r$ robots, i - the number of the current robot and n the covered distance.
Output: The parallel search strategy of $r - 2$ (middle) robots in a quadrant.
Initialization($r - i + 1, 5 * (i - 1)$).
for $v = 1$ **to** n **do**
    for $j = 1$ **to** $2(r - i)$ **do**
        Stairs($3 + 8(v - 1)$, horiz, NW).
        Stairs($5 + 8(v - 1)$, horiz, SE).
    end for
    Stairs($4 + 8(v - 1)$, horiz, NW).
    Stairs($6 + 8(v - 1)$, vert, SE).
    Stairs($8 + 8(v - 1)$, horiz, NW).
    for $j = 1$ **to** 2(i-1) **do**
        Stairs($7 + 8(v - 1)$, vert, SE).
        Stairs($9 + 8(v - 1)$, vert, NW).
    end for
    Stairs($10 + 8(v - 1)$, vert, SE).
end for

Algorithm 7. Robot-r(n)

Input: $k = 4r$ robots and n the covered distance.
Output: The parallel search strategy of the rth robot in a quadrant.
Initialization($1, 5(r - 1)$);
Stairs($8(v - 1) + 4$, horiz, NW).
Stairs($8(v - 1) + 6$, vert, SE).
Stairs($8(v - 1) + 8$, horiz, NW).
for $j = 1$ **to** $2(r - 1)$ **do**
    Stairs($8(v - 1) + 7$, vert, SE).
    Stairs($8(v - 1) + 9$, vert, NW).
end for
for $v = 2$ **to** n **do**
    for $k = 1$ **to** 2 **do**
        Stairs($8(v - 1) + 4k - 2$, vert, SE).
        Stairs($8(v - 1) + 4k$, horiz, NW).
    end for
    for $j = 1$ **to** $2(r - 1)$ **do**
        Stairs($8(v - 1) + 7$, vert, SE).
        Stairs($8(v - 1) + 9$, vert, NW).
    end for
end for

Theorem 5. *There exists an optimal asymptotic strategy for parallel search with k initial robots starting from a common origin and later adding new robots to the search.*

Proof (sketch). Given the distance to the origin for the additional robots, we can compute the exact time at which the additional searcher will meet up with the explored area. At this point the search agents switch from a k robot search pattern to a $k+1$ search pattern. The net cost of this transition effort is bounded by the diameter of the n ball at which the extra searcher joins, with no ill effect over the asymptotic competitive ratio. Hence, the search is asymptotically optimal. □

A parallel search with k robots with different speeds is another case which nicely illustrates how the abstract search strategy for robots with equal speed can be readily adapted to robots of varying speeds.

Theorem 6. *There exists an optimal strategy for parallel search with k robots with different speeds.*

Proof. Suppose we are given k robots with varying speeds. Let the speed of the k robots be $s_1, s_2, \ldots, s_k$ respectively. We consider the speeds to be integral, subject to proper scaling and rounding. Let $s = \sum_{i=1}^{k} s_i$. We use the strategy for $4s$ robots and we assign for each robot respectively: $4s_1, 4s_2, \ldots, 4s_k$ regions. It follows that every robot completes the exploration of its region at the same time as any other robot since the difference in area explored corresponds exactly to the difference in search speed and the search proceeds uniformly and optimally over the entire range as well. □

4.2 Probability of Detection

In real life settings there is a substantial probability that the search agent might miss the target even after exploring its immediate vicinity particularly in high or stormy seas. The search-and-rescue literature provides ready tables of probability of detection (POD) under various search conditions [6]. Figure 5(a) shows the initial probability of location map for a typical man overboard event. Figure 5(b) shows the probability of detection as a function of the width of the search area spanned. The unit search width magnitude is computed using location, time, target and search-agent specific information such as visibility, lighting conditions, size of target and height of search vessel. We consider then a setting in which a suitable POD distribution has been computed taking into account present visibility conditions and size of target (see Fig. 5(a)). Armed with this information, a robot must then make a choice between searching an unexplored cell in the lattice or revisiting a previously explored cell.

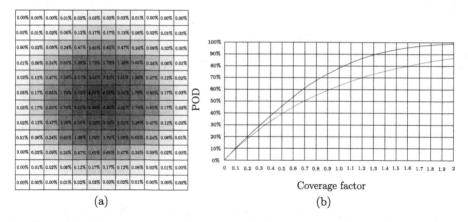

(a)   (b)

Fig. 5. (a) Initial probability map [6], (b) Average probabilities of detection (POD) over an area for visual searches using parallel sweeps (in blue: ideal search conditions, in red: normal search conditions) [6] (Color figure online).

Consider first an abstract model in which a robot can "teleport" from any given cell to another, ignoring any costs of movement related to this switch. The greedy strategy consists of robots moving to the cell with current highest probability of containing the target. Each cell is then searched using the corresponding pattern for the number of robots deployed in the cell.

Lemma 1. *Greedy is the optimal strategy for searching a probabilistic space under the teleportation model.*

Proof. Let p_i^j be the probability of discovering the target in cell i during the j visit, sorted in decreasing order. We now relabel them $p_1, p_2, \ldots$ The expected time of discovery is $\sum_{t=1}^{\infty} p_t \cdot t$ which is minimized when p_t are in decreasing order as follows. Assume by way of contradiction that the given minimal order is not in decreasing order, i.e., there exists i such that $p_i < p_{i+1}$. Now swap these two cells in the visiting order and we get that the cost before the swap was $\sum_{t=1}^{\infty} p_t \cdot t$, while after the swap is $\sum_{t=1, t\neq i}^{\infty} p_t \cdot t + i \cdot p_{i+1} + (i+1) \cdot p_i$. Subtracting these two quantities we get $i \cdot p_i + (i+1) \cdot p_{i+1} - i \cdot p_{i+1} - (i+1) \cdot p_i = p_{i+1} - p_i$ which is strictly positive from the assumption that $p_i < p_{i+1}$. This is a contradiction since $\sum_{t=1}^{\infty} p_t \cdot t$ was minimal.   □

Now we consider the case where moving from one search position to another happens at the same speed as searching. Observe that the probability of each cell evolves over time. It remains at its initial value so long as it is still unexplored and it becomes q^m times its initial value after m search passes, where $q = 1 - p$ is the probability of not detecting a target present in the current cell during a single search pass.

Probabilistic Search Algorithm. The algorithm creates *supercells* of size $h \times h$ unit cells, for some value of h, which depends on the total number of robots available to the search effort. The algorithm computes the combined probability of the target being found in a supercell which corresponds to the sum of the individual probabilities of the unit cells as given by the POD map.

At each time t, the algorithm considers the highest probability supercell and compares it to the lowest probability supercell being explored to determine a balance of robots to be assigned to each cell, in this case a transfer of robots from the low probability cell to the unexplored high probability cell. The search process continues until the target is found or the probability of finding it falls below a certain threshold. Once the probabilities have been rebalanced, we need to determine the source/destination pair for each robot. This is important since the distance between source and destination is dead search time, so we wish to minimize the amount of transit time. To this end, we establish a minimum-cost network flow [12] that computes the lowest total transit cost robot reassignment that satisfies the computed gains and losses.

More formally, let $C_1^t, C_2^t, \ldots, C_j^t$ be the areas being explored at time t by $r_1^t, r_2^t, \ldots, r_j^t$ robots respectively. The combined probability of a supercell is the sum of the probabilities of the cells inside it. When it is clear from the context what is the present t we will omit it from the superscript.

These combined probabilities are then sorted in decreasing order and the algorithm dispatches robots to the highest probability supercell until the marginal value of the robots is below that of an unexplored supercell. More precisely, let C_i and C_j be the two supercells of highest combined probability, p_i and p_j, respectively. The algorithm then assigns s robots to supercell C_i such that $p_i/s \geq p_j > p_i/(s+1)$. In other words the algorithm assigns robots to cells so that the expected gain per robot per cell is a approximately optimal. More specifically, the algorithm maintains two priority queues. One is a max priority queue (PQ) of supercells using the combined probability per robot as key. That is, supercell i appears with priority key equals to $p_i/(r_i + 1)$ where r_i is the present number of robots assigned to it by the algorithm. The other is a min PQ of supercells presently being explored with the residual probability key p_i/r_i.

The algorithm then compares the top element in the maxPQ with the top element in the minPQ. If the probability of the maxPQ is larger than the minPQ it transfers an additional robot to the maxPQ supercell, and decrements its key with updated priority. Similarly, the minPQ supercell losses a robot and its priority is incremented due to the loss of one robot. The algorithm continues transferring robots from minPQ supercells to maxPQ supercells. The algorithm however, does not remove the last robot from a supercell until all cells within it have been explored at least once.

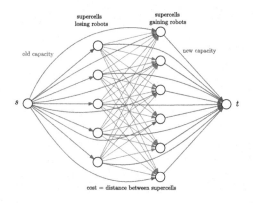

Fig. 6. Optimal robot reassignment via minimum cost network flow.

Once the algorithm has computed the number of robots gained/lost by each supercell, it establishes a minimum cost network flow problem to compute the lowest total transit cost robot-reassignment schedule that satisfies the computed gains and losses. This is modelled as a network flow in a complete bipartite graph (see Fig. 6). In this graph, nodes on the left side of the bipartite graph correspond to supercells losing robots, while nodes on the right correspond to supercells gaining robots. Every node (both losing and gaining) has an incoming arc from the source node with capacity equal to the old number of robots in the associated supercell and cost zero. Similarly, all nodes are connected to the sink with an edge of capacity exactly equal to the updated robot count of the associated supercell and cost zero as well. Lastly, the cross edges in the bipartite graph have infinite capacity and cost equal to the distance between the supercells represented by the end points.

From the construction it follows that the only way to satisfy the constraints is to reassign the robots from the losing supercell nodes to the gaining supercell nodes at minimum travel cost. This network flow problem can be solved in $O(E^2)$ time using the algorithm of Orlin [11]. In this case $E = O(n^2)$ and hence, in the worst case the minimum cost network flow algorithm runs in time $O(n^4)$, where n is the number of supercells.

Theorem 7. *The probabilistically weighted distributed scheduling strategy for the time interval $[0, t]$ can be computed in $O(t\, n^4)$ steps, where n is the size of the search grid.*

5   Conclusion

We present optimal strategies for robot swarm searches under both idealized and realistic considerations. We give pseudo-code showing that the search primitives are simple and can easily be implemented with minimal computational and navigational capabilities. We then give a heuristic to account for the probability of detection map often available in real life searches. The strategies proposed have a factor of k improved time to discovery as compared to a single searcher for the same total travel effort.

References

1. Alpern, S., Gal, S.: The Theory of Search Games and Rendezvous. Kluwer, London (2002)
2. Baeza-Yates, R., Culberson, J., Rawlins, G.: Searching in the plane. Inf. Comput. **106**, 234–252 (1993)
3. Baeza-Yates, R., Schott, R.: Parallel searching in the plane. Comput. Geom.: Theory Appl. **5**, 143–154 (1995)
4. Canadian Coast Guard/Garde Cotiere Canadienne. Merchant ship search and rescue manual (CANMERSAR) (1986)
5. Feinerman, O., Korman, A., Lotker, Z., Sereni, J.-S.: Collaborative search on the plane without communication. In: PODC, pp. 77–86 (2012)
6. IMO. IAMSAR Manual. Organization and Management, vol. I. IAMSAR. Mission Co-ordination, vol. II. IAMSAR. Mobile Facilities, vol. III (2010)
7. Koenig, S., Szymanski, B., Liu, Y.: Efficient, inefficient ANT coverage methods. Ann. Math. Artif. Intell. (Special Issue on ANT Robotics) **31**(1), 41–76 (2001)
8. Koopman, B.O.: Search and screening, Report No. 56 (ATI 64 627), Operations Evaluation Group, Office of the Chief of Naval Operation (1946)
9. López-Ortiz, A., Sweet, G.: Parallel searching on a lattice. In: Proceedings of the 13th Canadian Conference on Computational Geometry (CCCG) (2001)
10. National Search and Rescue Secretariat/Secrétariat national Recherche et sauvetage. CANSARP, SARScene, vol. 4 (1994)
11. Orlin, J.B.: A polynomial time primal network simplex algorithm for minimum cost flows. Math. Program. **78**, 109–129 (1997)
12. Papadimitriou, C.H., Steiglitz, K., Optimization, C.: Algorithms Complex. Dover Publications INC, New York (1998)
13. Stone, L.D.: Revisiting the SS central America search. In: International Conference on Information Fusion (FUSION), pp. 1–8 (2010)
14. U.S. Coast Guard Addendum to the U.S. National Search and Rescue Supplement (NSS) to the IAMSAR Manual. COMDTINST M16130.2F (2013)

Formation of General Position by Asynchronous Mobile Robots Under One-Axis Agreement

Subhash Bhagat[1]($\boxtimes$), Sruti Gan Chaudhuri[2], and Krishnendu Mukhopadhyaya[1]

[1] Indian Statistical Institute, Kolkata, India
{sbhagat_r,krishnendu}@isical.ac.in
[2] Jadavpur University, Kolkata, India
srutiganc@it.jusl.ac.in

Abstract. In the traditional model of autonomous, homogeneous mobile robots, the robots theoretically assumed to be transparent, i.e., they do not create any visual obstructions for the other robots. This paper strengthens this model, by incorporating the notion of obstructed visibility where the robots are considered to be opaque. Many of the existing algorithms require that each robot should have the complete knowledge of the positions of other robots. For this to happen in the new model no three robots can be collinear. This paper proposes a distributed algorithm for obtaining a general position (where no three robots are collinear) for the robots in finite time starting from an arbitrary configuration. The algorithm also assures collision free motion for each robot. The robots here are asynchronous, having no agreement in chirality. However, the robots agree on the direction of any one axis.

Keywords: Asynchronous · Oblivious · Obstructed visibility · General position · One-axis agreement · Swarm robots

1 Introduction

An interesting offshoot of research in robotics is the study of multi-robot systems, popularly known as *swarm robots*. A swarm of robots is a collection of identical, tiny mobile robots. The robots together perform a complex job, e.g., moving a big body, cleaning a big surface etc. A primary objective of a multi-robot system is to co-ordinate motions of the robots to form certain patterns. The traditional distributed model [12,14] for a swarm of robots or multi robot system, represents the mobile entities by distinct points located in the Euclidean plane. The robots are anonymous, indistinguishable, oblivious having no means of direct communication. They may not have common agreement in directions, orientation and unit distance. Each robot has sensing capability, by *vision*, which enables it to determine the positions (within its own coordinate system) of the other robots. The robots operate in rounds by executing *Look-Compute-Move* cycles. All robots may or may not be active in all the rounds. In a round, when a robot becomes active, it gets the positions (w.r.t its local coordinate system) of the

© Springer International Publishing Switzerland 2016
M. Kaykobad and R. Petreschi (Eds.): WALCOM 2016, LNCS 9627, pp. 80–91, 2016.
DOI: 10.1007/978-3-319-30139-6_7

other robots in its surroundings (Look) by its sensing capability. This snapshot is used to compute a destination point (Compute) for this robot. Finally, it moves towards this destination (Move).

Since the robots do not communicate with each other explicitly, the vision enables the robots to communicate with each other implicitly by sensing their relative positions and coordinate their actions accordingly.

1.1 Earlier Works

Majority of the investigations on pattern formation by [10,11,14,15] mobile robots assume that their visibility is unobstructed or full, i.e., if two robots A and B are located at a and b, they can see each other though other robots lie in between them on the line segment $\overline{ab}$ at that time. Very few observations on obstructed visibility (where A and B are not mutually visible if there exist other robots on the line segment $\overline{ab}$) have been made in different scenarios; such as,

- Ando et al. [4] first considers obstructed visibility for solving memoryless point convergence of mobile robots with limited visibility (a robot can see up to a certain fixed region around itself).
- the robots in the one dimensional space [6] - it presents a study on the uniform spreading of robots on a line, when the robots have obstructed visibility.
- the robots with visible lights [8,9] - in this model, each agent is provided with a local externally visible *light*, which is used as colors [2,8–10,13,14,16]. The robots implicitly communicate with each other using these colors as indicators of their states.
- the unit disc robot called *fat robots* [1,7] - in this model, the robots are not points but unit discs [1,5,7]. Unit disc robots obstruct the visibility of other robots. Bolla et al. [5] have presented a simulation for gathering such fat robots. [1,7] presents algorithm for gathering fat robots, where the robots obstruct the visibility of other robots. However, the algorithms in both [1,7] and do not assure collision free paths for the robots. The robots stop if they collide.

The mechanism of removing obstructed visibility, have been addressed recently in [2,3]. In [2] the authors have proposed algorithm for asynchronous robots in light model. Here, the robots starting from any arbitrary configuration form a circle which is itself an unobstructed configuration. The presence of a constant number of visible light (color) bits in each robot, implicitly helps the robots in communication and storing the past configuration. In [3], the robots obtain a obstruction free configuration for semi-synchronous robots by getting as close as possible.

1.2 Our Contribution

In this paper, we propose a distributed algorithm to remove obstructed visibility for asynchronous, oblivious robots by making a general configuration under one-axis agreement. The robots start from arbitrary distinct positions in the plane and reach a configuration where all of them can see each other. The algorithm presented in this paper also assures that the robots do not collide. The obstructed visibility model is no doubt improves the traditional model of multi robot system by incorporating real-life like characteristic. The algorithm presented in this paper can also be a preliminary step for any subsequent tasks which require complete visibility.

The organization of the paper is as follows: Sect. 2, defines the assumptions of the robot model used in this paper and presents the definitions and notations used in the algorithm. Section 3 presents an algorithm for obtaining general position by asynchronous robots. We also furnish the correctness of our algorithm in this section. Finally in Sect. 4, we conclude by providing the future directions of this work.

2   Model and Definitions

Let $\mathcal{R} = \{r_1, \ldots, r_n\}$ be a set of n homogeneous robots represented by points. All the robots may not be active at the same time. When active, the robots operate by executing cycles consisting of three states *look-compute-move* repeatedly. In *Look* state, a robot observes other robots and mark their positions in its local coordinate system. Each robot can observe 360° around itself up to an unlimited radius. However, they obstruct the visibility of other robots. In *Compute* state it computes a destination to move to. In *Move* state it moves to the computed destination. This cycle is executed *asynchronously* by the robots, i.e., the robots may not be in same state at any point of time. Thus a robot can observe other robots in motion and compute its destination based on obsolete locations of the moving robots. The robots are *oblivious*, in the sense that, after completion of a cycle they forget the computed data of that cycle and start a new cycle. The robots can not communicate through explicit massages passing. The movement of the robots are *non-rigid*, i.e., a robot may stop before reaching its destination. However, if a robot moves, it travels at least a finite minimum distance $\delta > 0$ towards its destination. The value of δ is not known to the robots. The total number of robots i.e., n is known to all the robots. The robots do not share a global coordinate system. Each of the robots has its own local coordinate system which may differ from the others. However, the robots only agree on the direction of one-axis. We consider the agreement on Y-axis which is conventionally north-south direction (one can also take the agreement on X-axis).

Initially the robots are positioned in distinct locations and are stationary. The robots are assumed to be fault free. Now we present some definitions and notations which will be used throughout the paper.

- **Position of a Robot:** $r_i(t)$ denotes the position of robot r_i at time t. A configuration $\mathcal{C}(t) = \{r_1(t), \ldots, r_n(t)\}$ is the set of positions occupied by the robots at time t. We denote the set of all such configurations by $\widetilde{C}$. We partition $\widetilde{C}$ into two classes: $\widetilde{C}_L$ and $\widetilde{C}_{NL}$, where $\widetilde{C}_L$ contains all the configurations in which all the robots in $\mathcal{R}$ lie on a straight line and $\widetilde{C}_{NL}$ contains all the

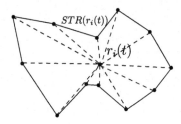

Fig. 1. An example showing the construction of $STR(r_i(t))$.

configurations in which there exist at least three non-collinear robot positions occupied by the robots in $\mathcal{R}$. We say that a configuration $\mathcal{C}(t)$ of R is in **general position** if no three robot positions in $\mathcal{C}(t)$ are collinear. By $\widetilde{C}_{GP}$, we denote the set of all configurations of $\mathcal{R}$ which are in general position. Clearly $\widetilde{C}_{GP} \subset \widetilde{C}_{NL}$.

- **Measurement of Angles:** By *an angle between two line segments*, if not stated otherwise, we mean the angle which is less than or equal to π.
- $\mathcal{V}(r_i(t))$: The vision, $\mathcal{V}(r_i(t))$, of robot r_i at time t is the set of positions occupied by the robots visible to r_i (r_i is not included). Note that if $r_j(t) \in \mathcal{V}(r_i(t))$, then $r_i(t) \in \mathcal{V}(r_j(t))$ and vice versa.

  We sort the robot positions in $\mathcal{V}(r_i(t))$ angularly in anti clockwise direction w.r.t. $r_i(t)$. Starting point can be any robot position in $\mathcal{V}(r_i(t))$. We connect them in that order to get a polygon $STR(r_i(t))$ (Fig. 1).
- A robot r_i is called an *non-terminal robot* at time t if it lies between two other robot positions i.e., on the line segment joining two other robot positions. The point $r_i(t)$ is called the non-terminal robot position. Otherwise, we call r_i a *terminal* robot.
- $DISP(r_i(t), r_j(t))$: When a robot r_i moves from $r_i(t)$ to $r_i(t')$, we call $\angle r_i(t)r_j(t)r_i(t')$ as the angle of displacement of r_i w.r.t. $r_j(t)$ and denote it by $DISP(r_i(t), r_j(t))$ (Fig. 2). It gives a measure of change of vision of the robots when they move.
- $d_{ij}^k(t)$: The straight line joining $r_i(t)$ and $r_j(j)$ is denoted by $\mathcal{L}_{ij}(t)$. The perpendicular distance of the line $\mathcal{L}_{ij}(t)$ from the point $r_k(t)$ is denoted by $d_{ij}^k(t)$.
- $D(r_i(t))$: Let $D(r_i(t))$ be minimum distance of any two robots in $\mathcal{V}(r_i(t))$.

3   Algorithm for Making of General Position

Let $\mathcal{C}(t_0)$ be the initial configuration of the robots. If $\mathcal{C}(t_0) \in \widetilde{C}_{GP}$, then it is done. Otherwise, the movements of the robots are planned in such a way that after a finite number of movements they reach a configuration in $\widetilde{C}_{GP}$. If an initial configuration $\mathcal{C}(t_0)$ is in $\widetilde{C}_L$, it is converted into a configuration in $\widetilde{C}_{NL}$. We describe our strategies in details as follows.

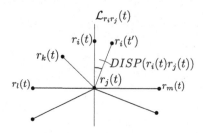

Fig. 2. An example illustrating $DISP(r_i(t), r_j(t))$.

3.1   Eligible Robots for Movements

Our approach selects a terminal robot r_i for movement at time t only if it satisfies any one of the following three conditions:

- $\mathcal{V}(r_i(t)) < n - 1$ and Y-axis of r_i does not contain any other robot position on it.
- $\mathcal{V}(r_i(t)) < n-1$, Y-axis of r_i contains at least one robot position and $r_i(t)$ has the highest $y-$coordinate value among all the robot positions on its Y-axis.
- $\mathcal{V}(r_i(t)) = n-1$, Y-axis of r_i contains at least one robot position and $r_i(t)$ has the highest $y-$coordinate value among all the robot positions on its Y-axis. This is essential to avoid deadlocks in the system.

The robots which are currently in non-terminal positions, do not move.

3.2   Computing Destination Point

Consider an arbitrary terminal robot r_i which finds itself eligible to move at time t. Note that, if there is collinearity in the configuration such a robot always exists. While computing the destination points for the robots, our approach takes care of the followings (i) the movements of the robots do not block the visibility of the other robots (ii) they do not collide with each other. The destination point for r_i depends on whether $\mathcal{C}(t)$ is in $\widetilde{C}_L$ or in $\widetilde{C}_{NL}$. Our algorithm decides two things regarding the movement of a robot: (i) the amount of movement, and (ii) the direction of movement.

- **Case-1: $\mathcal{C}$(t) $\in \widetilde{C}_{\mathbf{NL}}$**
  Three robots become collinear if the triangle, formed by their respective positions, collapses into a line due to their movements. In other words, one of the angles of the triangle must become zero. Our approach avoids the occurrences of such situations by considering two sets of angles. Let $\Gamma(r_i(t))$ denote the set of all angles made at r_i by two consecutive robot positions in $STR(r_i(t))$, i.e.,

$$\Gamma(r_i(t)) = \{\angle r_j r_i r_k : r_j, r_k \text{are two consecutive vertices on} STR(r_i(t))\}.$$

Let $\Gamma'(r_i(t))$ be the set of angles created by the neighbors of $r_i(t)$ at the other robot positions on $STR(r_i(t))$, i.e.,

$$\Gamma'(r_i(t)) = \{\angle r_u(t)r_j(t)r_i(t) : r_j \in \mathcal{V}(r_i(t)) \text{ and } r_u(t) \text{is a neighbor of } r_i(t)$$
$$\text{on } STR(r_j(t))\}.$$

The set $\Gamma(r_i(t))$ reflects the view of r_i about the robots in $\mathcal{V}(r_i(t))$ whereas $\Gamma'(r_i(t))$ gives the view the robots in $\mathcal{V}(r_i(t))$ about r_i.

The Direction of Movement: Our approach also decides the direction of the movement of r_i. For a robot r_j, consider the angle made by $\mathcal{L}_{ij}(t)$ with Y^+ and denote it by $\theta_{ij}(t)$. Let $\alpha(r_i(t)) = minimum\{\theta_{ij}(t) : r_j \in \mathcal{V}(r_i(t))$ and r_j does not lie on Y-axis of r_i } (tie, if any, is broken arbitrarily). Let $Bisec(r_i(t))$ denote the $\frac{1}{n^{2n}}$th bisector of $\alpha(r_i(t))$ closest to Y^+. It is a ray from $r_i(t)$ and the angle made by it with Y^+ is strictly less than $\frac{\pi}{2}$ since $n \geq 3$. When there is no other robot position on Y-axis of r_i, the robot r_i moves along Y^+. Otherwise r_i moves along $Bisec(r_i(t))$.

The Amount of Displacement: The maximum amount of displacement of r_i should be restricted in such a way that it does not create any new collinearity during or after the executions of their movements. If $r_j(t)$ and $r_k(t)$ are in $\mathcal{V}(r_i(t))$, the destination point of r_i should lie far enough from $\mathcal{L}_{jk}(t)$ so that even if all the three robots r_i, r_j and r_k move, they do not become collinear. Let $d(r_i(t)) = minimum\{d_{ij}^k(t), d_{ik}^j(t), d_{jk}^i(t) : \forall r_j, r_k \in \mathcal{V}(r_i(t))\}$. Our algorithm computes the destination point of r_i in such way that $DISP(r_i(t), r_j(t))$ would be small enough to avoid creation of new collinearities. Let $\beta(r_i(t))$ be the minimum of $\Gamma(r_i(t)) \cup \Gamma'(r_i(t))$. Note that $\beta(r_i(t)) \leq \frac{\pi}{3}$. Our approach demands $DISP(r_i(t), r_j(t))$ to be smaller than $\frac{1}{n^{2n}}$th fraction of $\beta(r_i(t))$.
Let

$$\sigma(r_i(t)) = \frac{1}{n^{2n}}minimum\{d(r_i(t)), D(r_i(t))\} * sin(\frac{\beta(r_i(t))}{n^{2n}})$$

- **Case-2: $\mathcal{C}(t_0) \in \widetilde{\mathbf{C}}_{\mathbf{L}}$**
In this case, all the robots lie on a straight line, say $\hat{\mathcal{L}}$. There are $n-2$ intermediate robots. The movement of even one terminal robot converts the present configuration into a configuration in $\widetilde{C}_{NL}$. Suppose r_i is one of the two terminal robots which finds $\mathcal{C}(t_0) \in \widetilde{\mathbf{C}}_{\mathbf{L}}$.

The Direction of Movement: If the robots do not lie on Y-axis of r_i, the robot r_i moves along Y^+. Otherwise r_i moves along X^+.

The Amount of Displacement: In this case $|\mathcal{V}(r_i(t_0))| = 1$. Let

$$\sigma(r_i(t)) = \frac{1}{n^{2n}}D(r_i(t_0))$$

Algorithm 1. ComputeDestination()

Input: $r_i(t)$ and $\mathcal{C}(t)$.
Output: a distance
1. **Case-1:** $|\mathcal{V}(r_i(t))| \geq 2$,
   Compute $\sigma(r_i(t))$ as defined in Case-1 of Subsect. **3.1**;
2. **Case-2:** $|\mathcal{V}(r_i(t))| = 1$,
   Compute $\sigma(r_i(t))$ as defined in Case-2 of Subsect. **3.1**;
3. return $\sigma(r_i(t))$;

Algorithm 2. MakeGeneralPosition()

Input: $\mathcal{C}(t)$, a configuration of a set robots $\mathcal{R}$.
Output: $\mathcal{C}(\hat{t})$, which is in general position.

1 **while** $|\mathcal{V}(r_i(t))| < n - 1 \vee Y\text{-axis contains other robots}$ **do**
2   **if** r_i *is non-terminal* **then**
3 | $p \leftarrow 0$;
4   **else**
5   **if** Y*-axis contains no other robots* **then**
6 | $p \leftarrow ComputeDestination(r_i(t), \mathcal{C}(t))$;
7   **else**
8   **if** $r_i(t)$ *is on the top of its own* Y*-axis* **then**
9 | $p \leftarrow ComputeDestination(r_i(t), \mathcal{C}(t))$;
10   **else**
11                 $p \leftarrow 0$;

12   Compute $\hat{r}_i(t)$ along the direction of movement of r_i and p distance apart from $r_i(t)$;
13   Move to $\hat{r}_i(t)$;

Let $\hat{r}_i(t)$ be the point on the direction of movement of r_i (i.e., on Y^+ or $Bisec(r_i(t))$ or X^+) at distance $\sigma(r_i(t))$ from $r_i(t)$. The destination point of $r_i(t)$ is $\hat{r}_i(t)$.

3.3 Correctness

The algorithm assures that the robots will form general position in finite number of movements. The termination of the algorithm is established by following observation and lemmas.

Let r_i, r_j and r_k be three arbitrary robots which are not collinear. In the following lemmas, we prove that r_i, r_j and r_k, never become collinear during the execution of our algorithm. These three robots will become collinear when $\triangle_{ijk}(t)$ collapses into a line due to the movements of the robots at some time point $t \geq t_0$. In other words all of $d_{ij}^k(t)$, $d_{ij}^k(t)$ and $d_{ij}^k(t)$ must become zero. Without loss of generality we prove that $d_{ij}^k(t)$ will never vanish.

We estimate the maximum decrement in the value of $d_{ij}^k(t)$ i.e., minimum value of $d_{ij}^k(t), t \geq t_0$, while robots move to break the collinearities.

Lemma 1. *Let r_i, r_j and r_k be three arbitrary robots, which are not collinear at time t_0. If each robot moves at most once, they do not become collinear during the whole execution of our algorithm.*

Proof. The maximum decrement or increment in the value of $d_{ij}^k(t_0)$ would occur if the robots move along lines perpendicular to $\mathcal{L}_{ij}()$. Depending on the initial positions of the robots, the following cases may arise.

- r_i, r_j and r_k are **Mutually Visible** at t_0: In a single movement of the robot which moves first, the displacement of the robot would be bounded above by $\frac{1}{n^{2n}}d_{ij}^k(t_0)$. This could increase $d_{ij}^k(t_0)$ to a value which would be bounded above by $(1 + \frac{1}{n^{2n}})d_{ij}^k(t_0)$. So, the observed value of $d_{ij}^k(t'), t' > t_0$ for the second robot to move would be bounded above by $(1 + \frac{1}{n^{2n}})d_{ij}^k(t_0)$. For a single movement of the second robot, the displacement of the robot would be bounded above by $(1 + \frac{1}{n^{2n}})\frac{d_{ij}^k(t_0)}{n^{2n}} < \frac{2}{n^{2n}}d_{ij}^k(t_0)$. Similarly, in a single movement of the third robot, the displacement of the robot would be bounded above by $(1 + \frac{3}{n^{2n}})\frac{d_{ij}^k(t_0)}{n^{2n}} < \frac{2}{n^{2n}}d_{ij}^k(t_0)$ (since $\frac{3}{n^{4n}} < \frac{1}{n^{2n}}$ and $n \geq$ 3). This would convert $d_{ij}^k(t_0)$ to the one which would be bounded above by $(1 + \frac{5}{n^{2n}})d_{ij}^k(t_0)$. This implies that, in a single movement of a robot, the displacement of the robot would be bounded above by $(1 + \frac{5}{n^{2n}})\frac{d_{ij}^k(t_0)}{n^{2n}} < \frac{2}{n^{2n}}d_{ij}^k(t_0)$. Thus, in general, we can assume that in a single movement of a robot, the displacement of the robot would be bounded above by $\frac{2}{n^{2n}}d_{ij}^k(t_0)$. This bound holds irrespective of the scheduling of actions of the robots. Thus, if each of the three robots moves once, we have,

$$d_{ij}^k(t) > (1 - \frac{6}{n^{2n}})d_{ij}^k(t_0) \tag{1}$$

  where $t > t_0$.
- r_i, r_j and r_k are **Not Mutually Visible** at t_0 : Since the three robots are not collinear, they form a triangle $\triangle_{ijk}(t_0)$. Again, since they are not mutually visible, at least one side of $\triangle_{ijk}(t_0)$ contains at least one robot. Without loss of generality, we consider the following scenarios.
  - $r_j(t_0), r_k(t_0) \in \mathcal{V}(r_i(t_0))$ and $r_k(t_0) \notin \mathcal{V}(r_j(t_0))$: Let r_u and r_v be the robots (not necessarily distinct) on $\mathcal{L}_{jk}(t_0)$, closest to r_j and r_k respectively (Fig. 3). As seen earlier, the displacements of r_j and r_k, in single movements, would be bounded above by $\frac{2}{n^{2n}}d_{ij}^u(t_0)$ and $\frac{2}{n^{2n}}d_{iv}^k(t_0)$ respectively. Both of these values are less than $\frac{2}{n^{2n}}d_{ij}^k(t_0)$. The movement of r_i would be bounded above by $\frac{2}{n^{2n}}d_{ij}^k(t_0)$ and the Eq. (1) holds.
  - $r_i(t_0) \notin \mathcal{V}(r_k(t_0)) \cup \mathcal{V}(r_j(t_0))$ and $r_j(t_0) \in \mathcal{V}(r_k(t_0))$: Let r_{u_1} and r_{v_1} be the robots (not necessarily distinct) on $\mathcal{L}_{ik}(t_0)$, closest to r_i and r_k respectively, r_{u_2} and r_{v_2} be the robots (not necessarily distinct) on $\mathcal{L}_{ij}(t_0)$, closest to r_i and r_j respectively (Fig. 4). The displacements of r_i,

r_j and r_k, in single movements, would be bounded above by $\frac{2}{n^{2n}}d_{iu_2}^{u_1}(t_0)$, $\frac{2}{n^{2n}}d_{jv_2}^{k}(t_0)$ and $\frac{2}{n^{2n}}d_{jv_1}^{k}(t_0)$ respectively. All of these values are less than $\frac{2}{n^{2n}}d_{ij}^{k}(t_0)$ and the Eq. (1) holds.

- $\boldsymbol{r_i(t_0) \notin \mathcal{V}(r_k(t_0)) \cup \mathcal{V}(r_j(t_0))}$ **and** $\boldsymbol{r_j(t_0) \notin \mathcal{V}(r_k(t_0))}$: Let r_{u_1} and r_{v_1} be the robots (not necessarily distinct) on $\mathcal{L}_{ik}(t_0)$, closest to r_i and r_k respectively, r_{u_2} and r_{v_2} be the robots (not necessarily distinct) on $\mathcal{L}_{ij}(t_0)$, closest to r_i and r_j respectively and r_{u_3} and r_{v_3} be the robots (not necessarily distinct) on $\mathcal{L}_{jk}(t_0)$, closest to r_j and r_k respectively (Fig. 5). The displacements of r_i, r_j and r_k, in single movements, would be bounded above by $\frac{2}{n^{2n}}d_{iu_2}^{u_1}(t_0)$, $\frac{2}{n^{2n}}d_{jv_2}^{u_3}(t_0)$ and $\frac{2}{n^{2n}}d_{v_1v_3}^{k}(t_0)$ respectively. All of these values are less than $\frac{2}{n^{2n}}d_{ij}^{k}(t_0)$. Thus the Eq. (1) holds.

All the above bounds hold even if (i) robots are observed while they are in motion, and (ii) stop before reaching their destinations. From Eq. (1) the lemma follows.

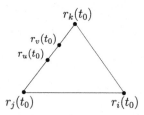

Fig. 3. An example where two pairs of robots are mutually visible.

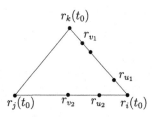

Fig. 4. An example where one pair of robots are mutually visible.

Lemma 2. *Let r_i, r_j and r_k be collinear robots at time t_0. If each robot moves at most once, they become non-collinear and remains non-collinear during whole execution of our algorithm.*

Proof. Without loss of generality, let r_j lie between r_i and r_k (there may be other robots on $\mathcal{L}_{ik}(t_0)$). Depending upon the initial positions of the robots, followings are the possible scenarios:

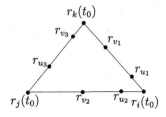

Fig. 5. An example where no pair of robots are mutually visible.

- **Case-1 ($\mathcal{L}_{ik}(t_0)$ is Not Coincident with Y-axis):** In this case, the directions of movements of all the three robots are on the same side of $\mathcal{L}_{ik}(t)$. At least one of r_i and r_k has to move before r_j finds itself as terminal robot. Thus, the movement of r_j would be bounded above by $\frac{1}{n^{2n}}d_{ik}^j(t)$ (even if $r_i \notin \mathcal{V}(r_j(t))$). Hence r_i, r_j and r_k do not remain collinear.
- **Case-2 ($\mathcal{L}_{ik}(t_0)$ is Coincident with Y-axis):** In this case, the directions of movements of the robots may be different sides of $\mathcal{L}_{ik}(t)$. However, the robots move one by one according to an ordering, depending upon their locations on $\mathcal{L}_{ik}(t_0)$. Exactly one of r_i and r_k, say r_i, moves first. Then r_j finds itself as terminal robot and moves. Finally, the r_k moves. The displacement of r_k is bounded above by $\frac{1}{n^{2n}}d_{ij}^k(t)$. Thus, they become non-collinear.

In a single move, robot r_i leaves all the initial lines of collinearity. Hence, it become visible to all those robots which were initially not visible to it provided no new collinearities are created with r_i. By Lemma 1 and Case-1, Case-2 above, r_i does not create any new collinearity in a single move. Hence it becomes visible to all other robots and remains so. Thus once r_i, r_j and r_k become collinear they never again become collinear.

Lemma 3. *Each robot moves at most once during the whole execution of our algorithm.*

Proof. According to our algorithm, a terminal robot moves if it can not see all other robots or one of the lines joining its position with other robot positions is coincident with its Y-axis. In both the cases, one move would be sufficient provided it does not create new collinearities. By Lemmas 1 and 2, in a single move, a robot r_i (i) does not create collinearities with the robots with which it was not collinear initially, and (ii) breaks the initial collinearities. Once r_i becomes visible to all the robots, in our approach, the other robots avoid creating collinearities with it. The robots which are still moving and have not noticed r_i, also do not create collinearities according to our approach. Hence, each robot in the system moves at most once during the whole execution of our algorithm.

Lemma 4. *The movements of the robots are collision free.*

Proof. Let r_i and r_j be two arbitrary robots and at least one of them move. Consider a robot r_k visible to at least one of r_i and r_j. By Lemma 3, each robot moves at most once. If r_i and r_j collide, then r_i, r_j and r_k would become collinear or remain collinear which are contradictions to Lemmas 1 and 2. Hence in our algorithm, the movements of the robots are collision free.

From the above results, we can state the following theorem:

Theorem 1. *A set of asynchronous, oblivious robots, placed in distinct location (initially not in general position) can form general position in finite time without any collision under one-axis agreement.*

4   Conclusion

In this paper we have presented a distributed algorithm for obtaining general position in finite time by a set of autonomous, homogeneous, oblivious, asynchronous robots having one-axis agreement. The algorithm assures that the robots will have collision free movements. Once the robots obtain general position, the next job could be to form any pattern maintaining the general position. Most of the existing pattern formation algorithms have assumed that the robots are transparent. Thus, designing algorithms for forming patterns by maintaining general position of the robots, may be a direct extension of this work.

References

1. Agathangelou, C., Georgiou, C., Mavronicolas, M.: A distributed algorithm for gathering many fat mobile robots in the plane. In: Proceedings of the 32nd ACM Symposium on Principles of Distributed Computing (PODC), pp. 250–259 (2013)
2. Di Luna, G.A., Flocchini, P., Gan Chaudhuri, S., Santoro, N., Viglietta, G.: Robots with lights: overcoming obstructed visibility without colliding. In: Felber, P., Garg, V. (eds.) SSS 2014. LNCS, vol. 8756, pp. 150–164. Springer, Heidelberg (2014)
3. Di Luna, G.A., Flocchini, P., Poloni, F., Santoro, N., Viglietta, G.: The mutual visibility problem for oblivious robots. In: Proceedings of 26th Canadian Conference on Computational Geometry (CCCG 2014) (2014)
4. Ando, H., Oasa, Y., Suzuki, I., Yamashita, M.: Distributed memoryless point convergence algorithm for mobile robots with limited visibility. IEEE Trans. Robot. Autom. **15**(5), 818–828 (1999)
5. Bolla, K., Kovacs, T., Fazekas, G.: Gathering of fat robots with limited visibility and without global navigation. In: Rutkowski, L., Korytkowski, M., Scherer, R., Tadeusiewicz, R., Zadeh, L.A., Zurada, J.M. (eds.) EC 2012 and SIDE 2012. LNCS, vol. 7269, pp. 30–38. Springer, Heidelberg (2012)
6. Cohen, R., Peleg, D.: Local spreading algorithms for autonomous robot systems. Theor. Comput. Sci. **399**, 71–82 (2008)
7. Czyzowicz, J., Gasieniec, L., Pelc, A.: Gathering few fat mobile robots in the plane. Theor. Comput. Sci. **410**(6–7), 481–499 (2009)
8. Das, S., Flocchini, P., Prencipe, G., Santoro, N., Yamashita, M.: The power of lights: synchronizing asynchronous robots using visible bits. In: Proceedings of the 32nd International Conference on Distributed Computing Systems (ICDCS), pp. 506–515 (2012)

9. Das, S., Flocchini, P., Prencipe, G., Santoro, N.: Synchronized dancing of oblivious chameleons. In: Ferro, A., Luccio, F., Widmayer, P. (eds.) FUN 2014. LNCS, vol. 8496, pp. 113–124. Springer, Heidelberg (2014)
10. Efrima, A., Peleg, D.: Distributed models and algorithms for mobile robot systems. In: van Leeuwen, J., Italiano, G.F., van der Hoek, W., Meinel, C., Sack, H., Plášil, F. (eds.) SOFSEM 2007. LNCS, vol. 4362, pp. 70–87. Springer, Heidelberg (2007)
11. Flocchini, P., Prencipe, G., Santoro, N., Widmayer, P.: Arbitrary pattern formation by asynchronous, anonymous, oblivious robots. Theor. Comput. Sci. **407**(1–3), 412–447 (2008)
12. Flocchini, P., Prencipe, G., Santoro, N.: Distributed Computing by Oblivious Mobile Robots. Morgan & Claypool, San Rafeal (2012)
13. Flocchini, P., Santoro, N., Viglietta, G., Yamashita, M.: Rendezvous of two robots with constant memory. In: Moscibroda, T., Rescigno, A.A. (eds.) SIROCCO 2013. LNCS, vol. 8179, pp. 189–200. Springer, Heidelberg (2013)
14. Peleg, D.: Distributed coordination algorithms for mobile robot swarms: new directions and challenges. In: Pal, A., Kshemkalyani, A.D., Kumar, R., Gupta, A. (eds.) IWDC 2005. LNCS, vol. 3741, pp. 1–12. Springer, Heidelberg (2005)
15. Suzuki, I., Yamashita, M.: Distributed anonymous mobile robots: formation of geometric patterns. SIAM J. Comput. **28**(4), 1347–1363 (1999)
16. Viglietta, G.: Rendezvous of two robots with visible bits. In: Flocchini, P., Gao, J., Kranakis, E., der Heide, F.M. (eds.) ALGOSENSORS 2013. LNCS, vol. 8243, pp. 291–306. Springer, Heidelberg (2014)

Graphs Algorithms

On Aligned Bar 1-Visibility Graphs

Franz J. Brandenburg$^{(\boxtimes)}$, Alexander Esch, and Daniel Neuwirth

University of Passau, 94030 Passau, Germany
{brandenb,eschalex,neuwirth}@fim.uni-passau.de

Abstract. A graph is called a bar 1-visibility graph, if its vertices can be represented as horizontal vertex-segments, called bars, and each edge corresponds to a vertical line of sight which can traverse another bar. If all bars are aligned at one side, then the graph is an *aligned bar 1-visibility graph*, *AB1V* graph for short.

We investigate *AB1V* graphs from different angles. First, there is a difference between maximal and optimal *AB1V* graphs, where optimal *AB1V* graphs have the maximum of $4n - 10$ edges. We show that optimal *AB1V* graphs can be recognized in $\mathcal{O}(n^2)$ time and prove that an *AB1V* representation is fully determined by either an ordering of the bars or by the length of the bars. Moreover, we explore the relations to other classes of beyond planar graphs and show that every outer 1-planar graph is a weak *AB1V* graph, whereas *AB1V* graphs are incomparable, e.g., with planar, k-planar, outer-fan-planar, $(1, j)$-visibility, and RAC graphs. For the latter proofs we also use a new operation, called path-addition, which distinguishes classes of beyond planar graphs.

1 Introduction

There is recent interest in beyond planar graphs, which comprise all classes of graphs that extend the planar graphs and are defined by restrictions on crossings. Particular examples are 1-planar graphs [4,18], fan planar graphs [2,3], quasi-planar graphs [20], right angle crossing (RAC) graphs [11], bar visibility graphs [10], bar $(1, j)$-visibility graphs [7], rectangle visibility graphs [15], and map graphs [8,23]. Besides, there are specializations, such as outer 1-planar graphs [1,14], IC-planar graphs [5], outer-fan-planar graphs [2] and *AB1V* graphs. The latter were introduced by Felsner and Massow [13] who called them semi bar 1-visibility graphs. The alignment of the bars is the important property to distinguish *AB1V* from general visibility representations.

Visibility is a major topic in graph drawing and computational geometry. A bar visibility representation displays each vertex by a horizontal bar and each edge by a vertical line of sight between the bars of the endvertices. There are several versions of visibility including *distinct, strong, ϵ* and *weak*. In the distinct, strong and ϵ-versions there is an edge if and only if there is a visibility by a line of sight. In the distinct case the endpoints of bars must have different x-coordinates [13,16]. Bars can be (half) open intervals in the strong case with a line of sight

Supported by the Deutsche Forschungsgemeinschaft (DFG), grant Br835/18-2.

M. Kaykobad and R. Petreschi (Eds.): WALCOM 2016, LNCS 9627, pp. 95–106, 2016.
DOI: 10.1007/978-3-319-30139-6_8

of width zero. In the ϵ-version, the bars are closed intervals and the lines of sight have width $\epsilon > 0$. This version is most commonly used and is sometimes called the strong version. In the weak version there is an edge if there is a visibility. Thus edges can be omitted if there is a visibility. Clearly, graphs in the weak version are exactly the subgraphs of graphs in the other versions. This assumption is relevant for a comparison with other classes of graphs, which are generally closed under taking subgraphs. In the planar case with non-transparent bars, distinct, strong, ϵ and weak visibility graphs have been fully characterized [16, 19, 22, 24]. In particular, for 2-connected graphs, the weak and the ϵ-visibility graphs are exactly the planar graphs, whereas $K_{2,3}$ has no strong visibility representation. All versions coincide for triangulated or maximal planar graphs.

Aligned bar 1-visibility representations and graphs were introduced by Felsner and Massow [13]. They used the distinct version of visibility and allow a line of sight to traverse up to k bars. A distinct $AB1V$ representation, $dAB1V$ for short, is characterized by two permutations, called t-order and r-order, respectively. The t-order is the top-down (or after a rotation the left-to-right) order of the bars of the vertices and the r-order is an ordering of the bars by length. Felsner and Massow established important properties of $dAB1V$ graphs. Every $dAB1V$ graph has a vertex of degree four and has at most $4n - 10$ edges. This bound is tight for all even $n \geq 4$. The graphs are 5-colorable, have clique number five and geometric thickness two. Last but not least they showed that for a given graph an r-order can be computed from a t-order.

$AB1V$ graphs are a proper subclass of bar 1-visibility graphs, which were introduced by Dean et al. [10] and further investigated by Sultana et al. [21] and by Evans et al. [12], who also compared them with other classes of beyond planar graphs. Bar $(1, j)$-visibility graphs [7] and 1-visibility graphs [4] specialize bar 1-visibility graphs and restrict the number of edges that may pass a bar to j and one passes, respectively.

In this work, we extend the research on $AB1V$ graphs. First, we show that $AB1V$ graphs are closed under path-addition. This is a novel operation that arbitrarily allows the addition of vertex-disjoint paths of sufficient length. The operation is later used to distinguish $AB1V$ from other classes of graphs and may be of interest on its own. Then we show that there are maximal $AB1V$ graphs with less than $4n - 10$ edges. A main result is the recognition of optimal $AB1V$ graphs in quadratic time. In addition, we complement a result of Felsner and Massow and show that the t-order can be computed from the r-order of a $dAB1V$ graph in linear time. Finally, we show that every outer 1-planar graph is a weak $AB1V$ graph such that an $AB1V$ representation can be constructed in linear time. However, $AB1V$ graphs are incomparable to e.g., planar, k-planar, outer-fan-planar, bar $(1, j)$-visibility, and RAC graphs.

The paper is organized as follows. In Sect. 2 we introduce basic concepts. Maximal graphs and path-additions are studied in Sects. 3 and 4. In Sect. 5 we investigate recognition problems. The relationship to other graph classes is discussed in Sect. 6 and we conclude with some open problems.

2   Preliminaries

We consider simple, undirected graphs $G = (V, E)$ with vertices listed in some arbitrary order. Edges are denoted by $e = (u, v)$. Let $N(v)$ denote the set of neighbors of a vertex v including v and $G[U]$ the subgraph induced by $U \subseteq V$.

We represent graphs by *aligned bar 1-visibility representations*, *AB1V* for short. As suggested by Felsner and Massow [13], we rotate the drawings, which due to the alignment are more intuitive and compact than ordinary visibility representations. In an *AB1V* representation, each vertex is represented by a vertical bar with bottom at $y = 0$. Each edge $e = (u, v)$ corresponds to a horizontal line of sight between the bars of u and v which can traverse another bar. We distinguish between *distinct*, *strong* and *weak* visibility. In the distinct and strong versions there is an edge if and only if there is a visibility, where in the distinct case all bars have a different length. In the weak version, there is an edge if there is a visibility. We denote this distinction by d, s or w, e.g., $dAB1V$.

A *partial AB1V* representation is the *AB1V* representation of an induced subgraph $G[U]$. An *extension* of a partial *AB1V* representation is an *AB1V* representation of $G[U \cup W]$ such that the restriction to $G[U]$ is the *AB1V* representation of $G[U]$.

From Felsner and Massow [13] we adopt the t- and r-orders for the description of $dAB1V$ representations which are the orderings of the vertices by the position from left to right and by the length of the bars, respectively. Then a partial *AB1V* representation is the restriction of the t- and r-orders to the vertices of U.

From the t-order we immediately obtain that every $sAB1V$ graph has a Hamilton path and every $wAB1V$ graph is sub-Hamiltonian (with a Hamilton cycle).

3   Maximality

AB1V graphs have a density of at most $4n - 10$ [13], however, there are sparser maximal *AB1V* graphs. An *AB1V* graph G is *maximal* if there is no proper *AB1V* supergraph H on the same set of vertices with more edges. Then the addition of any new edge violates *AB1V*. If there is no proper *AB1V* supergraph of the same size, then G is *optimal*, i.e., G has $4n - 10$ edges. The respective classes of graphs are denoted by $oAB1V$ and $mAB1V$, respectively.

Optimal and maximal coincide for some classes of graphs and they differ for others. A planar graph is maximal if and only if it is optimal and has $3n - 6$ edges. However, for outer 1-planar [1], 1-planar [6], and $(1, j)$-visibility graphs [7] there are maximal graphs which are not optimal.

First, we show that there are *AB1V* graphs with a unique *AB1V* representation: a rare result (and generally hard to prove) for beyond planar graphs.

Lemma 1. *There are AB1V graphs with a unique dAB1V representation, i.e., a unique t-order of the bars up to reflection.*

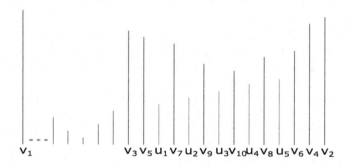

Fig. 1. An $AB1V$ graph with a unique $AB1V$ representation

Proof. (Sketch). Our graphs G_k, for $k \in \mathbb{N}$, consist of a base B and k K_5 subgraphs, which are each connected to three distinguished vertices at the left. The base is composed of three subgraphs with sets of vertices V_0, U_0 and W_0, respectively, see Fig. 1.

The subgraph $G[V_0]$ with $V_0 = \{v_1, \dots, v_{10}\}$ consists of six consecutive K_5. $G[U_0]$ with $U_0 = \{u_1, \dots, u_5\}$ is a path and adds a new path of length six from v_3 to v_4. Finally, $G[W_0]$ is K_5 with edges from some vertices of $G[W_0]$ to v_1, v_3, and v_5 and not to other vertices of V_0 or U_0. The uniqueness of the $AB1V$-representation is based on Observation 1. In consequence, the vertices of $G[\{v_1, v_2, v_3, v_4, v_5\}]$ and $G[\{v_6, v_7, v_8, v_9, v_{10}\}]$ must nest in t-order, vertices v_1 and v_{10} must be the first and last in any t-order, and the vertices of $G_0[W_0]$ must be places between v_1 and v_3 in any t-order.   □

Observation 1. If $(a_1, \dots, a_5)$ is the t-order of a K_5 in an $AB1V$ representation, then a_3 has the shortest bar of $a_1, \dots, a_5$ and thus is the first in the r-order of the K_5. In addition, there is no other vertex b with $a_1 < b < a_5$ in t-order with a bar of length at least the bar of a_3.

Theorem 1. *There are maximal $AB1V$ graphs with density $\frac{19}{5}n - 7$ for $n = 5k + 20, k \in \mathbb{N}$.*

Proof. Consider the graphs G_k from Lemma 1. Each K_5 subgraph $G_k[W_k]$ adds 19 edges, which finally results in $\frac{19}{5}n - 7$ edges for graphs of size $n = 5k + 20$. The $AB1V$ representation from Fig. 1 is maximal, and the graphs are maximal since the $AB1V$ representation is unique.   □

In consequence, a strict hierarchy among $AB1V$ graphs can be obtained.

Corollary 1. $oAB1V \subset mAB1V \subset dAB1V \subset sAB1V \subset wAB1V$.

4   Path-Addition

Next, we introduce a new operation, called *path-addition*, which can be used to distinguish classes of beyond planar graphs. Some classes are closed under path-addition, such as $AB1V$, bar 1-visibility and RAC graphs, whereas the planar, k-planar and $(1, j)$-visibility graphs are not closed.

Definition 1. *For a graph $G = (V, E)$, two vertices $u, v \in V$ and an internally vertex disjoint path $P = (u, w_1, \ldots, w_t, v)$ with $w_i \notin V$ for $1 \leq i \leq t$ from u to v, the* path-addition *results in a graph $G' = (V \cup W, E \cup Q \cup F)$ such that $W = \{w_1, \ldots, w_t\}$ is the set of internal vertices of P, Q consists of the edges of P and F is a set of edges with at least one endpoint in W. We denote G' by $G \oplus P \oplus F$.*

Definition 2. *A* class of graphs $\mathcal{G}$ is closed under path-addition *if for every graph G in $\mathcal{G}$ and for every internally vertex disjoint path P of length at least $|G| - 1$ between two vertices u and v of G there is a set of edges F such that $G \oplus P \oplus F$ is in $\mathcal{G}$.*

Path-addition is an extraordinary and powerful operation. If paths $P_1, \ldots, P_r$ are added successively to graph G, then the length of the paths increases at least exponentially such that $|P_i| \geq |G| - 1 + |P_1| + \cdots + |P_{i-1}|$ and $|P_i| \geq 2^{i-1}(G-1) + 1$. The edges of F_i are only constrained by the fact that one endvertex is from the new path, and they are added to preserve the class of graphs $\mathcal{G}$.

Theorem 2. *The $\{w, d, s\}AB1V$ graphs are closed under path-addition.*

Proof. Consider an $AB1V$ representation of G_i after paths $P_1, \ldots, P_i$ have been added to an $AB1V$ graph G. Increase the length of the bars of the vertices of G_i by $|G_i|$, such that they are long, and let the bars of the internal vertices of P_{i+1} have length $1, \ldots, |P_{i+1}|$, respectively, such that they are short. Insert the bars of the internal vertices of P_{i+1} between the bars of G_i, such that there is a 1-visibility between two consecutive vertices of P_{i+1}. This can be done in many ways if P_{i+1} is longer than the distance of its endvertices in the t-order of G_i.

The set of edges F_{i+1} is determined by the obtained $AB1V$ representation. If the bars of two consecutive vertices a, b of P_{i+1} are placed immediately to the left and right of the bar of vertex c of G_i, then F_{i+1} includes the edges (a, c) and (b, c) and if a and b are placed immediately to the left and right of the bar of another vertex c of P_{i+1}, then F_{i+1} includes the edges $(a, c), (b, c)$, and probably more edges between vertices of P_{i+1} according to the rules of $AB1V$. In any case, an appropriate set of edges F_{i+1} can be found to make G_{i+1} an $AB1V$ graph. □

Note that we add paths one at a time and each path has sufficient length in an $AB1V$ representation to connect the distinguished vertices u and v. A path may make some turns if it is too long. If an $AB1V$ representation is given then a path of length k would suffice, where k is the distance between the bars of u and v.

In the theory of minors, disjoint paths are used for edge-contraction. A path-addition introduces a disjoint path between any two vertices of a graph G. Hence, we can fill any graph G to a graph H that includes K_n as a minor, where $n \geq |G|$. In particular, the discrete graph with five vertices can be filled by 15 paths to a graph with a K_5 minor.

Corollary 2. *The planar graphs are not closed under path-addition.*

Suppose there is a cycle C of length r separating a graph G into an inner and an outer component such that the components are nonempty and C can be traversed only $c \cdot r$ times. Then we can add $c \cdot r + 1$ paths to violate this property. In consequence, we obtain:

Lemma 2. *The 1-planar (k-planar) and the $(1, j)$ bar 1-visibility graphs are not closed under path-addition.*

5 Recognition of Optimal $AB1V$ Graphs

The recognition problem of beyond planar graphs is $\mathcal{NP}$-hard, in general. From the aforementioned classes it is known that outer 1-planar graphs can be recognized in linear time [1, 14], map graphs in cubic time [9], and maximal outer-fan-planar graphs in polynomial time [2]. Here we solve the recognition problem for optimal $AB1V$ graphs. In addition, we show that the t-order can be computed in linear time from the r-order of a d$AB1V$ graph. This complements a result of Felsner and Massow [13] who computed the r-order from the t-order.

For our algorithms we use the following observation:

Observation 2.

- Every optimal $AB1V$ graph has a vertex of degree at most four, namely, the vertex with the shortest bar.
- The maximum complete subgraphs have size at most five [13].
- If G is an optimal $AB1V$ graph, then all vertices have degree at least four.
- If G is an optimal $AB1V$ graph, then for every $i = 5, \ldots, n$ the vertex with the i-th longest bar in a distinct $AB1V$ representation forms a K_5 with vertices with longer bars and it must be placed between these neighbors.
- The vertices with the four longest bars are in pairs at the left and right sides and they form a K_4.

Our recognition algorithm for optimal $AB1V$ graphs proceeds in two phases. First, it computes all K_5 subgraphs. Assuming the vertices of the actual K_5 have the longest bars in an $AB1V$ representation, it then takes an $AB1V$ representation and incrementally extends the partial $AB1V$ representation by adding a vertex of degree four.

Lemma 3. *If G is an optimal $AB1V$ graph, then G has at most $n - 4$ K_5 subgraphs, which can be computed in linear time.*

Proof. Every optimal $AB1V$ graph of size at least five has a vertex of degree four and several such vertices are possible. Our algorithm gets a vertex v of degree four, checks whether $G[N(v)]$ induces a K_5 and if so records the K_5. The algorithm removes v and proceeds on $G - v$, which is optimal, too.

Clearly, the access to v and the steps take $\mathcal{O}(1)$ time. Each K_5 subgraph can be assigned a vertex v of degree four in the actual graph, which results in at most $n - 4$ K_5 and a linear running time.

The correctness is clear by Observation 2. $\qquad\square$

Lemma 4. *There is a linear time algorithm which checks whether the partial AB1V representation of a K_5 subgraph of an optimal AB1V graph G can be extended to an AB1V representation of G, and if so computes an AB1V representation, i.e., the t- and r-orders.*

Proof. The algorithm implements Observation 2. Let $v_1, \ldots, v_5$ be the vertices of the K_5 with $v_1 > v_2 > v_3 > v_4 > v_5 > v_i$ for $i > 5$ in r-order and v_5 between two pairs from v_1, v_2, v_3, v_4 in the t-order of a partial AB1V representation.

If there is a partial AB1V representation of $G[V_i]$ and the vertices of $V_i = \{v_1, \ldots, v_i\}$ are assigned the $n + 1 - i$ longest bars, then the algorithm gets a vertex v_{i+1} which has four neighbors in V_i, assigns the bar of length $n - i$ and places the bar between (at the median) the bars of its four neighbors from V_i. This is a partial AB1V representation of $G[V_{i+1}]$ with $V_{i+1} = V_i \cup \{v_{i+1}\}$. If such a vertex does not exist or the placement is invalid, then the algorithm stops and rejects.

Processing a vertex takes $\mathcal{O}(1)$ time. If the algorithm succeeds it has computed the t- and r-orders of G. □

Theorem 3. *There is a quadratic time algorithm which checks whether a graph G is an optimal AB1V graph, and if so computes an AB1V representation.*

Proof. The algorithm takes each K_5 subgraph of G and successively tests all 120 partial AB1V representations for an extension. The algorithm succeeds if it finds an extension and rejects otherwise. If G is an optimal AB1V graph, the algorithm must succeed and it cannot succeed otherwise. There are $\mathcal{O}(n)$ many K_5 and each test takes $\mathcal{O}(n)$ time. □

Next, we show that the left-to-right order of the bars can be computed from an AB1V graph together with the knowledge on the length of the bars. The converse was shown in [13].

Theorem 4. *There is a linear time algorithm which computes a t-order given an r-order of a dAB1V graph G such that the t- and r-orders describe a dAB1V representation of G.*

Proof. Let $(r_1, \ldots, r_n)$ be the given r-order of G in decreasing order. Thus, vertex r_1 has a bar of length n and r_n has a bar of length one.

Consider a partial AB1V representation of $G[\{r_1, r_2, r_3, r_4\}]$, i.e., a left-to-right ordering of the vertices with the four longest bars. The algorithm successively extends $G[\{r_1, r_2, r_3, r_4\}]$ and in the i-th step adds vertex r_i. Let $V_{i-1} = \{r_1, \ldots, r_{i-1}\}$ for $i = 5, \ldots, n$.

For $i = 5, \ldots, n$, if r_i has four neighbors in V_{i-1}, then the partial AB1V representation is extended by placing r_i between (at the median) its four neighbors in V_{i-1}. If r_i has three neighbors in V_{i-1}, then one of its neighbors is the first (resp. last) vertex in the actual partial AB1V representation, which is extended by placing r_i in second (last but one) place. Finally, if r_i has only two neighbors in V_{i-1}, then r_i is an extreme vertex and is placed to the left (right) of the vertices of V_i, if the former leftmost (rightmost) vertex is its neighbor. The algorithm stops if this placement is invalid.

The correctness is due to the fact that the bars of $r_i, \ldots, r_n$ are shorter than the bar of r_{i-1}. Hence, the bar of r_i can see the bars of its neighbors from V_{i-1}.

Each run with an extension of a partial $AB1V$ representation of $G[V_{i-1}]$ takes linear time, and there are 24 such representations.   □

Together with a result from [13] we can conclude:

Corollary 3. *If G is a dAB1V graph, then it suffices to know (i) the t-order or (ii) the r-order to completely specify a AB1V representation of a dAB1V graph.*

6   Relationship to Other Classes of Graphs

For the comparison of $AB1V$ graphs with other classes of graphs we switch to weak $AB1V$ graphs, since distinct and strong $AB1V$ graphs are too restrictive, e.g., by a degree four vertex. Moreover, the other classes of graphs are generally closed under taking subgraphs. It is not difficult to transfer the later incomparability results to the o, m, d, s versions of $AB1V$ graphs.

Common drawings and visibility representations are equivalent in the planar case (and with 2-connectivity) when crossings are excluded. However, with singleton crossings, visibility representations are more powerful than common drawings, since the 1-planar graphs are a proper subclass of the 1-visibility graphs, which are the bar 1-visibility graphs where each bar is passed by at most one edge [4]. A similar result holds for $AB1V$ graphs.

Theorem 5. *There is a linear time algorithm that constructs a weak distinct $AB1V$ representation of an outer 1-planar graph.*

Proof. There is a linear time algorithm of Auer et al. [1] which tests whether a graph G is outer 1-planar and if so augments G to H by adding edges and constructs a (planar) maximal outer 1-planar embedding of H. The embedding of H consists of planar triangles and kites, which are K_4 that are embedded with a pair of crossing edges.

Let X be the set of crossing edges in the embedding of H and let $H - X$ be the graph after their removal. Then $H - X$ is an outerplanar graph with a Hamilton cycle where all inner faces are triangles or quadrangles. The dual of $H - X$ is a tree with t- and q-vertices corresponding to triangles and quadrangles, respectively. The outer face is ignored. We mark a leaf of T as root r, which is kept to the end.

First, we construct a (planar) bar visibility representation of $H - X$. Consider the root $r = (v_1, \ldots, v_k)$ with $k = 3, 4$ and with the edge (v_1, v_2) in the outer face. Then the t-order is the ordering of the vertices on the Hamilton cycle of H with $v_1 = 1$ and $v_2 = n$. Now, two vertices u, v span an interval.

The length of each bar, i.e., the r-order of $H - X$, is computed by successively removing a leaf from T. This extends Mitchell's algorithm for the recognition of maximal outerplanar graphs by successively removing a vertex of degree two [17]. A leaf b of T one-to-one corresponds to a vertex v of degree two if b is a triangle, and to two vertices u, v of degree two with an edge (u, v) if b is a quadrangle. Suppose that $i - 1$ vertices have been removed so far, where $i = 1, \ldots, n - i - j$ for $j = 0, 1$ and $j = 0$ if and only if the root is a triangle. Assign v a bar of length i if b is removed and v is a degree two vertices in the triangle b. The edges (u, v) and (v, w) incident to v are lines of sight at level i, i.e., at the top end of the bar of v. Increase i by one.

Accordingly, assign bars of length i and $i + 1$ to u and v (in any order) if b is a quadrangle and u, v are the degree two vertices corresponding to b and draw the lines of sight at the top of the bars. Increase i by two. Finally, assign bars of length n and $n - 1$ to the vertices v_1 and v_2 corresponding to the root and bars of length $n - 2$ resp. $n - 3$ to the other vertices of r.

The algorithm preserves the invariants that all bars have a distinct length and that the bars of all vertices in the interval between u and v have bars that are shorter than the bars of u and v if there is an edge (u, v). The latter is due to the fact that the vertices between u and v have been processed before u and v and it guarantees that an edge at level i is unobstructed.

Finally, we reinsert the crossing edges of X. For each pair of crossing edges (a, c) and (b, d) there is a quadrangle (a, b, c, d) with $a < b < c < d$ in t-order and $\{b, c\} < \{a, d\}$ in r-order. The bars of b and c have length i and $i + 1$, respectively. Then the bar of a vertex w in the interval from a to d has length at most $i - 1$ if $w \neq b, c$. Suppose the bar of b has length i, the case where the bar of c has length i is similar. Then draw the edge (a, c) as a planar line of sight at level $i + 1$ and draw the edge (b, d) as a line of sight at level $(i - 0.5)$ which traverses the bar of c. No other bars are affected. Hereby, we adapt the technique of Brandenburg [4] for the bar 1-visibility representation of 1-planar graphs to $AB1V$ representations. For integer coordinates scale by two.

In total we have obtained a weak distinct $AB1V$ representation of G where lines of sight are ignored if there is no edge. All stages of the algorithm take linear time. □

To establish inclusion relations we use Theorem 5 and the facts that outer 1-planar graphs are planar [1], that K_5 is a $wAB1V$ graph, and that K_6 is a bar 1-visibility graph and not a $wAB1V$ graph.

Corollary 4. *The outer 1-planar graphs are a proper subclass of the $wAB1V$ graphs, which in turn, are a proper subclass of the weak bar 1-visibility graphs.*

Next, we address the incomparability of $AB1V$ graphs with other classes of beyond planar graphs.

Theorem 6. *The classes of planar resp. k-planar graphs and of $\{w, d, s\}AB1V$ graphs are incomparable.*

Proof. As stated in Sect. 4, $AB1V$ graphs are closed under path-addition and planar and k-planar graphs are not. For the converse direction, note that every

AB1V graph has a vertex of degree at most four, however, there are planar graph of degree five, such as the dual of the football graph C_{60}, which has faces of size five and six.   □

Moreover, it is not difficult to find graphs which are *AB1V* graphs and are not planar (k-planar).

Clearly, *AB1V* graphs are bar visibility graphs, however, there is an incomparability if there is a bound on the number of lines of sight passing a bar, as in bar $(1, j)$-visibility graphs. The proof follows the reasoning behind Theorem 6.

Theorem 7. *The classes of bar $(1, j)$-visibility graphs and of $\{w, d, s\}AB1V$ graphs are incomparable.*

RAC graphs [11] are closed under path-addition. Nevertheless, they are incomparable to *AB1V* graphs, which can be proved by appropriate examples. Similarly, there are counterexamples for outer-fan-planar graphs.

Theorem 8. *The classes of $\{w, d, s\}AB1V$ and of RAC graphs are incomparable.*

Theorem 9. *The classes of $\{w, d, s\}AB1V$ and of outer-fan-planar graphs are incomparable.*

Proof. The graph $K_6 - e$ consists of K_5 and a vertex v of degree four and admits an *AB1V* representation with v at the first (or last) place in the t-order. Its density exceeds the maximum density of outer-fan-planar graphs [3].

Conversely, consider the outer-fan-planar graph G from Fig. 2, which is built in three stages. First, G has a central K_5 with vertices $v_1, \ldots, v_5$. For each edge (v_i, v_{i+1}) with $i = 1, \ldots, 5$ and $v_6 = v_1$ there is a $K_{2,4}$ with four new vertices $u_1, \ldots, u_4$ each, which are connected by edges (u_j, u_{j+1}) for $j = 1, 2, 3$ such that there is a path $(v_i, u_1, \ldots, u_4, v_{i+1})$. Finally, each edge $(v_i, u_1), \ldots, (u_4, v_{i+1})$ on this path belongs to a new K_5 with three new vertices w_a, w_b, w_c in each case. In total G has 50 vertices.

G is outer-fan-planar as shown by Fig. 2.

Assume that G has an *AB1V* representation. For each K_5 subgraph, Observation 1 applies. Suppose $(v_{i_1}, \ldots, v_{i_5})$ is the t-order of the vertices of the central K_5. Then there are vertices v_p and v_q with $v_{i_2} \leq v_p < v_q \leq v_{i_4}$ in the t-order which are consecutive in the central K_5 and have a $K_{2,4}$. The vertices $u_1, \ldots, u_4$ of this $K_{2,4}$ with the vertices v_p, v_q cannot be placed between the bars of v_p and v_q, since their bars must be shorter than the ones of v_p and v_q by Observation 1, and the shortest bar from $u_1, \ldots, u_4$ were not 1-visible from the bars of v_p and v_q.

Next, the bars of $u_1, \ldots, u_4$ must all be placed to the left of the bar of v_p or to the right of the bar of v_q. This is due to the fact that the vertices u_i, u_{i+1} together with their three new vertices w_a, w_b, w_c form K_5 and Observation 1 applies. Simply speaking, the bars of u_i and u_{i+1} with $i = 1, 2, 3$ must be close together and not separated by a long bar of $v_1, \ldots, v_5$. These arguments also apply to the K_5 with v_p, u_1, and with u_4, v_q, respectively. Assume the bars of u_1, u_2, u_3, u_4 are placed to the left of the bar of v_p; the other case is symmetric.

Then the (long) bar of v_p is placed between the bars of the K_5 with the vertices u_4 and v_q, which violates Observation 1. Hence, there is no $AB1V$ representation for G. $\square$

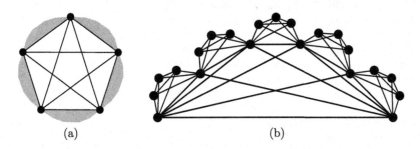

(a)   (b)

Fig. 2. An outer-fan-planar graph which is not an $AB1V$ graph. Gray areas in (a) are the subgraphs displayed in (b).

7 Conclusion

In this work we have investigated $AB1V$ representations and graphs, which are an interesting class of beyond planar graphs. We have deepened the studies of Felsner and Massow [13] and have added new properties and relationships.

Our studies have revealed many new problems, such as the closure of classes of graphs under path-addition and the relation of $AB1V$ graphs to other known classes of beyond planar graphs. Of interest are recognition problems, in particular for weak $AB1V$ graphs.

References

1. Auer, C., Bachmaier, C., Brandenburg, F.J., Gleißner, A., Hanauer, K., Neuwirth, D., Reislhuber, J.: Outer 1-planar graphs. Algorithmica (to appear, 2016)
2. Bekos, M.A., Cornelsen, S., Grilli, L., Hong, S.-H., Kaufmann, M.: On the recognition of fan-planar and maximal outer-fan-planar graphs. In: Duncan, C., Symvonis, A. (eds.) GD 2014. LNCS, vol. 8871, pp. 198–209. Springer, Heidelberg (2014)
3. Binucci, C., Di Giacomo, E., Didimo, W., Montecchiani, F., Patrignani, M., Symvonis, A., Tollis, I.G.: Fan-planarity: properties and complexity. Theor. Comput. Sci. **589**, 76–86 (2015)
4. Brandenburg, F.J.: 1-visibility representation of 1-planar graphs. J. Graph Algorithms Appl. **18**(3), 421–438 (2014)
5. Brandenburg, F.J., Didimo, W., Evans, W.S., Kindermann, P., Liotta, G., Montecchianti, F.: Recognizing and drawing IC-planar graphs. In: Di Giacomo, E., Lubiw, A. (eds.) GD 2015. LNCS, vol. 9411, pp. 295–308. Springer, Heidelberg (2016)

6. Brandenburg, F.J., Eppstein, D., Gleißner, A., Goodrich, M.T., Hanauer, K., Reisl-huber, J.: On the density of maximal 1-planar graphs. In: Didimo, W., Patrignani, M. (eds.) GD 2012. LNCS, vol. 7704, pp. 327–338. Springer, Heidelberg (2013)

7. Brandenburg, F.J., Heinsohn, N., Kaufmann, M., Neuwirth, D.: On bar $(1, j)$-visibility graphs. In: Rahman, M.S., Tomita, E. (eds.) WALCOM 2015. LNCS, vol. 8973, pp. 246–257. Springer, Heidelberg (2015)

8. Chen, Z., Grigni, M., Papadimitriou, C.H.: Map graphs. J. ACM **49**(2), 127–138 (2002)

9. Chen, Z., Grigni, M., Papadimitriou, C.H.: Recognizing hole-free 4-map graphs in cubic time. Algorithmica **45**(2), 227–262 (2006)

10. Dean, A.M., Evans, W., Gethner, E., Laison, J.D., Safari, M.A., Trotter, W.T.: Bar k-visibility graphs. J. Graph Algorithms Appl. **11**(1), 45–59 (2007)

11. Eades, P., Liotta, G.: Right angle crossing graphs and 1-planarity. Discrete Appl. Math. **161**(7–8), 961–969 (2013)

12. Evans, W.S., Kaufmann, M., Lenhart, W., Mchedlidze, T., Wismath, S.K.: Bar 1-visibility graphs vs. other nearly planar graphs. J. Graph Algorithms Appl. **18**(5), 721–739 (2014)

13. Felsner, S., Massow, M.: Parameters of bar k-visibility graphs. J. Graph Algorithms Appl. **12**(1), 5–27 (2008)

14. Hong, S., Eades, P., Katoh, N., Liotta, G., Schweitzer, P., Suzuki, Y.: A linear-time algorithm for testing outer-1-planarity. Algorithmica **72**(4), 1033–1054 (2015)

15. Hutchinson, J.P., Shermer, T., Vince, A.: On representations of some thickness-two graphs. Comput. Geom. **13**, 161–171 (1999)

16. Luccio, F., Mazzone, S., Wong, C.K.: A note on visibility graphs. Discrete Math. **64**(2–3), 209–219 (1987)

17. Mitchell, S.L.: Linear algorithms to recognize outerplanar and maximal outerplanar graphs. Inform. Process. Lett. **9**(5), 229–232 (1979)

18. Ringel, G.: Ein Sechsfarbenproblem auf der Kugel. Abh. aus dem Math. Seminar der Univ. Hamburg **29**, 107–117 (1965)

19. Rosenstiehl, P., Tarjan, R.E.: Rectilinear planar layouts and bipolar orientations of planar graphs. Discrete Comput. Geom. **1**, 343–353 (1986)

20. Suk, A.: k-quasi-planar graphs. In: Speckmann, B. (ed.) GD 2011. LNCS, vol. 7034, pp. 266–277. Springer, Heidelberg (2011)

21. Sultana, S., Rahman, M.S., Roy, A., Tairin, S.: Bar 1-visibility drawings of 1-planar graphs. In: Gupta, P., Zaroliagis, C. (eds.) ICAA 2014. LNCS, vol. 8321, pp. 62–76. Springer, Heidelberg (2014)

22. Tamassia, R., Tollis, I.G.: A unified approach a visibility representation of planar graphs. Discrete Comput. Geom. **1**, 321–341 (1986)

23. Thorup, M.: Map graphs in polynomial time. In: Proceedings of 39th FOCS, pp. 396–405. IEEE Computer Society (1998)

24. Wismath, S.: Characterizing bar line-of-sight graphs. In: Proceedings of 1st ACM Symposium Computational Geometry, pp. 147–152. ACM Press (1985)

A Necessary Condition and a Sufficient Condition for Pairwise Compatibility Graphs

Md. Iqbal Hossain[1(✉)], Sammi Abida Salma[1], and Md. Saidur Rahman[1]

Graph Drawing and Information Visualization Laboratory,
Department of Computer Science and Engineering,
Bangladesh University of Engineering and Technology, Dhaka, Bangladesh
{mdiqbalhossain,saidurrahman}@cse.buet.ac.bd, sammiq@gmail.com

Abstract. In this paper we give a necessary condition and a sufficient condition for a graph to be a pairwise compatibility graph (PCG). Let G be a graph and let G^c be the complement of G. We show that if G^c has two disjoint chordless cycles then G is not a PCG. On the other hand, if G^c has no cycle then G is a PCG. Our conditions are the first necessary condition and the first sufficient condition for pairwise compatibility graphs in general.

1 Introduction

Let T be an edge-weighted tree and let d_{min} and d_{max} be two non-negative real numbers such that $d_{min} \leq d_{max}$. A *pairwise compatibility graph (PCG)* of T for d_{min} and d_{max} is a graph $G = (V, E)$, where each vertex $u' \in V$ represents a leaf u of T and there is an edge $(u', v') \in E$ if and only if the distance between u and v in T, denoted by $d_T(u, v)$, lies within the range from d_{min} to d_{max}. We denote a pairwise compatibility graph of T for d_{min} and d_{max} by $PCG(T, d_{min}, d_{max})$. A graph G is a *pairwise compatibility graph (PCG)* if there exists an edge-weighted tree T and two non-negative real numbers d_{min} and d_{max} such that $G = PCG(T, d_{min}, d_{max})$. An edge-weighted tree T is called a *pairwise compatibility tree (PCT)* of a graph G if $G = PCG(T, d_{min}, d_{max})$ for some d_{min} and d_{max}. Figure 1(a) depicts a pairwise compatibility graph G and Fig. 1(b) of depicts a pairwise compatibility tree T of G for $d_{min} = 4$ and $d_{max} = 5$. Evolutionary relationships among a set of organisms can be modeled as pairwise compatibility graphs [5]. Moreover, the problem of finding a maximal clique can be solved in polynomial time for pairwise compatibility graphs if one can find their pairwise compatibility trees in polynomial time [5].

Constructing a PCT of a given graph is a challenging problem. It is interesting that there are some classes of graphs with very restricted structural properties whose PCT are unknown. For example, it is unknown that whether sufficiently large wheel graphs and grid graphs are $PCGs$ or not. It is known that some specific graphs of 8 vertices, 9 vertices, 15 vertices, and 20 vertices are not PCGs [4,8]. On the other hand, some restricted subclasses of graphs like, cycles, paths, trees, interval graphs, triangle free outerplanar 3-graphs are known as

© Springer International Publishing Switzerland 2016
M. Kaykobad and R. Petreschi (Eds.): WALCOM 2016, LNCS 9627, pp. 107–113, 2016.
DOI: 10.1007/978-3-319-30139-6_9

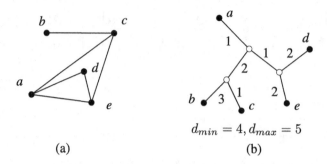

$d_{min} = 4, d_{max} = 5$

(a)   (b)

Fig. 1. (a) A pairwise compatibility graph G and (b) a pairwise compatibility tree T of G.

PCGs [7–9]. It is also known that any graph of at most seven vertices [2] and any bipartite graph with at most eight vertices [6] are PCGs. However the complete characterization of a PCG is not known.

In this paper we give a necessary condition and a sufficient condition for a graph to be a pairwise compatibility graph based on the complement of the given graph. Let G be a graph and let G^c be the complement of G. We prove that if G^c has two disjoint chordless cycles then G is not a PCG. On the other hand, if G^c has no cycle then G is a PCG.

The rest of the paper is organized as follows. In Sect. 2 we give some definitions that are used in this paper. In Sect. 3 we prove a necessary condition for PCG. We give a sufficient condition for PCG in Sect. 4. Finally, Sect. 5 presents some interesting open problems.

2   Preliminaries

Let $G = (V, E)$ be a simple graph with vertex set V and edge set E. Let V' and E' be subsets of V and E, respectively. The graph $G' = (V', E')$ is called a *subgraph* of G, and G' is an *induced* subgraph of G if E' is the set of all edges of G whose end vertices are in V'. The *complement* of G is the graph G^c with the vertex set V but whose edge set consists of the edges not present in G. A *chord* of a cycle C is an edge not in C whose endpoints lie in C. A *chordless cycle* of G is a cycle of length at least four in G that has no chord. For a vertex v of a graph G, $N(v) = \{u | (u, v) \in E\}$ denotes the *open neighborhood*. Let $G_1 = (V_1, E_1)$ and $G_2 = (V_2, E_2)$ be two induced subgraphs of G. We call the subgraphs G_1 and G_2 *disjoint* if they do not share a vertex and there is no edge $(u, v) \in E$ such that $u \in V_1$ and $v \in V_2$. The cycles C_1 and C_2 drawn by thick lines in the graph in Fig. 2 are disjoint cycles. On the other hand, the cycle C_1 drawn by thick line and the cycle C_3 indicated by dotted lines are not disjoint since the edge (c, g) has one end vertex c in C_1 and the other end vertex g in C_3.

Let $G = PCG(T, d_{min}, d_{max})$, and let v be a leaf of T. Then we denote the corresponding vertex of v in G by v' and vice versa. The following lemma is known on a forbidden structure of PCG.

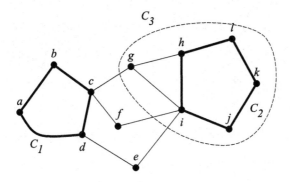

Fig. 2. A graph G with two disjoint chordless cycles whose complement is not a PCG.

Lemma 1 *[4].   Let C be the cycle a', b', c', d' of four vertices. If $C = PCG(T, d_{min}, d_{max})$ for some tree T and values d_{min} and d_{max}, then $d_T(a, c)$ and $d_T(b, d)$ cannot be both greater than d_{max}.*

A graph $G(V, E)$ is an *LPG (leaf power graph)* if there exists an edge-weighted tree T and a nonnegative number d_{max} such that there is an edge (u, v) in E if and only if for their corresponding leaves u', v' in T, we have $d_T(u', v') \leq d_{max}$. We write $G = LPG(T, d_{max})$ if G is an LPG for a tree T with a specified d_{max}. Again G is an *mLPG (minimum leaf power graph)* if there exists an edge-weighted tree T and a nonnegative number d_{min} such that there is an edge (u, v) in E if and only if for their corresponding leaves u', v' in T, we have $d_T(u', v') \geq d_{min}$. We write $G = mLPG(T, d_{min})$ if G is an mLPG for a tree T with a specified d_{min}. Both LPG and mLPG are subclasses of PCG [1]. The following lemmas are known on LPG and mLPG.

Lemma 2 *[3]. Let C_n be a cycle of length $n \geq 5$, then $C_n \notin mLPG$.*

Lemma 3 *[1]. The complement of every graph in LPG is in mLPG and conversely, the complement of every graph in mLPG is in LPG.*

3   Necessary Condition

In this section our aim is to prove that for a given graph G, if G^c has two disjoint chordless cycles of length four or more then G is not a PCG. We first prove the following lemma.

Lemma 4. *Let $G = (V, E)$ be a graph. Let $G_1 = (V_1, E_1)$ and $G_2 = (V_2, E_2)$ be two induced subgraphs of G with no common vertices in G_1 and G_2. Assume that $V_2 \subset N(u')$ and $V_1 \subset N(v')$ in G for every $u' \in V_1$ and every $v' \in V_2$. Let T_1 be a PCT of G_1 such that $G_1 = PCG(T_1, d_{min1}, d_{max1})$ and T_2 be a PCT of G_2 such that $G_2 = PCG(T_2, d_{min2}, d_{max2})$. If there exist two leaves a, c in T_1 such that $d_{T_1}(a, c) > d_{max1}$, and two leaves b, d in T_2 such that $d_{T_2}(b, d) > d_{max2}$, then G is not a PCG.*

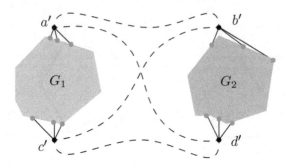

Fig. 3. Illustration for Lemma 4.

Proof. Assume for a contradiction that $G = PCG(T, d_{min}, d_{max})$. Then T has two subtrees T_1 and T_2 such that $G_1 = PCG(T_1, d_{min}, d_{max})$ and $G_2 = PCG(T_2, d_{min}, d_{max})$. Since there exists a pair of leaves a, c in T_1 such that $d_{T_1}(a, c) > d_{max}$ and exists a pair of leaves b, d in T_2 such that $d_{T_2}(b, d) > d_{max}$, both $d_T(a, c)$ and $d_T(b, d)$ are greater than d_{max} in T. Then G does not have the edges (a', c') and (b', d'). Hence the induced subgraph of vertices a', b', c', d' is a cycle in G because $V_2 \subset N(u')$ and $V_1 \subset N(v')$ in G for each $u' \in V_1$ and each $v' \in V_2$ (see Fig. 3), a contradiction to Lemma 1. $\qquad\square$

Since every vertex of G_1 is a neighbor of every vertex of G_2 in Lemma 4, there is no edge (u, v) in G^c where $u \in V_1$ and $v \in V_2$. That means G_1^c and G_2^c are disjoint for the graph G_1 and G_2 in Lemma 4. Thus the following lemma is immediate.

Lemma 5. *Let G be a graph. Let H_1 and H_2 be two disjoint subgraphs of G^c. For any $H_1^c = PCG(T_1, d_{min1}, d_{max1})$ and $H_2^c = PCG(T_2, d_{min2}, d_{max2})$, if T_1 has a pair of leaves whose weighted distance is greater than d_{max1} and T_2 has a pair of leaves whose weighted distance is greater than d_{max2}, then G is not a PCG.*

It would be interesting to investigate what could be the smallest subgraph H_1^c or H_2^c where there always exists a pair of leaves with weighted distance greater than d_{max} in its PCT. Our goal is to show that H_1^c or H_2^c could be a chordless cycle or the complement of a cycle.

It is known that every cycle C and its compliment C^c are PCGs [1,7]. One can easily observe that C_n^c $(n \geq 5)$ is not in $mLPG$. By Lemmas 2 and 3 we have the following lemma.

Lemma 6. *Let C_n be a cycle of length $n \geq 5$, then both C_n and C_n^c are in PCG but not in $\{mLPG, LPG\}$.*

Lemma 6 implies that in any valid PCT of C_n or C_n^c $(n \geq 5)$ there exist a pair of leaves whose weighted distance is greater than d_{max} and a pair of leaves whose weighted distance is smaller than d_{min}. We now prove the following lemma.

Lemma 7. *Let C_n be a cycle graph of vertices $n \geq 4$. Let $C_n^c = PCG(T, d_{min}, d_{max})$. Then there exist a pair of leaves in T whose weighted distance is greater than d_{max}.*

Proof. By Lemma 6 the claim is true for $n \geq 5$. We thus only prove for the case C_4. Let C_4 be the cycle a', b', c', d' of four vertices. Let $C_4^c = PCG(T_1, d_{min1}, d_{max1})$. We prove that there exist a pair of leaves in T_1 whose weighted distance is greater than d_{max1}. Note that C_4^c contains only two edges (a', c') and (b', d'). For contradiction assume that $d_{T_1}(a, b)$, $d_{T_1}(b, c)$, $d_{T_1}(c, d)$, $d_{T_1}(d, a)$ are smaller than d_{min1}. Then $C_4^c = mLPG(T_1, d_{min1})$. By Lemma 3, $C_4 = LPG(T_1, d_{max})$ or $PCG(T_1, 0, d_{max})$ for some d_{max}. Since (a', c') and (b', d') are the non-adjacent pair in C_4, both $d_{T_1}(a, c)$ and $d_{T_1}(b, d)$ are greater than d_{max}. But both $d_{T_1}(a, c)$ and $d_{T_1}(b, d)$ can not be greater than d_{max} by Lemma 1, a contradiction.   □

We are now ready to prove our main result of this section as in the following theorem.

Theorem 1. *Let G be a graph. Let H_1 and H_2 be two disjoint induced subgraphs of G^c. If each of H_1 and H_2 is either a chordless cycle of at least four vertices or C_n^c for $n \geq 5$, then G is not a PCG.*

Proof. Let H_1 be a chordless cycle or compliment of a cycle of at least five vertices. Let $H_1^c = PCG(T_1, d_{min1}, d_{max1})$ for any feasible T_1, d_{min1} and d_{max1}. By Lemma 7, T_1 has a pair of leaves a, b such that $d_{T_1}(a, b) > d_{max1}$. Similarly, if $H_2^c = PCG(T_2, d_{min2}, d_{max2})$ then T_2 has a pair of leaves c, d such that $d_{T_2}(c, d) > d_{max2}$. Then by Lemma 5, G is not a PCG.   □

4 Sufficient Condition

In this section we show that if the complement of a graph has no cycle then the graph is a PCG.

Salma *et. al.* [7] showed that every tree $\mathcal{T}$ is a PCG where $\mathcal{T} = PCG(T, 3, 3)$. They compute the edge-weighted tree T easily by taking a copy of $\mathcal{T}$ and attaching each vertex as a pendant vertex with its original one. Then set weight 1 to each edge. It can be easily verified that $\mathcal{T}$ is a PCT of T for $d_{max} = 3$ and $d_{min} = 3$. For the same settings it is also true that $\mathcal{T} = LPG(T, 3)$. We extend this technique and show that every forest is a LPG as in the following lemma.

Lemma 8. *Let $\mathcal{T}$ be a forest. Then $\mathcal{T} = LPG(T, 3)$. Furthermore the edge-weighted tree T can be found in linear time.*

Proof. Let $\mathcal{T}$ be a forest of trees $\mathcal{T}_1, \mathcal{T}_2, \cdots, \mathcal{T}_k$. We find $\mathcal{T}_i = LPG(T_i, 3)$ for each tree as described above. Then for $2 \geq i \geq k$ we join T_i with $\sum T_{i-1}$, where $\sum T_{i-1}$ is the resultant merged trees up to $i - 1$, as follows. Take two internal vertices u and v from T_i and $\sum T_{i-1}$, respectively. If no internal vertex exists in T_i or $\sum T_{i-1}$ then u or v could be a leaf. Then add an edge between u and v in $\sum T_i$ with weight 2. In this way we get $T = \sum T_k$. Figures 4(a) and 4(b) illustrate a forest $\mathcal{T}$ and an edge-weighted tree T for $\mathcal{T} = LPG(T, 3)$, respectively. It is easy to see that T for $\mathcal{T} = LPG(T, 3)$ can be constructed in linear time.   □

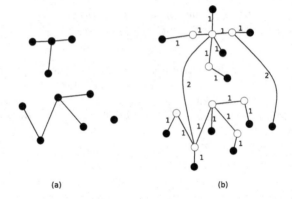

Fig. 4. (a) A forest $\mathcal{T}$ and (b) the edge-weighted tree T for $\mathcal{T} = LPG(T, 3)$.

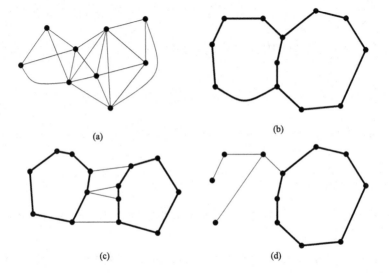

Fig. 5. Illustration for open problems

We now show that if the complement of a graph has no cycle then the graph is a PCG as in the follows theorem.

Theorem 2. *Let G be a graph. If G^c has no cycle then G is a PCG.*

Proof. Let G be a graph. If G^c has no cycle then G^c is a forest. By Lemma 8, $G^c = LPG(T, 3)$. It can be easily verified that $G = mLPG(T, 4)$. $\quad\square$

5   Conclusion

In this paper we have given a necessary condition and a sufficient condition for a pairwise compatibility graph. We have shown that if the complement of a given graph G contains two disjoint chordless cycles or two disjoint complements of cycles then G is not a PCG. On the other hand, if the complement of G do not have any cycle then G is a PCG. The results of this paper suggest several open problems which could be helpful to close the gap between our two conditions. For example: decide whether G is PCG or not (a) if G^c has no chordless cycle, i.e. length of each induced cycle is 3, (b) if G^c has two sufficiently large chordless cycles and share some vertices (see the example in Fig. 5(b)) or join by some edges (see the example in Fig. 5(c)), and (c) G^c has at most one sufficiently large cycle (see the example in Fig. 5(d)).

Acknowledgment. This work is done in the Department of Computer Science and Engineering (CSE) of Bangladesh University of Engineering and Technology (BUET). The authors gratefully acknowledge the support received from BUET.

References

1. Calamoneri, T., Montefusco, E., Petreschi, R., Sinaimeri, B.: Exploring pairwise compatibility graphs. Theor. Comput. Sci. **468**, 23–36 (2013)
2. Calamoneri, T., Frascaria, D., Sinaimeri, B.: All graphs with at most seven vertices are pairwise compatibility graphs. Comput. J. **56**(7), 882–886 (2012)
3. Calamoneri, T., Petreschi, R., Sinaimeri, B.: On relaxing the constraints in pairwise compatibility graphs. In: Rahman, M.S., Nakano, S. (eds.) WALCOM 2012. LNCS, vol. 7157, pp. 124–135. Springer, Heidelberg (2012)
4. Durocher, S., Mondal, D., Rahman, M.S.: On graphs that are not PCGs. Theor. Comput. Sci. **571**, 78–87 (2015)
5. Kearney, P.E., Munro, J.I., Phillips, D.: Efficient generation of uniform samples from phylogenetic trees. In: Benson, G., Page, R.D.M. (eds.) WABI 2003. LNCS (LNBI), vol. 2812, pp. 177–189. Springer, Heidelberg (2003)
6. Mehnaz, S., Rahman, M.: Pairwise compatibility graphs revisited. In: International Conference on Informatics, Electronics Vision (ICIEV), pp. 1–6, May 2013
7. Salma, S.A., Rahman, M.S., Hossain, M.I.: Triangle-free outerplanar 3-graphs are pairwise compatibility graphs. J. Graph Algorithms Appl. **17**(2), 81–102 (2013)
8. Yanhaona, M.N., Bayzid, M.S., Rahman, M.S.: Discovering pairwise compatibility graphs. Discrete Math. Algorithms Appl. **2**(4), 607–623 (2010). Springer
9. Yanhaona, M.N., Hossain, K.S.M.T., Rahman, M.S.: Pairwise compatibility graphs. J. Appl. Math. Comput. **30**, 479–503 (2009)

Mixing Times of Markov Chains
of 2-Orientations

Stefan Felsner[1][(✉)] and Daniel Heldt[1]

Institut Für Mathematik, Technische Universität Berlin, Berlin, Germany
felsner@math.tu-berlin.de

Abstract. We study Markov chains for α-orientations of plane graphs, these are orientations where the outdegree of each vertex is prescribed by the value of a given function α. The set of α-orientations of a plane graph has a natural distributive lattice structure. The moves of the up-down Markov chain on this distributive lattice corresponds to reversals of directed facial cycles in the α-orientation.

A 2-orientation of a plane quadrangulation is an orientation where every inner vertex has outdegree 2. We show that there is a class of plane quadrangulations such that the up-down Markov chain on the 2-orientations of these quadrangulations is slowly mixing. On the other hand the chain is rapidly mixing on 2-orientations of quadrangulations with maximum degree at most 4.

1 Introduction

Let $G = (V, E)$ be a graph and let $\alpha : V \to \mathbb{N}$ be a function, an α-orientation of G is an orientation with $\text{outdeg}(v) = \alpha(v)$ for all vertices $v \in V$. A variety of interesting combinatorial structures on planar graphs can be modeled as α-orientations. Examples are spanning trees, Eulerian orientations, Schnyder woods of triangulations, separating decompositions of quadrangulations. These and further examples are discussed in [5,10]. In this paper we are interested in Markov chains to sample uniformly from the α-orientations of a given planar graph G for a fixed α.

A uniform sampler may be used to get data for a statistical approach to typical properties of α-orientations. Under certain conditions it can also be used for approximate counting of α-orientations.

In [10, Sect. 6.2] it is shown that counting α-orientations can be reduced to counting perfect matchings of a related bipartite graph. The latter problem can be approximately solved using the permanent algorithm of Jerrum et al. [12]. This algorithm also builds on random sampling.

For sampling α-orientations of plane graphs, however, there is a more natural Markov chain. The reversal of the orientation of a directed cycle in an α-orientation yields another α-orientation. If G is a plane graph and $\vec{G}, \vec{G}'$ are α-orientations of G, then we define $\vec{G} < \vec{G}'$ whenever $\vec{G}'$ is obtained by reverting a clockwise cycle of $\vec{G}$. In [5] it has been shown that this order relation makes the set of α-orientations of G into a distributive lattice.

M. Kaykobad and R. Petreschi (Eds.): WALCOM 2016, LNCS 9627, pp. 114–127, 2016.
DOI: 10.1007/978-3-319-30139-6_10

A finite distributive lattice is the lattice of down-sets of some poset P. Let a 'step' consist in adding/removing a random element of P to/from the down-set. These step yield the *up-down Markov chain* on the distributive lattice. A nice feature of the up-down Markov chain is that it is monotone, see [16]. A monotone Markov chain is suited for using *coupling from the past*, see [17]. This method allows to sample exactly from the uniform distribution on the elements of a distributive lattice.

The challenge in applications of the up-down Markov chain is to analyze its mixing time. In [16] some examples of distributive lattices are described where this chain is rapidly mixing but there are examples where the mixing is slow. Miracle et al. [15] have investigated the mixing time of the up-down Markov chain for 3-orientations, a class of α-orientations intimately related to Schnyder woods. They show that there is a class of plane triangulations such that the up-down Markov chain on the 3-orientations of these triangulations is slowly mixing. They also show that the chain is rapidly mixing on 3-orientations of plane triangulations with maximum degree at most 6.

In this paper we present similar results for the up-down Markov chain on the 2-orientations of plane quadrangulations. These special 2-orientations are of interest because they are related to separating decompositions, a structure with many applications in floor-planning and graph drawing. We refer to [6,8,11] and references given there for literature on the subject. Specifically we show that there is a class of plane quadrangulations such that the up-down Markov chain on the 2-orientations of these quadrangulations is slowly mixing. On the other hand the chain is rapidly mixing on 2-orientations of quadrangulations with maximum degree at most 4.

In the full paper [7] we also revisit the case of 3-orientations, there we have somewhat simpler examples, compared to those from [15]. Our examples also have a smaller maximum degree, $O(\sqrt{n})$ instead of $O(n)$ on graphs with n vertices. There we also exhibit a function α and a class of plane graphs of maximum degree 6 such that the up-down Markov chain on the α-orientations of these graphs is slowly mixing.

2   Preliminaries

In the first part of this section we give some background on the up-down Markov chain on general α-orientations. Then we discuss 2-orientations and the associated separating decompositions. Finally we provide some background on mixing times for Markov chains.

2.1   The Up-Down Markov Chain of α-orientations

Let G be a plane graph and $\alpha : V \to \mathbb{N}$ be such that G admits α-orientations. For α-orientations $\vec{G}, \vec{G}'$ of G we define $\vec{G} < \vec{G}'$ when $\vec{G}'$ is obtained by reverting a simple clockwise cycle of $\vec{G}$. The transitive closure of this relation makes the set of α-orientations of G into a distributive lattice, see [5] or [9].

The steps of the up-down Markov chain on a distributive lattice $\mathcal{L} = (X, <)$ correspond to changes $x \leftrightarrow x'$ for covering pairs $x \prec x'$, i.e., pairs $x < x'$ such that there is no $y \in X$ with $x < y < x'$. In other words the up-down Markov chain performs a random walk on the diagram of the lattice. The transition probabilities are (usually) chosen uniformly with a nonzero probability for staying in a state. Since the diagram of a lattice is connected the chain is ergodic. It is also symmetric, hence, the unique stationary distribution is the uniform distribution.

The steps of the up-down Markov chain of α-orientations are given by certain reversals of cycles. For a clean description we need the notion of a *rigid edge*. An edge of $G = (V, E)$ is α-*rigid* if it has the same direction in every α-orientation of G. Let $R \subseteq E$ be the set of α-rigid edges. Since directed cycles of an α-orientation $\vec{G}$ can be reversed, rigid edges never belong to directed cycles. Define $r(v)$ as the number of rigid edges that have v as a tail and let $\alpha'(v) = \alpha(v) - r(v)$. Now α-orientations of G and α'-orientation of $G' = (V, E - R)$ are in bijection. And with the inherited plane embedding of G' the distributive lattices are isomorphic.

If G' is disconnected then we can shift connected components of G' to get a plane drawing $G^{\#}$ without nested components. Since the orientation, clockwise or counterclockwise, of a directed cycle in G' and $G^{\#}$ is identical the distributive lattices of α'-orientations are isomorphic. The steps of the up-down Markov chain of α'-orientations of $G^{\#}$ are easy to describe, they correspond to the reversal of cycles that form the boundary of bounded faces, the face boundaries of $G^{\#}$ are the *essential cycles* for the up-down Markov chain of α-orientations of G. In slight abuse of notation we also refer to the up-down Markov chain of α-orientations of G as the *face flip Markov chain*, after all the essential cycles of G are faces in $G^{\#}$.

From the previous description it follows that the elements of the poset P_α whose down-sets correspond to elements of $\mathcal{L}_\alpha(G)$, i.e., to α-orientations of G, are essential cycles. It is important to keep the following in mind:

Fact A. An essential cycle can correspond to several elements of the poset P_α.

This fact is best illustrated with an example. Figure 1(left) shows the octahedron graph G_{oct} with an Eulerian orientation, this is an α orientation with $\alpha(v) = 2$ for all v. The orientation is the minimal one in the lattice, it has no counterclockwise oriented cycle. Figure 1(middle) depicts the poset P_α the labels of the elements of P_α refer to the corresponding faces of G_{oct}. The elements $1, 1', 1''$ all correspond to the same face of G_{oct}, this face has to be reversed three times in a sequence of face flips that transforms the minimal Eulerian orientation into the maximal.

The elements of P_α can be found as follows. Let $\vec{G}_{\text{min}}$ be the minimal α-orientation, i.e., the one without counterclockwise cycles. Starting from $\vec{G}_{\text{min}}$ perform *flips*, i.e., reversals of essential cycles from clockwise to counterclockwise, in any order until no further flip is possible. The unique α-orientation that admits no flip is the maximal one. The flips of a maximal flip-sequence S are the elements

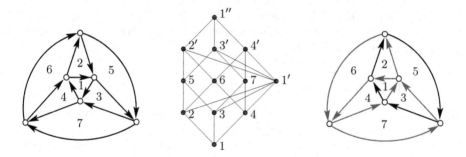

Fig. 1. Left: A minimal α-orientation. Middle: The poset P_α. Right: The α-orientation corresponding to the down set $\{1, 2, 3, 4, 1', 6, 7, 4'\}$ of P_α.

of P_α. Let $\hat{p}(f)$ be the number of times an essential cycle f has been flipped in S. Hence, the elements of P_α are $\{(f, i) : f \text{ essential cycle}, 1 \le i \le \hat{p}(f)\}$.

If essential cycles f and f' share an edge e then from observing the orientation of e we find that between any two appearances of f in a flip-sequence there is a appearance of f'. From this we obtain

Fact B. If essential cycles f and f' share an edge, then $|\hat{p}(f) - \hat{p}(f')| \le 1$.

The above discussion is based on [5] where α-orientations of G have been analyzed via α-potentials, an encoding of down-sets of P_α. If $\vec{G}$ is an α-orientation, then we say that an essential cycle f is at *potential level* i in $\vec{G}$ if (f, i) belongs to the down-set $D_{\vec{G}}$ of P_α corresponding to $\vec{G}$ but $(f, i + 1) \notin D_{\vec{G}}$.

2.2 2-Orientations and Separating Decompositions

A *quadrangulation* is a plane graphs whose faces are uniformly of degree 4. Equivalently quadrangulations are maximal bipartite plane graphs.

Let Q be a quadrangulation, we call the color classes of the bipartition white and black and name the two black vertices on the outer face s and t. A 2-*orientation* of Q is an orientation of the edges such that $\text{outdeg}(v) = 2$ for all $v \ne s, t$. Since a quadrangulation with n vertices has $2n - 4$ edges it follows that s and t are sinks.

A *separating decomposition* of Q is an orientation and coloring of the edges of Q with colors red and blue such that two conditions hold:

(1) All edges incident to s are ingoing red and all edges incident to t are ingoing blue.
(2) Every vertex $v \ne s, t$ is incident to a nonempty interval of red edges and a nonempty interval of blue edges. If v is white, then, in clockwise order, the first edge in the interval of a color is outgoing and all the other edges of the interval are incoming. If v is black, the outgoing edge is the last one of its color in clockwise order (Fig. 2).

Fig. 2. Edge orientations and colors (Color figure online).

Separating decompositions have been studied in [6,8,11]. Relevant to us is that the 'forget function' that associates a 2-orientation with a separating decomposition is a bijection, see [11].

2.3 Markov Chains and Mixing Times

We refer to [13] for basics on Markov chains. In applications of Markov chains to sampling and approximate counting it is critical to determine how quickly a Markov chain M converges to its stationary distribution π. Let $M^t(x, y)$ be the probability that the chain started in x has moved to y in t steps. The *total variation distance* at time t is $\|M^t - \pi\|_{TV} = \max_{x \in \Omega} \frac{1}{2} \sum_y |M^t(x, y) - \pi(y)|$, here The *mixing time* of M is defined as $\tau_{\text{mix}} = \min(t : \|M^t - \pi\|_{TV} \leq 1/4)$. The state space Ω of the Markov chains considered by us consists of sets of graphs on n vertices. Such a chain is *rapidly mixing* if τ_{mix} is upper bounded by a polynomial of n.

A key tool for lower bounding the mixing time of a Markov chain is the *conductance* defined as $\Phi_M = \min_{S \subseteq \Omega, \pi(S) \leq 1/2} \frac{1}{\pi(S)} \sum_{s_1 \in S, s_2 \notin S} \pi(s_1) \cdot M(s_1, s_2)$. The connection with τ_{mix} is given by

Fact T. $\tau_{\text{mix}} \geq (4\Phi_M)^{-1}$.

This is Theorem 7.3 from [13]. A similar result was already shown in [19]. Consider a partition $\Omega^-, \Omega^0, \Omega^+$ of the state space with the property that all paths of the transition graph of the Markov chain that connect Ω^- and Ω^+ contain a vertex from Ω^0, i.e., $M(s_1, s_2) = 0$ for all $s_1 \in \Omega^-$ and $s_2 \in \Omega^+$. An easy computation based on the definition of conductance shows:

Fact C. If $\Omega^-, \Omega^0, \Omega^+$ is such a partition of Ω, then $\Phi_M \leq \frac{\pi(\Omega^0)}{\min\{\pi(\Omega^-), \pi(\Omega^+)\}}$.

A state space has *hour-glass* shape if it has a partition that leads to a super-polynomial lower bound on Φ_M.

3   Markov Chains for 2-Orientations

In this section we study the Markov chain M_2 for 2-orientations of plane quadrangulations. This is a special instance of an up-down Markov chain. A step of the chain consists in the reversal of directed essential cycle.

Lemma 1. *The essential cycles for the Markov chain M_2 of a plane quadrangulation are the four-cycles that contain no rigid edge.*

The *Markov chain M_2* can now be readily described. In each step it chooses a four-cycle C and $p \in [0, 1]$ uniformly at random. If C is directed in the current orientation $\vec{Q}$ and $p \leq 1/2$, then C is reversed, otherwise the new state equals the old one. The stationary distribution of M_2 is the uniform distribution.

In Subsect. 3.2 we prove that M_2 is rapidly mixing for quadrangulations of maximum degree ≤ 4. First, however, we show an exponential lower bound for the mixing time of M_2 on a certain family of quadrangulations.

3.1   Slow Mixing for 2-Orientations

Theorem 1. *Let Q_n be the quadrangulation on $5n + 1$ vertices shown in Fig. 3. The Markov chain M_2 on 2-orientations of Q_n has $\tau_{mix} > 3^{n-3}$.*

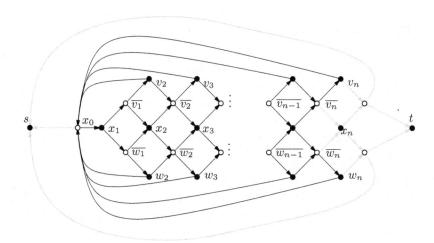

Fig. 3. The graph Q_n with the unique 2-orientation containing the edge (x_0, x_1). Rigid edges are shown gray (Color figure online).

Proof. Let Ω be the set of 2-orientations of Q_n. We define a partition $\Omega_L, \Omega_c, \Omega_R$ of this set. The edge (x_0, s) is rigid, the second out-edge (x_0, a) of x_0 is called *left* if $a \in \{v_2, \dots, v_n\}$, it is *right* if $a \in \{w_2, \dots, w_n\}$ and it is *central* if $a = x_1$. Now $\Omega_L, \Omega_c, \Omega_R$ are the sets 2-orientations where the second out-edge of x_0 is left, central, and right respectively. The next claim shows that we can apply Fact C.

Claim 1. If $\vec{Q}_1 \in \Omega_L$ and $\vec{Q}_2 \in \Omega_R$, the $M_2(\vec{Q}_1, \vec{Q}_2) = 0$.

If $\vec{Q} \to \vec{Q}'$ is a step of M_2 which changes the second out-edge $\vec{e}$ of x_0, then the step corresponds to the reversal of a four-cycle containing $\vec{e}$. There is no four-cycle of Q_n that contains a left and a right edge of x_0.   △

Claim 2. $|\Omega_c| = 1$ and Fig. 3 shows the unique 2-orientation in this set.

Claim 3. $|\Omega_L| = |\Omega_R| \geq \frac{1}{2}(3^{n-1} - 1)$.

From the symmetry of Q_n we get $|\Omega_L| = |\Omega_R|$. Now let P_k be the set of directed path from x_0 to v_k in $\vec{Q}$ from Fig. 3. If $p \in P_k$ then (v_k, x_0) together with p forms a directed cycle in $\vec{Q}$. Reverting this cycle yields a 2-orientation that contains the edge (x_0, v_k). This 2-orientation belongs to Ω_L. Different paths in P_k yield different orientations. Therefore, $|\Omega_L| \geq \sum_k |P_k|$.

It remains to evaluate $|P_k|$. With induction we easily obtain that in $\vec{Q}$ there are exactly 3^{i-1} directed paths from x_0 to either of $\overline{v_i}$ and $\overline{w_i}$. Hence $|P_k| = 3^{k-2}$ and $|\Omega_L| \geq \sum_{2 \leq k \leq n} 3^{k-2} = \frac{1}{2}(3^{n-1} - 1)$. △

The three claims together with Fact C yield $\Phi_{M_2(Q_n)} \leq \frac{2}{3^{n-1}-1}$. Which implies the theorem via Fact T. □

3.2   The Tower Chain for Low Degree Quadrangulations

Following ideas originating from [14] we define a tower Markov chain M_{2T} that extends M_2. A single step of M_{2T} can combine several steps of M_2. Using a coupling argument we show that M_{2T} is rapidly mixing on quadrangulations of degree at most 4. With the comparison technique this positive result will then be extended to M_2.

The basic approach for our analysis of M_{2T} on low degree quadrangulations is similar to what Fehrenbach and Rüschendorf [4] did on certain subgraphs of the quadrangular grid. In the context of 3-orientations of triangulations similar methods were applied by Creed [1] to certain subgraphs of the triangular grid and later by Miracle et al. [15] to general triangulations. As Creed [1] noted these is an inaccurate claim in the proof of [4]. Later [15] stepped into the same trap. In 3.2.1 below we discuss these issues and show how to repair the proofs.

Let $\vec{Q}$ be a 2-orientation and C be a simple cycle. With $e^+(C)$ we denote the number of clockwise edges of C and with $e^-(C)$ the number of counterclockwise edges. If f is a four-cycle and $\nu(f) = |e^+(f) - e^-(f)|$, then $\nu(f)$ can take the values 0, 2, and 4. The face f is *oriented* if $\nu(f) = 4$, it is *scrambled* if $\nu(f) = 0$, and it is *blocked* is $\nu(f) = 2$. If f is blocked, then three edges have the same orientation and one edge does not. We call this the *blocking edge* of f.

A *tower* of length k is a sequence $(f_1, f_2, \ldots, f_k)$ of four-cycles of $\vec{Q}$ such that each f_i for $i = 1, .., k - 1$ is blocked and f_k is oriented. Moreover, in f_i the blocked edge of f_{i-1} is opposite to the blocked edge of $f_{i-1}i$ for $i = 2, .., k - 1$. Figure 4 shows a tower of length 5.

If T is a tower, then all edges that belong to some $f \in T$ but are non-blocking have the same orientation relative to f. This shows that non-blocking edges belong to a unique $f \in T$. Hence after removing all blocking edges from a tower T of length k we obtain a connected region whose boundary ∂T is an oriented cycle with $2k+2$ edges. This is the *boundary cycle* of the tower. The boundary cycle need not be simple but each edge of ∂T only belongs to a single face $f_i \in T$. Therefore, we can also obtain the effect of reverting ∂T by reverting $f_k, f_{k-1}, \ldots, f_1$ in this order.

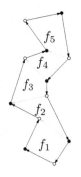

Lemma 2. *If f is a four-cycle, then there is at most one tower starting with $f = f_1$.*

Fig. 4. A tower of length 5.

We are ready to describe the *tower Markov chain* M_{2T}. If M_{2T} is in state $\vec{X}$ then it performs the transition to the next step as follows: an essential four-cycle f, and a $p \in [0,1]$ are each chosen uniformly at random. If in $\vec{X}$ there is a tower T_f of length k starting with f then revert ∂T_f if

- ∂T_f is clockwise and either $k = 1$ and $p \leq 1/2$ or $k > 1$ and $p \leq 1/(4k)$,
- ∂T_f is counterclockwise and either $k = 1$ and $p > 1/2$ or $k > 1$ and $p \geq 1 - 1/(4k)$.

In all other cases the new state is again $\vec{X}$.

Since the steps of M_2 are also steps of M_{2T} the chain is connected. In the orientation obtained by reverting the tower $T = (f_1, \ldots, f_k)$ there is the tower $T' = (f_k, \ldots, f_1)$ whose reversal leads back to the original orientation. Since both towers have the same length the chain is symmetric and its stationary distribution is uniform. For the next lemma the degree condition is indispensable.

Lemma 3. *Let Q have maximum degree ≤ 4 and let $T = (f_1, \ldots, f_k)$ be a tower and $\hat{f} \neq f_k$ be an oriented face in a 2-orientation of Q. If T and $\hat{f}$ share an edge e but $\hat{f}$ and f_1 share no edge, then e is the edge of f_k opposite to the blocking edge of f_{k-1}.*

Proof. Let (u_i, v_i) be the blocking edge of f_i. We extend the labeling of vertices of T such that ∂T is the directed cycle $v_0, v_1, \ldots, v_{k-1}, v_k, u_k, u_{k-1}, \ldots, u_1, u_0$.

If (u_{i+1}, u_i) with $i \geq 1$ is an edge of $\hat{f}$ and $u_{i-1} \notin \hat{f}$, then $\hat{f}$ contains an out-edge of u_i which is not part of T. However, u_i contains the out-edges (u_i, v_i) and (u_i, u_{i-1}). This contradicts $\text{outdeg}(u_i) = 2$.

If (v_i, v_{i+1}) with $i \geq 1$ is an edge of $\hat{f}$ and $v_{i-1} \notin \hat{f}$, then $\hat{f}$ contains an in-edge of v_i which is not part of T. Vertex v_i also contains the in-edges (u_i, v_i) and (v_{i-1}, v_i). Now v_i has in-degree ≥ 3, since $\text{outdeg}(v_i) = 2$ the degree is at least 5. A contradiction.

We are not interested in edges shared by $\hat{f}$ and f_1, i.e., in edges containing u_0 or v_0. Therefore, the only remaining candidate for e is the edge (v_k, u_k).   □

Theorem 2. *Let Q be a plane quadrangulation with n vertices so that each inner vertex is adjacent to at most 4 edges. The mixing time of M_{2T} on 2-orientations of Q satisfies $\tau_{mix} \in O(n^5)$.*

The proof of Theorem 2 is based on the path coupling theorem of Dyer and Greenhill [3]. The following simple version of the theorem suits our needs.

Theorem 3 (Dyer–Greenhill). *Let M be a Markov chain with state space Ω. If there is a graph $\mathcal{G}_M$ with vertex set Ω and a coupling (X_t, Y_t) of M such that with the graph distance $d : \Omega \times \Omega \to \mathbb{N}$ based on $\mathcal{G}_M$ we have:*

$$\mathbb{E}[d(X_{t+1}, Y_{t+1})] \le d(X_t, Y_t) \quad and \quad \Pr(d(X_{t+1}, Y_{t+1}) \ne d(X_t, Y_t)) \ge \rho$$
$$then \ \tau_{mix}(M) \le 2\lceil e/\rho\rceil \, diam(\mathcal{G}_M)^2.$$

The coupling of M_{2T} used for the proof of Theorem 2 is the trivial one, i.e., we run chains X_t and Y_t with the same choices of f and p in each step.

The graph $\mathcal{G}$ will be the transition graph of M_2, i.e., the distance between 2-orientations $\vec{X}$ and $\vec{Y}$ equals the number of four-cycles that have to be reversed to get from $\vec{X}$ to $\vec{Y}$.

Lemma 4. *The maximum potential* $\hat{p}_{max} = \max_f \hat{p}(f)$ *of an essential cycle is less than* n.

Proof. Let Q be the quadrangulation whose 2-orientations are in question. It is convenient to replace Q by $Q^{\#}$ so that essential cycles are just faces. Recall that $\hat{p}$ of the outer face is 0 and $|\hat{p}(f) - \hat{p}(f')| \le 1$ for any two adjacent faces (Fact B). Since a quadrangulation has $n - 2$ faces we obtain $(n - 3)$ as an upper bound for $\hat{p}_{max}$. $\square$

Lemma 5. *The diameter of* $\mathcal{G}$ *is at most* $n^2/2$.

Proof. The height of the lattice $\mathcal{L}_\alpha(Q^{\#})$ is the length of a maximal flip sequence, i.e., $\sum_f \hat{p}(f)$. Using (Fact B) as in the proof of the previous lemma we find that $\sum_f \hat{p}(f) \le 0 + 1 + \ldots + (n - 3)$. This is $< n^2/2$.

In the diagram of a distributive lattice the diameter is attained by the distance between the zero and the one, i.e., between the global minimum and the global maximum. This distance is exactly the height of the lattice. Since $\mathcal{G}$ is the cover graph (undirected diagram) of the distributive lattice $\mathcal{L}_\alpha(Q)$ we obtain that the diameter of $\mathcal{G}$ is at most $n^2/2$. $\square$

3.2.1   Finding an Appropriate ρ.

To get a reasonable ρ the following argument is tempting and was actually used in [4,15]: For given $\vec{X}$ and $\vec{Y}$ there is always at least one essential cycle f whose reversal in $\vec{X}$ reduces the distance to $\vec{Y}$. If $(X_t, Y_t) = (X, Y)$ and this cycle f is chosen by M_{2T}, then with probability 1/2 the distance is reduced. There are at most $n - 3$ essential cycles. Hence we may set $\rho = 1/(2n)$.

Indeed for up-down Markov chains on distributive such a statement holds. If I and J are down-sets of the poset P, then there is an $x \in P$ whose addition to or removal from I decreases the distance to J. In the context of α-orientations, however, an f whose reversal in $\vec{X}$ reduces the distance to $\vec{Y}$ may be oriented in $\vec{Y}$ with the same orientation as in $\vec{X}$. In that case if f is chosen by M_{2T} the reversal of f is applied to both or to none.

Figure 5 shows that there are cases where a pairs (X_t, Y_t) exists such that $\Pr(d(X_{t+1}, Y_{t+1}) \neq d(X_t, Y_t)) = 0$.

To overcome this problem we now define the *slow tower Markov chain* M_{S2T}. If M_{S2T} is in state $\vec{X}$ then it performs the transition to the next step as follows: an essential four-cycle f, a value i with $0 \leq i < n$ and a $p \in [0,1]$ are each chosen uniformly at random. If f is not at potential level i in $\vec{X}$, then nothing is done and $\vec{X}$ is the new state. Otherwise, the step of M_{S2T} equals the step of M_{2T} with the pair (f, p).

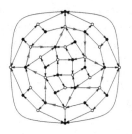

Fig. 5. A 2-orientation $\vec{Q}$ such that the 2-orientation $\vec{Q}'$ obtained by reverting the blue cycle has the same oriented faces (Color figure online).

Lemma 6. *If* (X_t, Y_t) *is a trivial coupling of the slow chain* M_{S2T}, *then* $1/(2n^2)$ *is a lower bound on* $\Pr(d(X_{t+1}, Y_{t+1}) \neq d(X_t, Y_t))$.

Proof. For given $\vec{X}$ and $\vec{Y}$ there is always at least one essential cycle f_1 whose reversal in $\vec{X}$ reduces the distance to $\vec{Y}$. If f_1 appears in $\vec{X}$ and $\vec{Y}$ with the same orientation then the potential level of f_1 in $\vec{X}$ and $\vec{Y}$ is different. Hence, if for the step of M_{S2T} the triple (f, i, p) is chosen such that $f = f_1$ and i is the potential level of f in $\vec{X}$ and p is such that f is actually reversed, then the distance decreases.

The probability for choosing f is at least $1/n$. For i and p the probabilities are $1/n$ and $1/2$ respectively. Together this yields the claimed bound.   □

3.2.2 Completing the proof of Theorem 2.

In Lemma 7 we show that if $(\vec{X}, \vec{Y})$ is an edge of $\mathcal{G}$ and $(\vec{X}^+, \vec{Y}^+)$ is the pair obtained after a single coupled step of the tower chain M_{2T}, then $\mathbb{E}[d(\vec{X}, \vec{Y}) - d(\vec{X}^+, \vec{Y}^+)] \leq 0$. Note that a step of the coupled slow chain M_{S2T} moves the pair $(\vec{X}, \vec{Y})$ to one of $(\vec{X}, \vec{Y})$ or $(\vec{X}^+, \vec{Y}^+)$. Hence Lemma 7 also applies to M_{S2T}.

Assuming Lemma 7 we get the following:

Proposition 1. *Let* Q *be a plane quadrangulation with* n *vertices so that each inner vertex is adjacent to at most 4 edges. The mixing time of* M_{S2T} *on 2-orientations of* Q *satisfies* $\tau_{mix}(M_{S2T}) \in O(n^6)$

Proof. The condition $\mathbb{E}[d(X_{t+1}, Y_{t+1})] \leq d(X_t, Y_t)$ needed for the application of Theorem 3 follows from Lemma 7 because on a single edge M_{S2T} is just a slowed down version of M_{2T} a linearity of expectation. Applying Theorem 3 with parameters $\rho = \frac{1}{2n^2}$ (Lemma 6) and $\text{diam}(\mathcal{G}) \leq n^2/2$ (Lemma 5) yields $\tau_{mix}(M_{S2T}) \leq en^6$.   □

The mixing time of the slow chain could thus be proven with a coupling that allows an application of the theorem of Dyer and Greenhill. Now consider a single state $\vec{X}_t$ evolving according to the slow chain M_{S2T}. Note that this is exactly as if we would run the tower chain M_{2T} but only allow a transition to be conducted if an additional uniform random variable $q \in \{0, \ldots, n-1\}$ takes the value $q = 0$. It follows that the mixing times of M_{S2T} and of M_{2T} deviate by a factor of n. Therefore, $\tau_{\mathrm{mix}}(M_{2T}) \leq en^5$.

To complete the proof of Theorem 2 it remains to prove Lemma 7.

Lemma 7. *If $(\vec{X}, \vec{Y})$ is an edge of $\mathcal{G}$ and $(\vec{X}^+, \vec{Y}^+)$ is the pair obtained after a single coupled step of M_{2T}, then $\mathbb{E}[d(\vec{X}, \vec{Y}) - d(\vec{X}^+, \vec{Y}^+)] \leq 0$.*

Proof. Since $(\vec{X}, \vec{Y})$ is an edge of $\mathcal{G}$ they differ in the orientation of exactly one face $\hat{f}$. We assume w.l.o.g that $\hat{f}$ is oriented clockwise in $\vec{X}$ and counterclockwise in $\vec{Y}$.

Let f be the face chosen for the step of M_{2T}. Depending on f we analyze $d(\vec{X}^+, \vec{Y}^+)$ in three cases.

A. If $f = \hat{f}$, then depending on the value of p face f is reversed either in $\vec{X}$ or in $\vec{Y}$. After the step the orientations $\vec{X}^+$, $\vec{Y}^+$ coincide. The expected change of distance in this case is -1.

B. If f and $\hat{f}$ share an edge and $f \neq \hat{f}$ there are three options depending on the type of f in $\vec{Y}$.

   1. Face f is oriented in $\vec{Y}$, necessarily clockwise. It follows that in $\vec{X}$ face f starts the clockwise tower $(f, \hat{f})$ of length two. In $\vec{Y}$ a face f is a clockwise tower of length 1. If $p \leq 1/8$ both towers are reversed so that $\vec{X}^+$ and $\vec{Y}^+$ coincide. If $1/8 < p \leq 1/2$, then f is reversed in $\vec{Y}$ while $\vec{X}^+ = \vec{X}$, in this case the distance increases by 1. If $p > 1/2$ both orientations remain unchanged. The expected change of distance in this case is $\frac{1}{8} \cdot (-1) + (\frac{1}{2} - \frac{1}{8}) \cdot (+1) + \frac{1}{2} \cdot 0 = \frac{1}{4}$.

   2. Face f is scrambled in $\vec{Y}$. In this case f is blocked in $\vec{X}$ and it may start a tower of length k. If $p \leq 1/(4k)$ this tower is reverted which results in a increase of distance by k. In all other cases the distance remains unchanged. hence, the expected change of distance in this case is $\leq 1/4$.

   3. Face f is blocked in $\vec{Y}$. Then it is either oriented or scrambled in $\vec{X}$. After changing the role of $\vec{X}$ and $\vec{Y}$ we can use the analysis of the other two cases to conclude that the expected change of distance is again $\leq 1/4$.

C. Finally, suppose that f and $\hat{f}$ have no edge in common.

   1. If f starts a tower in $\vec{X}$ which has no edge in common with $\hat{f}$, then f starts the very same tower in $\vec{Y}$ and the coupled chain will either revert both towers or none of them. The distance remains unchanged.

   2. Now let f start a tower $T = (f_1, \ldots, f_k)$ in $\vec{X}$ which has an edge in common with $\hat{f}$. The case where $\hat{f}$ and $f_1 = f$ share an edge was considered in **B**. Now Lemma 3 implies that either $\hat{f} = f_k$ or $\hat{f} \neq f_k$ and the shared edge is such that $(f_1, \ldots, f_k, \hat{f})$ is a tower in $\vec{Y}$. Hence, with T there is a tower T' in $\vec{Y}$ that starts in f and has length $k \pm 1$, moreover T and T' have the same orientation. Let ℓ be the larger of the lengths of T and T'. With a probability of $1/(4\ell)$ both towers are reversed and the distance decreases by 1. With a probability of $1/(4(\ell-1)) - 1/(4\ell)$ only the shorter of the two towers is reversed and the distance increases by $\ell - 1$. With the remaining probability both orientations remain unchanged. The expected change of distance in this case is $\frac{1}{4\ell} \cdot (-1) + (\frac{1}{4(\ell-1)} - \frac{1}{4\ell}) \cdot (\ell - 1) = 0$.

Let m be the number of essential four-cycles, i.e., the number of options for f. Combining the values for the change of distance in cases **A, B, C** and the probability of these cases we obtain: $\mathbb{E}[d(\vec{X}, \vec{Y}) - d(\vec{X}^+, \vec{Y}^+)] \leq \frac{1}{m}(-1) + \frac{4}{m}(1/4) + \frac{m-5}{m}0 = 0$. $\qquad\square$

3.3   Comparison of M_{2T} and M_2

The comparison of the mixing times of M_{2T} and M_2 is based on a technique developed by Diaconis and Saloff-Coste [2]. We will use Theorem 4 a variant due to Randall and Tetali [18].

Let M and $\widetilde{M}$ be two reversible Markov chains on the same state space Ω such that M and $\widetilde{M}$ have the same stationary distribution π. With $E(M)$ we denote the edges of the directed transition graph of M, i.e., $(x, y) \in E(M)$ whenever $M(x, y) > 0$. Define $E(\widetilde{M})$ alike. For each $(x, y) \in E(\widetilde{M})$ define a canonical path γ_{xy} as a sequence $x = v_0, v_1, \ldots, v_k = y$ of transitions of M, i.e. $(v_i, v_{i+1}) \in E(M)$ for all i. Let $|\gamma_{xy}|$ be the length of γ_{xy} and for $(x, y) \in E(M)$ let $\Gamma(x, y) := \{(u, v) \in E(\widetilde{M}) : (x, y) \in \gamma_{uv}\}$. Further let

$$\mathcal{A} := \max_{(x,y) \in E(M)} \left\{ \frac{1}{\pi(x) M(x, y)} \sum_{(u,v) \in \Gamma(x,y)} |\gamma_{uv}| \pi(u) \widetilde{M}(u, v) \right\}$$

and let $\pi_\star := \min_{x \in \Omega} \pi(x)$.

Theorem 4. *In the above setting* $\tau_{mix}(M) \leq 4 \log(4/\pi_\star)\, \mathcal{A}\, \tau_{mix}(\widetilde{M})$.

We are going to apply this theorem with $M = M_2$ and $\widetilde{M} = M_{2T}$. Both chains are symmetric, hence reversible, and have the uniform distribution π as stationary distribution.

The definition of the canonical paths comes quite natural. A transition $(\vec{U}, \vec{V})$ of M_{2T} corresponds to the reversal of ∂T for some tower T of $\vec{U}$. Suppose that $T = (f_1, \ldots, f_k)$ and recall that the effect of reverting ∂T can also be obtained by reverting $f_k, f_{k-1}, \ldots, f_1$ in this order. Reverting them one by one yields a path in $E(M)$, this path is chosen to be $\gamma_{\vec{U}\vec{V}}$.

If $|\gamma_{\vec{U}\vec{V}}| = k$, i.e., the transition $(\vec{U}, \vec{V})$ corresponds to a tower of length k, then $M_{2T}(\vec{U}, \vec{V}) = 1/(4k)$, hence, $|\gamma_{\vec{U}\vec{V}}| M_{2T}(\vec{U}, \vec{V}) = 1/4$. Also π is constant so that $\pi(\vec{U})/\pi(\vec{X}) = 1$. For an upper bound on $\mathcal{A}$ we therefore only have to estimate the number of tower moves that have a canonical path that contains the face flip at f that moves $\vec{X}$ to $\vec{Y}$. If $T = (f_1, \ldots, f_k)$ is such a tower with $f = f_i$, then $(f_1, \ldots, f_{i-1}, f)$ is a tower in $\vec{X}$ and $(f_k, \ldots, f_{i+1}, f)$ is a tower in $\vec{Y}$. Since a tower is defined by its initial face each of $\vec{X}$ and $\vec{Y}$ has at most n towers, all the more each has at most n towers ending in f. This shows $|\Gamma(\vec{X}, \vec{Y})| \leq n^2$ and $\mathcal{A} \leq n^2/4$.

It remains to find $\pi_\star = \frac{1}{|\Omega|}$. Since a quadrangulation has $2n - 4$ edges it has at most 2^{2n} orientations this would suffice for our purposes. However, a better upper bound of 1.9^n for the number of 2-orientations was obtained in [10].

Given the above ingredients for the comparison theorem and the mixing time of $\tau_{\mathrm{mix}}(M_{2T}) \in O(n^5)$ from Theorem 2 we have the theorem:

Theorem 5. *Let Q be a plane quadrangulation with n vertices so that each inner vertex is adjacent to at most 4 edges. The mixing time of the face reversal Markov chain M_2 on 2-orientations of Q satisfies $\tau_{mix}(M_2) \in O(n^8)$.*

4   Concluding Remarks and Open Problems

In this work we have studied 2-orientations, a special class of α-orientations. We have obtained some of the few known results about the complexity of sampling α-orientations of planar graphs. It would be of interest to extend the study to other interesting instances, e.g. to the sampling of Eulerian orientations. Another challenge is to better understand the dependence on the degree. So far we have rapid mixing for degree ≤ 4 and slow mixing examples with maximum degree $O(n)$, this is a huge gap.

References

1. Creed, P.J.: Sampling eulerian orientations of triangular lattice graphs. J. Discr. Alg. **7**, 168–180 (2009)
2. Diaconis, P., Saloff-Coste, L.: Comparison theorems for reversible Markov chains. An. Appl. Prob. **3**, 696–730 (1993)
3. Dyer, M., Greenhill, C.: A more rapidly mixing Markov chain for graph colourings. Rand. Struct. Alg. **13**, 285–317 (1998)
4. Fehrenbach, J., Rüschendorf, L.: Markov chain algorithms for Eulerian orientations and 3-Colourings of 2-Dimensional cartesian grids. Statistics Decisions **22**, 109–130 (2004)
5. Felsner, S.: Lattice structures from planar graphs. Electr. J. Combin. **11**(1), 24 (2004)
6. Felsner, S., Fusy, É., Noy, M., Orden, D.: Bijections for Baxter families and related objects. J. Combin. Theory Ser. A **18**, 993–1020 (2011)
7. Felsner, S., Heldt, D.: Mixing times of markov chains on degree constrained orientations of planar graphs (2015). http://page.math.tu-berlin.de/~felsner/Paper/mix-alpha.pdf
8. Felsner, S., Huemer, C., Kappes, S., Orden, D.: Binary labelings for plane quadrangulations and their relatives. Discr. Math. Theor. Comp. Sci. **12**(3), 115–138 (2010)
9. Felsner, S., Knauer, K.: ULD-lattices and Δ-bonds. Comb. Probab. Comput. **18**(5), 707–724 (2009)
10. Felsner, S., Zickfeld, F.: On the number of planar orientations with prescribed degrees. Electr. J. Combin. **15**, 41p (2008)

11. de Fraysseix, H., de Mendez, O.P.: On topological aspects of orientations. Discr. Math. **229**, 57–72 (2001)
12. Jerrum, M., Sinclair, A., Vigoda, E.: A polynomial-time approximation algorithm for the permanent of a matrix with nonnegative entries. J. ACM **51**, 671–697 (2004)
13. Levin, D., Peres, Y., Wilmer, E.: Markov Chains and Mixing Times. AMS, Providence (2009)
14. Luby, M., Randall, D., Sinclair, A.: Markov chain algorithms for planar lattice structures. In: 36th FOCS, pp. 150–159 (1995)
15. Miracle, S., Randall, D., Streib, A.P., Tetali, P.: Mixing times of Markov chains on 3-orientations of planar triangulations. In: Proceeding AofA 2012, pp. 413–424, Proceeding AQ, Discrete Mathematics and Theory Computer Science (2012). arxiv:1202.4945
16. Propp, J.: Generating random elements of finite distributive lattices. Electr. J. Combin. **4**(2), R15 (1997)
17. Propp, J.G., Wilson, D.B.: Exact sampling with coupled Markov chains and applications to statistical mechanics. Rand. Struct. Alg. **9**(1&2), 223–252 (1996)
18. Randall, D., Tetali, P.: Analyzing Glauber dynamics by comparison of Markov chains. J. Math. Phys. **41**, 1598–1615 (1997)
19. Sinclair, A., Jerrum, M.: Approximate counting, uniform generation and rapidly mixing Markov chains. Inf. Comput. **82**, 93–133 (1989)

Computational Geometry

Computing a Minimum-Width Square Annulus in Arbitrary Orientation

[Extended Abstract]

Sang Won Bae$^{(\boxtimes)}$

Department of Computer Science, Kyonggi University, Suwon, South Korea
swbae@kgu.ac.kr

Abstract. In this paper, we address the problem of computing a minimum-width square annulus in arbitrary orientation that encloses a given set of n points in the plane. A square annulus is the region between two concentric squares. We present an $O(n^3 \log n)$-time algorithm that finds such a square annulus over all orientations.

1 Introduction

An annulus informally depicts a ring-shaped region in the plane. More specifically, an annulus of a simple closed curve C, such as a circle, with a reference point inside C can be regarded as the region between two concentric homothets of C. Given a set P of n points in the plane, finding geometric shapes that best fits P is an important variant of shape matching problems. If the shape is restricted to C under a certain family of transformations, then this problem is equivalent to finding the *minimum-width annulus* that contains P. Among others, the case when C is chosen as a circle has been most intensively studied. The minimum-width circular annulus problem has been first addressed independently by Wainstein [16] and by Roy and Zhang [13], resulting in $O(n^2)$-time algorithms. The same time bound can be achieved by using the observation that the center of a minimum-width circular annulus corresponds to a vertex of the nearest-site Voronoi diagram of P, a vertex of the farthest-site Voronoi diagram of P, or an intersection point of two edges of the two diagrams [9]. The currently best exact algorithm takes $O(n^{\frac{3}{2}+\epsilon})$ time by Agarwal and Sharir [3]. Linear-time approximation schemes are also known by Agarwal et al. [2] and by Chan [7].

The minimum-width circular annulus problem has applications in facility location in a sense that the center of the optimal annulus minimizes the difference between the maximum and the minimum distances from the center to input points with respect to the Euclidean metric. Of course, in some applications, other metrics like the L_1 or L_∞ metric would be more appropriate. In this sense, the *square annulus* or *rectangular annulus* problem naturally arises. Abellanas

This research was supported by Basic Science Research Program through the National Research Foundation of Korea (NRF) funded by the Ministry of Science, ICT & Future Planning (2013R1A1A1A05006927) and by the Ministry of Education (2015R1D1A1A01057220).

© Springer International Publishing Switzerland 2016
M. Kaykobad and R. Petreschi (Eds.): WALCOM 2016, LNCS 9627, pp. 131–142, 2016.
DOI: 10.1007/978-3-319-30139-6_11

et al. [1] considered minimum-width rectangular annuli that are axis-parallel, and presented two algorithms taking $O(n)$ or $O(n \log n)$ time: one minimizes the width over rectangular annuli with arbitrary aspect ratio and the other does over rectangular annuli with a prescribed aspect ratio, respectively. Gluchshenko et al. [10] presented an $O(n \log n)$-time algorithm that computes a minimum-width axis-parallel square annulus, and proved a matching lower bound, while the second algorithm by Abellanas et al. can do the same in the same time bound. The $\log n$ gap between the rectangular and the square annulus problems could be understood in a geometric point of view. In both cases, the outer boundary of an optimal annulus can be chosen as a smallest axis-parallel rectangle or square enclosing P, as shown in [1,10], but the smallest enclosing rectangle is unique while there are in general infinitely many smallest enclosing squares. If one considers rectangular or square annuli in arbitrary orientation, the problem gets more difficult. Mukherjee et al. [12] presented an $O(n^2 \log n)$-time algorithm that computes a minimum-width rectangular annulus in arbitrary orientation and arbitrary aspect ratio. However, to our best knowledge, there is no known algorithm for the minimum-width square annulus in arbitrary orientation. We aim to give the first algorithmic results to this variant of the problem.

A variant of the problem where the outer or inner boundary of the resulting annulus is fixed has also been studied. Duncan et al. [8] and De Berg et al. [6] independently showed that the minimum-width circular annulus can be computed in $O(n \log n)$ time in this case. Barequet et al. [4] and Barequet and Goryachev [5] considered the case when the prescribed shape C is given as any convex or simple polygon for this variant of the problem. When C is a square and its orientation can be chosen arbitrarily, their results imply that the minimum-width square annulus can be computed in $O(n^4 \log n)$ time, provided that the side length of its outer, inner or middle square is given.

In this work, we consider the minimum-width square annulus problem in arbitrary orientation, and present an $O(n^3 \log n)$-time exact algorithm. Note that this is the first algorithm for the problem. Comparing to the results of Barequet and Goryachev [5], our algorithm is more efficient while dropping the restriction on the size of the resulting annulus.

The omitted proofs and additional figures will be provided in a full version.

2   Preliminaries

For any square in the plane $\mathbb{R}^2$, its *center* is the intersection point of its two diagonals and its *radius* is half its side length. Two squares are called *concentric* if they share a common center and any pair of their sides are either parallel or orthogonal. A *square annulus* A is the region between two concentric squares, including its boundary. The *width* of a square annulus A is the difference of radii of the two concentric squares determining A.

The *orientation* of a line or line segment ℓ in the plane is a nonnegative value $\theta \in [0, \pi)$ such that the rotated copy of the x-axis by θ counter-clockwise is parallel to ℓ. If the orientation of a line or line segment is θ, then we say that the line or line segment is θ-*aligned*. A rectangle, a square, or a square annuls is

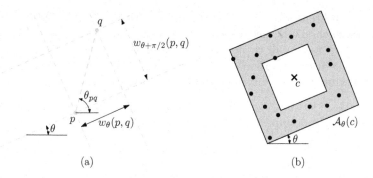

Fig. 1. (a) $w_\theta(p,q)$ and $w_{\theta+\pi/2}(p,q)$. (b) The minimum-width θ-aligned square annuls $\mathcal{A}_\theta(c)$ with center c enclosing points P.

also called θ-aligned for some $\theta \in [0, \pi/2)$ if each of its sides is either θ-aligned or $(\theta + \pi/2)$-aligned.

For any two points $p, q \in \mathbb{R}^2$, let $\overline{pq}$ denote the line segment joining p and q, and $|\overline{pq}|$ denote the Euclidean length of $\overline{pq}$. We will often discuss the distance between the orthogonal projections of p and q onto any θ-aligned line, denoted by $w_\theta(p,q)$. It is not difficult to see that $w_\theta(p,q) = |\overline{pq}| \cdot |\cos(\theta_{pq} - \theta)|$, where θ_{pq} denotes the orientation of $\overline{pq}$. See Fig. 1(a). Also, we define $d_\theta(p,q) :=$ $\max\{w_\theta(p,q), w_{\theta+\pi/2}(p,q)\}$ to be the convex distance between p and q with its unit disk being a unit θ-aligned square. Note that $d_\theta(p,q)$ is exactly the radius of the smallest θ-aligned square with center p that contains q in its boundary.

In a specific orientation $\theta \in [0, \pi/2)$, we regard any θ-aligned line to be *horizontal* and directed from left to right, and any $(\theta + \pi/2)$-aligned line to be *vertical* and directed from bottom to above. For any $p, q \in \mathbb{R}^2$, we say that p is to the *left* of q, or q is to the *right* of p, in θ if the orthogonal projection of p onto a θ-aligned line is prior to that of q. Similarly, p is *below* q or equivalently q is *above* p in θ if the orthogonal projection of p onto a $(\theta + \pi/2)$-aligned line is prior to that of q. For example, in Fig. 1(a), p is to the left of and below q in θ.

Let P be a set of points in $\mathbb{R}^2$. In orientation $\theta \in [0, \pi/2)$, let l_θ^*, r_θ^*, t_θ^* and b_θ^* be the leftmost, rightmost, topmost, and bottommost points in θ among those in P. Then, the smallest θ-aligned rectangle $\mathcal{R}_\theta$ enclosing P is uniquely determined by these four extreme points l_θ^*, r_θ^*, t_θ^* and b_θ^*. The *height* of $\mathcal{R}_\theta$ is the length of a vertical side of $\mathcal{R}_\theta$ and the *width* of $\mathcal{R}_\theta$ is the length of its horizontal side. That is, the height of $\mathcal{R}_\theta$ is equal to $w_{\theta+\pi/2}(t_\theta^*, b_\theta^*)$ and its width is equal to $w_\theta(l_\theta^*, r_\theta^*)$.

In this paper, we are interested in square annuli enclosing P. If we fix an orientation $\theta \in [0, \pi/2)$ and a center $c \in \mathbb{R}^2$, then there is a unique minimum-width θ-aligned square annulus containing P, which is determined by the smallest square that encloses P and the largest square whose interior contains no point of P. We denote this annulus by $\mathcal{A}_\theta(c)$. See Fig. 1(b). Thus, the minimum-width square annulus problem in a fixed orientation θ is to find an optimal center c^* that minimizes the width of $\mathcal{A}_\theta(c)$ over $c \in \mathbb{R}^2$. If we further consider arbitrary orientation, then the problem asks to find an optimal pair (θ^*, c^*) of orientation and center that minimize the width of $\mathcal{A}_\theta(c)$ over $c \in \mathbb{R}^2$ and $\theta \in [0, \pi/2)$.

We will often face with functions of a particular form $a\sin(\theta + b)$ for some $a, b \in \mathbb{R}$. Such a function is called *sinusoidal functions of period* 2π. Through out this paper, we call them shortly *sinusoidal functions*, when their period is 2π. Obviously, the equation $a\sin(\theta + b) = 0$ has at most one zero over $\theta \in [0, \pi)$. The following property of sinusoidal functions is well known and easily derived.

Lemma 1. *The sum of two sinusoidal functions is also sinusoidal. Therefore, the graphs of two sinusoidal functions cross at most once over* $[0, \pi)$. □

Note that $\cos(\theta)$ is also sinusoidal as $\cos(\theta) = \sin(\theta + \pi/2)$. Observe that for a fixed pair of points $p, q \in \mathbb{R}^2$, the function $w_\theta(p, q)$ is a piecewise sinusoidal function of $\theta \in [0, \pi)$ with at most one breakpoint.

3   Square Annuli in Fixed Orientation

In this section, we discuss square annuli in a fixed orientation that contain a given set P of n points. Throughout this section, we fix an orientation $\theta \in [0, \pi/2)$. Let $H = w_{\theta+\pi/2}(t^*_\theta, b^*_\theta)$ and $W = w_\theta(l^*_\theta, r^*_\theta)$ be the height and the width of $\mathcal{R}_\theta$, and let o be the center of $\mathcal{R}_\theta$. Here, we assume without loss of generality that $H \geq W$. Let m be the midpoint of line segment $\overline{t^*_\theta b^*_\theta}$ and L_θ be the θ-aligned line through m. Consider smallest θ-aligned squares enclosing P. The trace of the centers of all such squares forms a line segment $C_\theta \subset L_\theta$. Note that the length of C_θ is exactly equal to $H - W$ and its midpoint coincides with $o \in L_\theta$, and that $w_\theta(o, m) = w_\theta(m', m)$ where m' denotes the midpoint of $\overline{l^*_\theta r^*_\theta}$. Gluchshenko et al. [10] proved the following lemma.

Lemma 2 (Gluchshenko et al. [10]). *There exists a minimum-width θ-aligned square annulus that contains P whose center lies on C_θ.* □

Hence, a minimum-width square annulus in orientation θ is $\mathcal{A}_\theta(c^*)$ such that $c^* \in C_\theta$ minimizes the width of $\mathcal{A}_\theta(c)$ over $c \in C_\theta$. Observe further that, for all $c \in C_\theta$, the outer square of $\mathcal{A}_\theta(c)$ is always determined by t^*_θ and b^*_θ, that is, its radius is always $\frac{1}{2}H$, a constant. On the other hand, the inner square of $\mathcal{A}_\theta(c)$ is determined by a point $p \in P$ that is the closest from c with respect to the distance function $d_\theta(p, c)$, that is, the radius of the inner square of $\mathcal{A}_\theta(c)$ is exactly $\min_{p \in P} d_\theta(p, c)$. Thus, the width of $\mathcal{A}_\theta(c)$ is equal to $\frac{1}{2}H - \min_{p \in P} d_\theta(p, c)$, and our goal is to maximize $\min_{p \in P} d_\theta(c, p)$ over $c \in C_\theta$.

For convenience, we introduce a parameter $\lambda \in \mathbb{R}$ for points on L_θ such that those points $c \in L_\theta$ to the left of m are parameterized by $\lambda = -w_\theta(m, c)$ while the others $c \in L_\theta$ are parameterized by $\lambda = w_\theta(m, c)$. We denote by $c(\lambda)$ the point $c \in L_\theta$ parameterized by a specific $\lambda \in \mathbb{R}$. Observe that the endpoints of C_θ are parameterized by $\lambda = w_\theta(m, m') \pm \frac{1}{2}(H - W)$. Let $f_p(\lambda) := d_\theta(p, c(\lambda))$ be a real-valued function for each $p \in P$. Our problem can thus be solved by computing the lower envelope $\mathcal{F}$ of the n functions f_p over $\lambda \in \mathbb{R}$ and searching for a highest point on $\mathcal{F}$ over $\lambda \in [w_\theta(m, m') - \frac{1}{2}(H-W), w_\theta(m, m') + \frac{1}{2}(H-W)]$.

As also observed by Gluchshenko et al. [10], the function f_p is continuous and piecewise linear such that its graph consists of exactly three linear pieces whose slopes are -1, 0, and 1 in order. We call the piece with slope 0 of the graph of

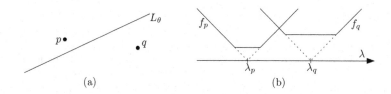

Fig. 2. (a) Two points $p, q \in P$ and the line L_θ. (b) The graphs of f_p and f_q.

f_p the *plateau* of p, and the two other pieces with slopes -1 and 1 the *left* and *right wings* of p, respectively. The plateau of p has length exactly $2w_{\theta+\pi/2}(p, m)$ and is at height $w_{\theta+\pi/2}(p, m)$. Also, the extensions of the two wings of p always cross at $\lambda_p \in \mathbb{R}$ such that $w_\theta(p, c(\lambda_p)) = 0$. See Fig. 2.

Lemma 3. *For $p, q \in P$, the graphs of f_p and f_q intersect exactly once, unless $w_\theta(m, p) = w_\theta(m, q)$ or $w_{\theta+\pi/2}(m, p) = w_{\theta+\pi/2}(m, q)$.*

Lemma 3 implies that the lower envelope $\mathcal{F}$ consists of $O(n)$ pieces, as it corresponds to a Davenport-Schinzel sequence of order 1 [14], and can be computed in $O(n \log n)$ time [11]. The following lemma immediately follows from the fact that the function f_p for each $p \in P$ is a convex function.

Lemma 4. *A highest point of $\mathcal{F}$ over C_θ corresponds to a breakpoint of $\mathcal{F}$ or an endpoint of C_θ, $\lambda = w_\theta(m', m) - \frac{1}{2}(H - W)$ or $w_\theta(m', m) + \frac{1}{2}(H - W)$.* □

Hence, for a fixed orientation $\theta \in [0, \pi/2)$, we can compute a minimum-width θ-aligned square annulus containing P in $O(n \log n)$ time by computing the lower envelope $\mathcal{F}$ and checking every breakpoint of $\mathcal{F}$ and each endpoint of C_θ.

4   Square Annuli over Arbitrary Orientations

In this section, we present an algorithm that finds a minimum-width square annulus containing P over all orientations $\theta \in [0, \pi/2)$. For the purpose, as θ continuously increases from 0 to $\pi/2$, we maintain the lower envelope $\mathcal{F}$ in orientation θ, collect all information about candidates of highest points of the envelope and find an optimal orientation using the collected information.

As done in [12], we start by decomposing $[0, \pi/2)$ into *primary intervals*. A primary interval is determined by a maximal interval $I \subseteq [0, \pi/2)$ satisfying that the four extreme points l_θ^*, r_θ^*, t_θ^*, and b_θ^* stay constant for all $\theta \in I$ and there is no $\theta \in I$ such that the height and the width of $\mathcal{R}_\theta$ are equal. In order to have all primary intervals, we apply the rotating caliper algorithm by Toussaint [15] after computing the convex hull of P, and then specify all $\theta \in [0, \pi/2)$ such that the height and the width of $\mathcal{R}_\theta$ are equal.

Lemma 5. *There are $O(n)$ primary intervals, and the decomposition of $[0, \pi/2)$ into the primary intervals can be specified in $O(n \log n)$ time.*

Our algorithm works for each primary interval $I \subseteq [0, \pi/2)$, separately, in an event-driven way; it maintains its invariants by handling events.

4.1   The Invariants

In the following, let I be any fixed primary interval, and $H(\theta)$ and $W(\theta)$ be the height and the width of $\mathcal{R}_\theta$ as functions of $\theta \in I$. Then, by definition, the four extreme points $l^* = l^*_\theta, r^* = r^*_\theta, t^* = t^*_\theta, b^* = b^*_\theta$ are fixed, and $H(\theta) - W(\theta)$ is either always positive or always negative for all $\theta \in I$. Without loss of generality, we assume that $H(\theta) - W(\theta) > 0$ for $\theta \in I$. We then define L_θ and C_θ, and use the same parameterization λ of L_θ as done in Sect. 3. Note that C_θ corresponds to interval $[\beta_\theta - \alpha_\theta, \beta_\theta + \alpha_\theta]$ of $\mathbb{R}$, where $\alpha_\theta = \frac{1}{2}(H(\theta) - W(\theta))$ and $\beta_\theta = w_\theta(m', m)$, where m' denotes the midpoint of $\overline{l^* r^*}$. For each $p \in P$, let $f_p(\theta, \lambda) := d_\theta(p, c(\lambda))$ be a function of two variables $\theta \in I$ and $\lambda \in \mathbb{R}$. Then the lower envelope $\mathcal{F}_\theta$ at $\theta \in I$ corresponds to $\min_{p \in P} f_p(\theta, \lambda)$, that is, $\mathcal{F}_\theta(\lambda) = \min_{p \in P} f_p(\theta, \lambda)$.

We will maintain a few data structures as the *invariants* of the algorithm: the combinatorial structure of $\mathcal{F}_\theta$, a balanced binary search tree $\mathcal{T}$ for points P, and the set E of pieces on $\mathcal{F}_\theta$ corresponding to the two endpoints of C_θ.

- The combinatorial structure of $\mathcal{F}_\theta$ is encoded by a sequence of the pieces appearing on $\mathcal{F}_\theta$ in the λ-increasing order. Each piece can be represented by a pair (p, σ) for $p \in P$ and $\sigma \in \{-1, 0, 1\}$ if it is the piece of p whose slope is σ. We store the combinatorial structure of $\mathcal{F}_\theta$ into a balanced binary search tree that supports the following operations in logarithmic time: (1) inserting a new piece or removing an existing piece, and (2) finding the relevant piece at a given $\lambda \in \mathbb{R}$. By an abuse of notation, we denote by $\mathcal{F}$ this data structure storing the envelope $\mathcal{F}_\theta$ over $\theta \in I$.
- For each $p \in P$, let λ_p and ζ_p be two sinusoidal functions of $\theta \in I$ associated with p such that $\lambda_p(\theta) = -w_\theta(m, p)$ if p is to the left of m, or $\lambda_p(\theta) = w_\theta(m, p)$, otherwise, and $\zeta_p(\theta) = w_{\theta + \pi/2}(m, p)$. The binary search tree $\mathcal{T}$ at θ stores points in P indexed by $(\lambda_p(\theta), \zeta_p(\theta))$ for $p \in P$ as follows. The leaves of $\mathcal{T}$ correspond to all points of P in the increasing order of $\lambda_p(\theta)$ and also store the descriptions of λ_p and ζ_p for each $p \in P$. Each internal node v of $\mathcal{T}$ is associated with two points $p_v, q_v \in P$ such that $\lambda_{p_v}(\theta) \geq \lambda_p(\theta)$ for all $p \in P(v)$ and $\zeta_{q_v}(\theta) \leq \zeta_p(\theta)$ for all $p \in P(v)$, where $P(v)$ denotes the set of $p \in P$ stored in a leaf contained in the subtree of $\mathcal{T}$ rooted at v. Thus, $\mathcal{T}$ acts as a binary search tree on P with respect to the first index $\lambda_p(\theta)$ of $p \in P$, and in addition supports the following operations in logarithmic time: (1) inserting or removing a point, (2) given $p, q \in P$, finding $r \in P$ such that $\lambda_p(\theta) < \lambda_r(\theta) < \lambda_q(\theta)$ and $\zeta_r(\theta) \leq \zeta_{r'}(\theta)$ for all $r' \in P$ with $\lambda_p(\theta) < \lambda_{r'}(\theta) < \lambda_q(\theta)$.
- The set E consists of exactly two pieces appearing in $\mathcal{F}$ that correspond to the two endpoints of C_θ at θ.

4.2   Combinatorial Changes of the Invariants and Events

As θ continuously increases over I, the invariants $\mathcal{F}, \mathcal{T}$, and E suffer combinatorial changes. To capture those changes, we handle *events* of several types defined to be a specific value of $\theta \in I$ describing a degenerate scene:

(a) An *alignment event* occurs at $\theta \in I$ such that $\lambda_p(\theta) = \lambda_q(\theta)$ or $\zeta_p(\theta) = \zeta_q(\theta)$ for some $p, q \in P$ with $p \neq q$.

(b) A *midline event* occurs when $\zeta_p(\theta) = 0$, that is, p lies on L_θ for some $p \in P$.

(c) A *wing event* occurs when the left (or right) endpoint of the plateau of p lies on the right (or left, respectively) wing of q, for some $p, q \in P$ with $p \neq q$.

(d) A *triple event* occurs when the intersection of the right wing of p and the left wing of p' appears in $\mathcal{F}$ and lies on the plateau of q for some $p, p', q \in P$.

(e) An *endpoint event* occurs when a breakpoint of $\mathcal{F}$ corresponds to an endpoint of C_θ.

A combinatorial change in $\mathcal{F}$ happens when a piece (p, σ) is about to appear onto it or disappear from it. If the piece is the plateau of p, that is, $\sigma = 0$, then either an alignment or a triple event occurs, while if the piece is a wing of p, then either an alignment or a wing event occurs. See Fig. 3. Each structural change of $\mathcal{T}$ happens exactly when an alignment event occurs, while the function description of ζ_p for each $p \in P$ may also be changed if and only if a midline event occurs. Finally, each change of the set E corresponds to either an endpoint event or an alignment event. Hence, by predicting all those events and handling them correctly, we can maintain the structures $\mathcal{F}$, $\mathcal{T}$ and E over $\theta \in I$.

(a)   (b)   (c)   (d)

Fig. 3. (a) Alignment events. (b) Midline event. (c) Wing event. (d) Triple event.

4.3   Computing Events

Here, we focus on how to compute events before they occur. The events of the first three types are relatively easy to predict.

Lemma 6. *All of the alignment, midline, and wing events can be computed explicitly in $O(n^2)$ time.*

By the above lemma, we can precompute all the alignment, midline, and wing events before running the main loop of our algorithm. However, triple and endpoint events are dependent on the current envelope $\mathcal{F}$. They will be computed while the main loop is being executed by exploiting the invariants.

For any $p, p', q \in P$, let TRI(p, q, p') be $\theta' \in I$ when the intersection of $(p, 1)$ and $(p', -1)$ lies on $(q, 0)$. At $\theta' = $ TRI(p, q, p'), we have $\lambda_{p'}(\theta') - \lambda_p(\theta') = 2\zeta_q(\theta')$, so given p, p', q, one can compute TRI(p, q, p') in $O(1)$ time by solving the sinusoidal equation. Note that TRI(p, q, p') is uniquely determined by Lemma 1. This triple event TRI(p, q, p') indeed occurs at θ' only if the intersection point of $(p, 1)$ and $(p', -1)$ appears in $\mathcal{F}_{\theta'}$. There are two cases: whether the plateau $(q, 0)$ of q is about to appear or disappear at θ'. For the latter case, the three pieces $(p, 1)$,

$(q,0)$, and $(p',-1)$ must be consecutive in $\mathcal{F}$ just before θ', while for the former case, $(p,1)$ and $(p',-1)$ are consecutive in $\mathcal{F}$ but q is not seen from $\mathcal{F}$ just before θ'. For $p,p' \in P$ such that $(p,1)$ and $(p',-1)$ are consecutive in $\mathcal{F}$ at $\theta \in I$, let $\mathrm{TRI}_\theta(p,p')$ be a triple event at $\theta' > \theta$ associated with p and p' that indeed occurs, assuming that there is no alignment or wing event between θ and θ'.

Lemma 7. *Given $p,p' \in P$ such that $(p,1)$ and $(p',-1)$ are consecutive in $\mathcal{F}$ at $\theta \in I$, $\mathrm{TRI}_\theta(p,p')$ can be computed in $O(\log n)$ time.*

In a similar fashion, we predict the next endpoint event. Let ENDPT_θ^+ and ENDPT_θ^- be the next endpoint event at $\theta' > \theta$ that occurs at the endpoints $\beta_{\theta'} + \alpha_{\theta'}$ and $\beta_{\theta'} - \alpha_{\theta'}$ of $C_{\theta'}$, respectively, assuming that there is no change locally at the piece corresponding to each endpoint in $\mathcal{F}$ between θ and θ'.

Lemma 8. *Each of ENDPT_θ^+ and ENDPT_θ^- can be computed in $O(1)$ time.*

Note that the predicted triple event $\mathrm{TRI}_\theta(p,p')$ or endpoint events ENDPT_θ^+ and ENDPT_θ^- may not really occur if another event that causes some changes in the invariants occurs before it. Thus, we treat those predicted triple and endpoint events as tentative events for a while and modify them whenever necessary while handling other events, so that finally every occurred triple and endpoint event is correctly handled.

4.4   The Main Loop: Handling the Next Event

Before the main loop, we initialize the invariants $\mathcal{F}$, $\mathcal{T}$ and E for the starting point θ_0 of the primary interval I. This can be done in $O(n \log n)$ time as discussed in Sect. 3.

We also maintain the *event queue* $\mathcal{Q}$ that is a priority queue storing events indexed by its occurring time so that the earliest upcoming event can be extracted from $\mathcal{Q}$ and adding or deleting a specific event in logarithmic time.

We compute all the alignment, midline and wing events in $O(n^2)$ time by Lemma 6 and insert them into $\mathcal{Q}$ in $O(n^2 \log n)$ time. For any two consecutive pieces in $\mathcal{F}$ that are of the form $(p,1)$ and $(p',-1)$, we compute $\mathrm{TRI}_{\theta_0}(p,p')$ by Lemma 7 and insert it into $\mathcal{Q}$. Also, for any three consecutive pieces in $\mathcal{F}$ that are of the form $(p,1)$, $(q,0)$, and $(p',-1)$, in order, we compute $\mathrm{TRI}(p,q,p')$ and insert it into $\mathcal{Q}$. Lastly, compute $\mathrm{ENDPT}_{\theta_0}^+$ and $\mathrm{ENDPT}_{\theta_0}^-$ by Lemma 8 and insert them into $\mathcal{Q}$. Note that $\mathcal{Q}$ will contain at most $O(n)$ triple and endpoint events at every time.

We are then ready to start the main loop. In the main loop, we extract the upcoming event from $\mathcal{Q}$ and handle it according to its type. While handling each event, (i) we update our invariants $\mathcal{F}$, $\mathcal{T}$ and E, (ii) modify some triple and endpoint events in $\mathcal{Q}$ that are turned to be invalid by the update by deleting them from $\mathcal{Q}$ and inserting newly computed ones into $\mathcal{Q}$.

We first describe how to update the invariants in the following.

(a) *(Alignment event).* Let $p,q \in P$ be associated with this alignment event at $\theta \in I$. We first discuss how to update $\mathcal{T}$. We have two cases: $\lambda_p(\theta) = \lambda_q(\theta)$

or $\zeta_p(\theta) = \zeta_q(\theta)$. In the former case, the order of p and q is about to be reversed with respect to λ. at θ. Thus, we apply the new order of p and q in $\mathcal{T}$. This can be done simply by deleting p and q from $\mathcal{T}$, and reinserting them with the new order of $\lambda_p(\theta + \epsilon)$ and $\lambda_q(\theta + \epsilon)$ for an arbitrarily small positive ϵ. In the latter case, the structure of $\mathcal{T}$ remains the same but the order of p and q with respect to ζ. is about to be reversed. Thus, we update all the internal nodes along the paths from p and q to the root by applying the new order of $\zeta_p(\theta + \epsilon)$ and $\zeta_q(\theta + \epsilon)$.

Next, we update the envelope $\mathcal{F}$. We also distinguish two cases: $\lambda_p(\theta) = \lambda_q(\theta)$ or $\zeta_p(\theta) = \zeta_q(\theta)$.

(i) For the former case, we assume without loss of generality that $\zeta_p(\theta) < \zeta_q(\theta)$, and $\lambda_p(\theta-\epsilon) > \lambda_q(\theta-\epsilon)$ for a sufficiently small positive real $\epsilon > 0$. So, we will have $\lambda_p(\theta + \epsilon) < \lambda_q(\theta + \epsilon)$ right after θ. If the plateau $(p, 0)$ of p does not appear in $\mathcal{F}$, then this alignment event causes no change in $\mathcal{F}$. We thus assume that $(p, 0)$ appears in $\mathcal{F}$. Since $(q, 0)$ and $(q, 1)$ are going to stick out from the right wing $(p, 1)$ of p, we decide if $(q, 0)$ will appear in the envelope $\mathcal{F}$ right after θ by checking the right neighbor of $(p, 0)$ or $(p, 1)$. If so, we add $(q, 0)$ and $(q, 1)$ into $\mathcal{F}$ accordingly. On the other hand, on the left wing $(p, -1)$ of p, $(q, -1)$ and $(q, 0)$ get removed from $\mathcal{F}$ if they appear in $\mathcal{F}$ to the left of $(p, -1)$.

(ii) For the latter case, where we have $\zeta_p(\theta) = \zeta_q(\theta)$, we assume that $\zeta_p(\theta - \epsilon) < \zeta_q(\theta - \epsilon)$ for a sufficiently small positive real $\epsilon > 0$ but we are going to have $\zeta_p(\theta + \epsilon) > \zeta_q(\theta + \epsilon)$ right after θ. We further assume without loss of generality that $\lambda_p(\theta) < \lambda_q(\theta)$. Let Z be the intersection between the plateaus $(p, 0)$ and $(q, 0)$ of p and q at θ. If Z has no intersection with the current envelope $\mathcal{F}_\theta$, then we do not have any change in its combinatorial structure $\mathcal{F}$. Thus, we assume that Z intersects $\mathcal{F}_\theta$.

If the right wing $(p, 1)$ of p also appears in $\mathcal{F}$, then so does the plateau $(q, 0)$ of q. In this case, $(p, 1)$ is about to disappear while $(q, -1)$ is about to appear in $\mathcal{F}$ right after θ. We thus delete $(p, 1)$ from $\mathcal{F}$ and insert $(q, -1)$ into $\mathcal{F}$ between $(p, 0)$ and $(q, 0)$.

Otherwise, if $(p, 1)$ is not in $\mathcal{F}$, then $(q, 0)$ is not in $\mathcal{F}$, either. Further, if Z contains the breakpoint of $\mathcal{F}_\theta$ between $(p, 0)$ and its left neighbor, then $(p, 0)$ is about to disappear and $(q, 0)$ is about to appear in $\mathcal{F}$ right after θ; so, we delete $(p, 0)$ from $\mathcal{F}$ and insert $(q, 0)$ into $\mathcal{F}$ instead. In the other case, where Z does not contain the breakpoint of $\mathcal{F}_\theta$ between $(p, 0)$ and its left neighbor, no piece is deleted while $(q, -1)$ and $(q, 0)$ should be inserted into $\mathcal{F}$ in this order.

After having updated $\mathcal{T}$ and $\mathcal{F}$, we update E by checking which piece of $\mathcal{F}$ corresponds to each endpoint $\beta_\theta - \alpha_\theta$ and $\beta_\theta + \alpha_\theta$ of C_θ.

(b) *(Midline event).* Let $p \in P$ be associated with this midline event at $\theta \in I$. We then have $\zeta_p(\theta) = 0$. In this case, there is no structural change in $\mathcal{F}$, $\mathcal{T}$, and E but the function description ζ_p of $w_{\theta+\pi/2}(p, m)$ should be updated; more precisely, the sinusoidal function ζ_p gets negated after θ.

(c) *(Wing event).* Let $p, q \in P$ be associated with this wing event at $\theta \in I$. Assume without loss of generality that the right endpoint of the plateau of p

lies on the left wing of q at θ. The other cases can be handled in a symmetric way. If not both of $(p,0)$ and $(q,-1)$ appear in $\mathcal{F}$ in a consecutive way, then we do not have any change in $\mathcal{F}$, so ignore this event. Otherwise, this is the case where the right wing $(p,1)$ is about to appear in $\mathcal{F}$ between $(p,0)$ and $(q,-1)$ or disappear from $\mathcal{F}$. According to the case, we insert $(p,1)$ into $\mathcal{F}$ between $(p,0)$ and $(q,-1)$ or delete it from $\mathcal{F}$. There is no change in $\mathcal{T}$ and E by any wing event.

(d) *(Triple event).* Let $\mathrm{TRI}(p,q,p')$ be this event at $\theta \in I$. Then, the plateau $(q,0)$ of q is about to appear in $\mathcal{F}$ between $(p,1)$ and $(p',-1)$ or disappear from $\mathcal{F}$. According to the case, we insert $(q,0)$ into $\mathcal{F}$ between $(p,1)$ and $(p',-1)$ or delete it from $\mathcal{F}$. In this case, there is no change in $\mathcal{T}$ and E.

(e) *(Endpoint event).* Let $(p,\sigma) \in E$ be the piece that is relevant to this endpoint event at $\theta \in I$. Let (p',σ') be a piece in $\mathcal{F}$ neighboring (p,σ) such that the breakpoint between (p,σ) and (p',σ') corresponds to this event. Then, we remove (p,σ) from E and insert (p',σ') into E. There is no change in $\mathcal{F}$ and $\mathcal{T}$ by any endpoint event.

After having updated the invariants as described above, we modify triple and endpoint events in $\mathcal{Q}$. For the two endpoint events in $\mathcal{Q}$, we simply recompute ENDPT_θ^+ and ENDPT_θ^- by Lemma 8 and replace the existing ones by the new ones. In order to modify triple events, we perform the following: For each (p,σ) disappeared from $\mathcal{F}$ by this update, we delete every triple event that involves p from $\mathcal{Q}$, while for each (p,σ) newly inserted into $\mathcal{F}$ by this update, we compute triple events that involves (p,σ) and its neighbors by Lemma 7, and insert them into $\mathcal{Q}$. Also, for the two points $p,q \in P$ that have changed their order in $\mathcal{T}$, we delete the triple events of the form $\mathrm{TRI}(r,p,r')$ or $\mathrm{TRI}(r,q,r')$ from $\mathcal{Q}$, if exist. We then recompute $\mathrm{TRI}_\theta(r,r')$ by Lemma 7 and insert it into $\mathcal{Q}$. Note that the update of the invariants $\mathcal{F}$, $\mathcal{T}$ and E by a single event is local and involves a constant number of pieces. Thus, we have the following lemma.

Lemma 9. *Each event can be handled in $O(\log n)$ time.*

While running the main loop of our algorithm, we create, delete, and handle a number of events. The following lemma bounds the total number of events we create by $O(n^2)$.

Lemma 10. *The number of events created by the algorithm for a primary interval I is at most $O(n^2)$.*

Therefore, the invariants $\mathcal{F}$, $\mathcal{T}$, and E are maintained over $\theta \in I$ in $O(n^2 \log n)$ time by our algorithm.

Lemma 11. *Our algorithm correctly maintain the invariants $\mathcal{F}$, $\mathcal{T}$ and E in $O(n^2 \log n)$ time over a primary interval I.*

4.5   Finding an Optimal Orientation over a Primary Interval

By maintaining $\mathcal{F}$ over $\theta \in I$ as described above, we now have enough information to trace candidates of highest points of $\mathcal{F}_\theta(\lambda)$ over $\lambda \in C_\theta$ and $\theta \in I$. Recall

Lemma 4 that there exists a highest point of $\mathcal{F}_\theta$ at fixed θ that corresponds to a breakpoint of $\mathcal{F}_\theta$ or an endpoint of C_θ. As $\theta \in I$ continuously increases, the height of each breakpoint of $\mathcal{F}_\theta$ or the height of $\mathcal{F}_\theta$ at each endpoint of C_θ changes continuously as a function of θ.

Each breakpoint of $\mathcal{F}_\theta$ is determined by two neighboring pieces (p, σ) and (p', σ') in $\mathcal{F}_\theta$ for $p, p' \in P$ and $\sigma, \sigma' \in \{-1, 0, 1\}$. Conversely, for any pair $((p, \sigma), (p', \sigma'))$ of two pieces, we say that the pair is *alive at* $\theta \in I$ if (p, σ) and $(p'\sigma')$ are consecutive in $\mathcal{F}_\theta$ and the breakpoint between them falls into C_θ. An *alive interval* of the pair $((p, \sigma), (p', \sigma'))$ is a maximal sub-interval I' of I such that the pair is alive at every $\theta \in I'$ and no midline event is contained in I'. Note that there may be several alive intervals for a pair of pieces. Let $g_{p\sigma p'\sigma'}(\theta)$ be the partial function defined over alive intervals of $((p, \sigma), (p', \sigma'))$ that maps θ to the height of the breakpoint between the two pieces in $\mathcal{F}_\theta$. Similarly, an *alive interval* of each piece (p, σ) is a maximal sub-interval I' of I such that $(p, \sigma) \in E$ corresponding to a common endpoint of C_θ at every $\theta \in I'$ and no midline event is contained in I'. Let $g_{p\sigma}$ be the partial function defined over alive intervals of (p, σ) that maps θ to the height of (p, σ) at the corresponding endpoint of C_θ.

Let $\mathcal{G}$ be the set of all those partial functions defined above. We first observe that each function in $\mathcal{G}$ is indeed sinusoidal.

Lemma 12. *Any function $g \in \mathcal{G}$ is sinusoidal in its alive interval.*

Another key observation on alive intervals is the following.

Lemma 13. *The endpoints of each alive interval correspond to two events handled by the above algorithm. Moreover, the number of alive intervals is $O(n^2)$.*

By Lemma 13, we can explicitly collect all alive intervals and the corresponding functions defined on them while maintaining the invariants $\mathcal{F}$, $\mathcal{T}$ and E. Thus, $\mathcal{G}$ together with the alive intervals can be obtained in the same time as in Lemma 11. Let Γ be the set of curved segments corresponding to the graphs of partial functions in $\mathcal{G}$. Lemmas 12 and 13 immediately imply the following.

Lemma 14. *The set Γ consists of $O(n^2)$ segments of sinusoidal curves. Moreover, any two segments in Γ cross at most once.*   □

Now, let $G(\theta) := \max_{g \in \mathcal{G}} g(\theta)$ be the pointwise maximum over $g \in \mathcal{G}$ at θ. Note that the function G corresponds to the upper envelope of curved segments in Γ. Lemma 14 implies that the upper envelope of curved segments in Γ forms a Davenport-Schinzel sequence of order 3 [14]. The upper envelope of Γ can be computed in $O(n^2 \log n)$ time by Hershberger [11], though its complexity can be as large as $\Theta(n^2 \alpha(n))$ [14]. Thus, the function G can be computed in $O(n^2 \log n)$ time and is piecewise sinusoidal with $O(n^2 \alpha(n))$ breakpoints.

Recall that our goal is to minimize $\min_{\lambda \in C_\theta} \{\frac{1}{2} H(\theta) - \mathcal{F}_\theta(\lambda)\}$ over $\theta \in I$ by Lemma 2, where $\mathcal{F}_\theta(\lambda) = \min_{p \in P} f_p(\theta, \lambda)$. Since $H(\theta)$ is independent from $\lambda \in C_\theta$, it is equivalent to minimize $\frac{1}{2} H(\theta) - \max_{\lambda \in C_\theta} \mathcal{F}_\theta(\lambda)$ over $\theta \in I$. By Lemma 4, we now have $\max_{\lambda \in C_\theta} \mathcal{F}_\theta(\lambda) = \max_{g \in \mathcal{G}} g(\theta) = G(\theta)$. Thus, our last task is to find a minimum of $\frac{1}{2} H(\theta) - G(\theta)$ over $\theta \in I$. Since $H(\theta)$ is sinusoidal, the function $\frac{1}{2} H(\theta) - G(\theta)$ is also piecewise sinusoidal by Lemma 1, so we can find an optimal orientation that minimizes $\frac{1}{2} H(\theta) - G(\theta)$ in $O(n^2 \alpha(n))$ time.

Therefore, in $O(n^2 \log n)$ time, we can find an optimal orientation over a primary interval. The corresponding annulus can be obtained simultaneously, or by running the algorithm for fixed orientation described in Sect. 3. Since there are $O(n)$ primary intervals by Lemma 5, we finally conclude our main theorem.

Theorem 1. *Given a set P of n points in $\mathbb{R}^2$, a minimum-width square annulus that contains P over all orientations can be computed in $O(n^3 \log n)$ time.* □

References

1. Abellanas, M., Hurtado, F., Icking, C., Ma, L., Palop, B., Ramos, P.: Best fitting rectangles. In: European Workshop on Computational Geometry (EuroCG 2003) (2003)
2. Agarwal, P.K., Har-Peled, S., Varadarajan, K.R.: Approximating extent measures of points. J. ACM **51**(4), 606–635 (2004)
3. Agarwal, P., Sharir, M.: Efficient randomized algorithms for some geometric optimization problems. Discrete Comput. Geom. **16**, 317–337 (1996)
4. Barequet, G., Briggs, A.J., Dickerson, M.T., Goodrich, M.T.: Offset-polygon annulus placement problems. Comput. Geom.: Theory Appl. **11**, 125–141 (1998)
5. Barequet, G., Goryachev, A.: Offset polygon and annulus placement problems. Comput. Geom.: Theory Appl. **47**(3, Part A), 407–434 (2014)
6. de Berg, M., Bose, P., Bremner, D., Ramaswami, S., Wilfong, G.: Computing constrained minimum-width annuli of point sets. Comput.-Aided Des. **30**(4), 267–275 (1998)
7. Chan, T.: Approximating the diameter, width, smallest enclosing cylinder, and minimum-width annulus. Int. J. Comput. Geom. Appl. **12**, 67–85 (2002)
8. Duncan, C., Goodrich, M., Ramos, E.: Efficient approximation and optimization algorithms for computational metrology. In: Proceedings of the 8th ACM-SIAM Symposium Discrete Algorithms (SODA 1997), pp. 121–130 (1997)
9. Ebara, H., Fukuyama, N., Nakano, H., Nakanishi, Y.: Roundness algorithms using the Voronoi diagrams. In: Abstracts 1st Canadian Conference on Computational Geometry (CCCG), pp. 41 (1989)
10. Gluchshenko, O.N., Hamacher, H.W., Tamir, A.: An optimal $o(n \log n)$ algorithm for finding an enclosing planar rectilinear annulus of minimum width. Oper. Res. Lett. **37**(3), 168–170 (2009)
11. Hershberger, J.: Finding the upper envelope of n line segments in $O(n \log n)$ time. Inf. Proc. Lett. **33**, 169–174 (1989)
12. Mukherjee, J., Mahapatra, P., Karmakar, A., Das, S.: Minimum-width rectangular annulus. Theor. Comput. Sci. **508**, 74–80 (2013)
13. Roy, U., Zhang, X.: Establishment of a pair of concentric circles with the minimum radial separation for assessing roundness error. Comput.-Aided Des. **24**(3), 161–168 (1992)
14. Sharir, M., Agarwal, P.K.: Davenport-Schinzel Sequences and Their Geometric Applications. Cambridge University Press, New York (1995)
15. Toussaint, G.: Solving geometric problems with the rotating calipers. In: Proceedings of the IEEE MELECON (1983)
16. Wainstein, A.: A non-monotonous placement problem in the plane. In: Software Systems for Solving Optimal Planning Problems, Abstract: 9th All-Union Symp. USSR, Symp., pp. 70–71 (1986)

A General Framework for Searching on a Line

Prosenjit Bose[1] and Jean-Lou De Carufel[2]([⊠])

[1] School of Computer Science, Carleton University, Ottawa, Canada
[2] School of Electrical Engineering and Computer Science,
University of Ottawa, Ottawa, Canada
jdecaruf@uottawa.ca

Abstract. Consider the following classical search problem: a target is located on the line at distance D from the origin. Starting at the origin, a searcher must find the target with minimum competitive cost. The classical competitive cost studied in the literature is the ratio between the distance travelled by the searcher and D. Note that when no lower bound on D is given, no competitive search strategy exists for this problem. Therefore, all competitive search strategies require some form of lower bound on D.

We develop a general framework that optimally solves several variants of this search problem. Our framework allows us to match optimal competitive search costs for previously studied variants such as: (1) where the target is fixed and the searcher's cost at each step is a constant times the distance travelled, (2) where the target is fixed and the searcher's cost at each step is the distance travelled plus a fixed constant (often referred to as the *turn cost*), (3) where the target is moving and the searcher's cost at each step is the distance travelled.

Our main contribution is that the framework allows us to derive optimal competitive search strategies for variants of this problem that do not have a solution in the literature such as: (1) where the target is fixed and the searcher's cost at each step is $\alpha_1 x + \beta_1$ for moving distance x away from the origin and $\alpha_2 x + \beta_2$ for moving back with constants $\alpha_1, \alpha_2, \beta_1, \beta_2$, (2) where the target is moving and the searcher's cost at each step is a constant times the distance travelled plus a fixed constant turn cost. Notice that the latter variant can have several interpretations depending on what the turn cost represents. For example, if the turn cost represents the amount of time for the searcher to turn, then this has an impact on the position of the moving target. On the other hand, the turn cost can represent the amount of fuel needed to make an instantaneous turn, thereby not affecting the target's position. Our framework addresses all of these variations.

1 Introduction

Consider the following classical search problem: a target is located on the line at distance D from the origin. Starting at the origin, a searcher must find the target

This work was supported by FQRNT and NSERC.

M. Kaykobad and R. Petreschi (Eds.): WALCOM 2016, LNCS 9627, pp. 143–153, 2016.
DOI: 10.1007/978-3-319-30139-6_12

with minimum competitive cost. The classical competitive cost studied in the literature is defined as the ratio between the distance travelled by the searcher and D. This problem and many of its variants have been extensively studied both in mathematics and computer science. For an encyclopaedic overview of the field, the reader is referred to the following books on the area [1,2,10]. Techniques developed to solve this family of problems have many applications in various fields such as robotics, scheduling, clustering, or routing to name a few [4,5,8,11–14]. In particular, solutions to these problems have formed the backbone of many competitive online algorithms (see [6] for a comprehensive overview). Note that when no lower bound on D is given, no competitive search strategy exists for this problem. If an adversary places the target at a distance $\epsilon > 0$ from the origin and the first step taken by the algorithm is $\xi > 0$ in the wrong direction, the ratio ξ/ϵ cannot be bounded. Therefore, all competitive search strategies require some form of lower bound on D.

We introduce a general framework to resolve several variations of this classical search problem. Our framework allows us to match optimal competitive search costs for previously studied variants. The first one is the classical problem where the target is fixed and the searcher's cost at each step is a constant times the distance travelled. This was first studied by Gal [9,10] and subsequently by Baeza-Yates et al. [3]. Given a lower bound of λ on D, an optimal strategy is the following: $x_i = 2^i \lambda$ $(i \geq 1)$. At step i, if i is even, move to x_i and then return to the origin. If i is odd, move to $-x_i$ and then return to the origin. All known search strategies exhibit this alternating behaviour. The competitive cost of this strategy is 9 in the worst case. With our framework, we find the strategy $x_i = (i+1)2^i \lambda$, which also has a competitive cost of 9 in the worst case (refer to Lemma 1).

The following variant we can solve with our framework is the one where the target is fixed and the searcher's cost at each step is the distance travelled plus a fixed constant (often referred to as the *turn cost*). This was first studied by Alpern and Gal [2, Sect 8.4]. They provided a strategy with expected competitive cost $9 + 2t/\lambda$. They left open the question of whether this is optimal. Demaine et al. [7] addressed a deterministic variant of the problem. Their strategy is $x_i = \frac{1}{2}(2^i - 1) t$ $(i \geq 1)$. The total cost of this strategy is $9D + 2t$ in the worst case. Their strategy does not require a lower bound on D, the distance from the origin to the target. Therefore, their search strategy cannot be competitive with respect to only D (by the adversarial argument presented above). However, their search strategy is competitive with respect to the worst case cost of any online search strategy. Notice that when there is a cost of $t > 0$ charged for each turn made by the searcher, in essence $t + D$ is a lower bound on the worst case cost of any online search strategy. This is because in the worst case, any online strategy may start in the wrong direction and have to make at least one turn. Therefore, their search strategy is competitive with respect to $t + D$ as opposed to just D. Surprisingly, with our framework, we prove that when $t/2\lambda \leq 1$, the optimal competitive cost is still 9 in the worst case (refer to Lemma 1). When $t/2\lambda \geq 1$, the optimal search cost is $\left(9 + 2\frac{\left(\frac{t}{2\lambda} - 1\right)^2}{\frac{t}{2\lambda}}\right) D$ (refer to Theorem 1).

Moreover, our framework allows us to resolve the more general variant where the target is fixed and the searcher's cost at each step is $\alpha_1 x + \beta_1$ for moving distance x away from the origin and $\alpha_2 x + \beta_2$ for moving back with positive constants $\alpha_1, \alpha_2, \beta_1, \beta_2$ such that $\alpha_1 + \alpha_2 > 0$ (refer to Sect. 4.1).

The third variant that is encompassed by our framework is the one where the target is moving and the searcher's cost at each step is a the distance travelled. This was first studied by Gal [10]. Suppose that the searcher travels at speed 1 and the target travels at speed $0 < w < 1$. Given a lower bound of λ on D, an optimal strategy is $x_i = \left(2 \frac{1+w}{1-w}\right)^i \lambda$ $(i \geq 1)$. The competitive cost of this strategy is $\frac{(3+w)^2}{(1-w)^3}$ in the worst case. With our framework, we find the strategy $x_i = \left(\frac{1+3w}{(1-w)(1+w)} i + \frac{1}{1+w}\right) \left(2 \frac{1+w}{1-w}\right)^i \lambda$, which also has a competitive cost of $\frac{(3+w)^2}{(1-w)^3}$ in the worst case (refer to Lemma 3). Moreover, our framework allows us to resolve the more general variant where the target is moving and the searcher's cost at each step is a constant times the distance travelled plus a fixed constant turn cost (refer to Sect. 3). Surprisingly, even in this setting, when $t/2\lambda$ is small compared to w, the turn cost has no effect on the optimal competitive cost (refer to Lemma 3). Notice that the latter variant can have several interpretations depending on what the turn cost represents. For example, if the turn cost represents the amount of time for the searcher to turn, then this has an impact on the position of the moving target. On the other hand, the turn cost can represent the amount of fuel needed to make an instantaneous turn, thereby not affecting the target's position. Our framework addresses all of these variations. To the best of our knowledge, the only general framework for solving line-searching problems is due to Gal [9,10]. A precise description of our more general framework is outlined in Sect. 4.

2   Searching on a Line with Turn Cost

A *search strategy* for the problem of searching on a line is a function $S(i) = (x_i, r_i)$ defined for all integers $i \geq 1$. At step i, the searcher travels a distance of x_i on ray $r_i \in \{\text{left}, \text{right}\}$. If he does not find the target, he goes back to the origin and proceeds with step $i + 1$. Let D be the distance between the searcher and the target at the beginning of the search. Traditionally, the goal is to find a strategy S that minimizes the *competitive ratio* $CR(S)$ (or *competitive cost*) defined as the total distance travelled by the searcher divided by D, in the worst case. If D is given to the searcher, any strategy S such that $S(1) = (D, \text{left})$ and $S(2) = (D, \text{right})$ is optimal with a competitive ratio of 3 in the worst case. If D is unknown, a lower bound $\lambda \leq D$ must be given to the searcher, otherwise the competitive ratio is unbounded in the worst case.

Let $Left = \{i \mid r_i = \text{left}\}$ and $Right = \{i \mid r_i = \text{right}\}$. To guarantee that, wherever the target is located, we can find it with a strategy S, we must have $\sup_{i \in Left} x_i = \sup_{i \in Right} x_i = \infty$. We say that S is *monotonic* if the sequences $(x_i)_{i \in Left}$ and $(x_i)_{i \in Right}$ are strictly increasing. The strategy S is said to be *periodic* if $r_1 \neq r_2$ and $r_i = r_{i+2}$ for all $i \geq 1$. We know from previous work (see [3,10] for instance) that there is an optimal strategy that is periodic and monotonic. Let us say that a strategy is *fully monotonic* if the sequence $(x_i)_{i \geq 1}$ is monotonic and non-decreasing. We can make the following assumption without loss of generality.

Periodic-Monotonic Assumption: *There exists an optimal search strategy that is periodic and fully monotonic.*

We now address the problem where we are given a turn cost t. If the searcher knows D, the optimal strategy is still $x_i = D$ and it has a competitive cost of $\sup_{D \geq \lambda} \frac{3D+t}{D} = 3 + \frac{t}{\lambda}$ in the worst case. When $t = 0$, where we get a competitive cost of 3 in the worst case. This corresponds to the case where D is known and there is no turn cost. For the rest of the section, we suppose that the searcher does not know D. Throughout this section, we let γ be the competitive cost of the optimal strategy for searching on a line with turn cost, in the worst case.

In the worst case, it takes at least 2 steps to find the target. We find

$$\frac{2x_1 + t + \lambda}{\lambda} = \sup_{D \geq \lambda} \frac{2x_1 + t + D}{D} \leq \gamma. \tag{1}$$

In general, if the searcher does not find the target at step n but finds it at step $n + 2$, we have $x_n < D \leq x_{n+2}$. Therefore, we find

$$\frac{\sum_{i=1}^{n+1}(2x_i + t) + x_n}{x_n} = \sup_{x_n < D \leq x_{n+2}} \frac{\sum_{i=1}^{n+1}(2x_i + t) + D}{D} \leq \gamma. \tag{2}$$

We first show that if $\frac{t}{2\lambda} \leq 1$, a competitive cost of 9 is still achievable in the worst case.

Lemma 1. *If $\frac{t}{2\lambda} \leq 1$, the strategy $x_i = \left(\left(\left(1 - \frac{t}{2\lambda}\right) i + \left(1 + \frac{t}{2\lambda}\right) \right) 2^i - \frac{t}{2\lambda} \right) \lambda$ is optimal and has a competitive cost of 9.*

Proof. Using (1) and (2), we can show that the strategy x_i has the prescribed competitive cost in the worst case. The strategy x_i is optimal since it cannot do better than the optimal strategy for searching on a line. □

When $t = 0$, we find $x_i = (i+1)2^i\lambda$, which has a competitive cost of 9 in the worst case. Before presenting the result for the case where $\frac{t}{2\lambda} > 1$, we need the following technical lemma whose proof is omitted due to lack of space.

Lemma 2. *If $\frac{t}{2\lambda} > 1$, then $\gamma > 9$.*

Theorem 1. *If $\frac{t}{2\lambda} \geq 1$, the strategy $x_i = \left(\left(1 + \frac{t}{2\lambda}\right) \left(1 + \left(\frac{t}{2\lambda}\right)^{-1}\right)^i - \frac{t}{2\lambda} \right) \lambda$ is optimal and has a competitive cost of $\frac{2x_1 + t + \lambda}{\lambda} = 9 + 2\frac{\left(\frac{t}{2\lambda} - 1\right)^2}{\frac{t}{2\lambda}}$.*

Proof. Using (1) and (2), we can show that the strategy x_i has the prescribed competitive cost in the worst case. Moreover, if $\frac{t}{2\lambda} = 1$, then $9 + 2\frac{\left(\frac{t}{2\lambda} - 1\right)^2}{\frac{t}{2\lambda}} = 9$. For the rest of the proof, we suppose that $\frac{t}{2\lambda} > 1$ (hence, $\gamma > 9$ by Lemma 2).

From (1), we find

$$0 < \lambda \leq x_1 \leq \frac{(\gamma - 1)\lambda - t}{2} \tag{3}$$

and from (2), we find

$$x_{n+1} \leq \tau_0 x_n - \mu_0 - \nu_0 \sum_{i=1}^{n-1} (x_i + \kappa_0) \tag{4}$$

for all $n \geq 1$, where $\tau_0 = \frac{\gamma - 3}{2}$, $\mu_0 = t$, $\nu_0 = 1$ and $\kappa_0 = \frac{1}{2} t$. We can prove by induction that for all $0 \leq m \leq n - 1$,

$$x_{n+1} \leq \tau_m x_{n-m} - \mu_m - \nu_m \sum_{i=1}^{n-1-m} (x_i + \kappa_0), \tag{5}$$

where $\tau_{m+1} = \tau_0 \tau_m - \nu_m$, $\mu_{m+1} = \mu_0 \tau_m + \mu_m + \kappa_0 \nu_m$ and $\nu_{m+1} = \nu_0 \tau_m + \nu_m$. From the theory of characteristic equations, we find[1]

$$\tau_m = \frac{\tau_0(\tau_0 - r_2) - \nu_0}{r_1 - r_2} r_1^m - \frac{\tau_0(\tau_0 - r_1) - \nu_0}{r_1 - r_2} r_2^m,$$

$$\mu_m = \frac{(\tau_0 + 1)\mu_0 + (\nu_0 - 1)\kappa_0 + (\kappa_0 - \mu_0)r_2}{r_1 - r_2} r_1^m$$
$$- \frac{(\tau_0 + 1)\mu_0 + (\nu_0 - 1)\kappa_0 + (\kappa_0 - \mu_0)r_1}{r_1 - r_2} r_2^m + \kappa_0,$$

$$\nu_m = \frac{\nu_0(\tau_0 + 1 - r_2)}{r_1 - r_2} r_1^m - \frac{\nu_0(\tau_0 + 1 - r_1)}{r_1 - r_2} r_2^m,$$

where $r_1 > r_2$ are the roots of the polynomial $X^2 - (\tau_0 + 1)X + (\tau_0 + \nu_0)$. Notice that since $\gamma > 9$, we have $\tau_0 > 3$ and hence, $1 < r_2 < r_1$.

From (5) with $m := n - 1$, we have $x_{n+1} \leq \tau_{n-1} x_1 - \mu_{n-1}$ for all $n \geq 1$. By the *periodic-monotonic assumption*, x_{n+1} is increasing with respect to n. Also, x_{n+1} is unbounded. Therefore, $\delta(n) = \tau_{n-1} x_1 - \mu_{n-1}$ must be unbounded. Let us study $\delta(n)$, which can be written as $\delta(n) = a r_1^n - b r_2^n - \kappa_0$, where

$$a = \frac{\tau_0(\tau_0 - r_2) - \nu_0}{r_1(r_1 - r_2)} x_1 - \frac{(\tau_0 + 1)\mu_0 + (\nu_0 - 1)\kappa_0 + (\kappa_0 - \mu_0)r_2}{r_1(r_1 - r_2)},$$

$$b = \frac{\tau_0(\tau_0 - r_1) - \nu_0}{r_2(r_1 - r_2)} x_1 - \frac{(\tau_0 + 1)\mu_0 + (\nu_0 - 1)\kappa_0 + (\kappa_0 - \mu_0)r_1}{r_2(r_1 - r_2)}.$$

We will prove that $a \geq 0$ by contradiction. Suppose that $a < 0$. We have

$$\frac{d\delta}{dn} = a \log(r_1) r_1^n - b \log(r_2) r_2^n = b \log(r_2) r_2^n \left(\frac{a \log(r_1)}{b \log(r_2)} \left(\frac{r_1}{r_2} \right)^n - 1 \right).$$

---

[1] Even though $\nu_0 - 1 = 0$, we do not simplify the expressions for τ, μ and ν since later in the paper, we re-use them with different values for τ_0, μ_0, ν_0 and κ_0.

If $b \geq 0$, then $\frac{d\delta}{dn} \leq 0$, which means that δ is not increasing. In other words, δ is bounded, which is a contradiction. If $b \leq 0$, since $1 < r_2 < r_1$, there exists a rank n_0 such that $\frac{a \log(r_1)}{b \log(r_2)} \left(\frac{r_1}{r_2}\right)^n > 1$ for all $n \geq n_0$. This implies that $\frac{d\delta}{dn} < 0$ for all $n \geq n_0$. Thus, $\delta(n)$ is decreasing for all $n \geq n_0$. In other words, δ is bounded. Consequently, we must have $a \geq 0$.

Thus, we can find a lower bound on γ by finding the smallest γ which satisfies

$$\begin{cases} 0 < \lambda \leq x_1 \leq \frac{(\gamma-1)\lambda - t}{2}, \\ 0 \leq \frac{\tau_0(\tau_0 - r_2) - \nu_0}{r_1(r_1 - r_2)} x_1 - \frac{(\tau_0+1)\mu_0 + (\nu_0 - 1)\kappa_0 + (\kappa_0 - \mu_0)r_2}{r_1(r_1 - r_2)}. \end{cases} \tag{6}$$

Equivalently,

$$\begin{cases} \gamma \geq \frac{t^2 + 3tx_1 + (t + 2x_1)\sqrt{t(t + 2x_1)}}{tx_1}, \\ \gamma \geq \frac{2x_1 + t + \lambda}{\lambda}, \\ x_1 \geq \lambda > 0. \end{cases}$$

This optimization problem solves to $\gamma = 9 + 2\frac{\left(\frac{t}{2\lambda} - 1\right)^2}{\frac{t}{2\lambda}}$. $\qquad\square$

3   Searching a Moving Target on a Line with Turn Cost

In this section, we characterize an optimal strategy for searching a moving target on a line with turn cost. We suppose that the searcher travels at speed 1. The speed w of the target, where $0 \leq w < 1$, is known. In the worst case, the target and the searcher start moving in opposite directions, the target never slows down and never changes direction. Suppose that the turn cost $t = 0$. If the searcher knows D and the side of the line where the target is, it has to travel at distance $\frac{D}{1-w}$ to reach the target. If only D is known, the optimal strategy is to walk $\frac{D}{1-w}$ in one direction, then go back to the origin and travel at distance $\frac{D + 2\frac{D}{1-w}w}{1-w} = \frac{1+w}{(1-w)^2} D$ in the opposite direction. In the worst case, we get a competitive cost of $\frac{2\frac{D}{1-w} + \frac{1+w}{(1-w)^2} D}{D} = \frac{3-w}{(1-w)^2}$.

The *competitive cost* of an online algorithm usually refers to the cost of the algorithm in the worst case divided by the cost of an optimal offline algorithm. When both a turn cost and a moving target are involved, this definition can be interpreted in different ways. Traditionally, for line-searching problems, the cost of the optimal offline algorithm is D. This is the framework that was used by Gal to find an optimal strategy for searching a moving target on a line (see [10, Sect 7.5]). However, the cost of the optimal offline algorithm is $\frac{1}{1-w}D$ since, even if the searcher knows the exact position of the target, this target is still moving. Therefore, both comparing to D or to $\frac{1}{1-w}D$ make sense. But then, the optimal cost for one is a factor of $(1-w)$ of the other. Therefore, in this section, we focus on the cases where we compare the cost of the online algorithm to D.

When we introduce a turn cost, there are at least two ways of calculating the cost of the online algorithm. The turn cost t can be seen as representing time or fuel, which are different in this context. If t represents time, then the searcher has to wait for t units of time before moving back to the origin. While the searcher is waiting, the target has time to move a bit further. If t represents fuel, then we can imagine that it takes no time for the searcher to turn. Hence, it takes 0 units of time for the searcher to change direction and the target has no time to move. In Sect. 3.1, we study the case where t is considered to be time. In Sect. 3.2, we study the case where t is considered to be fuel.

3.1   Turn Cost Is Time — Competitive with Respect to D

Suppose that t is considered to be time. That is, when the searcher changes direction, it takes time t and the target can take advantage of that extra time to escape. When D is known, the optimal strategy is to travel at distance $\frac{D}{1-w}$ in one direction, then go back to the origin and travel at distance $\frac{D+\left(2\frac{D}{1-w}+t\right)w}{1-w} =$ $\frac{1+w}{(1-w)^2}D + \frac{w}{1-w}t$. In the worst case, we get a competitive cost of

$$\sup_{D\geq\lambda} \frac{2\frac{D}{1-w}+t+\frac{1+w}{(1-w)^2}D+\frac{w}{1-w}t}{D} = \frac{3-w}{(1-w)^2} + \frac{t}{\lambda(1-w)}.$$

For the rest of this section, we suppose that the searcher does not know D. Let γ be the competitive cost of the optimal strategy for searching a moving target on a line with turn cost, in the worst case.

In the worst case, it takes at least 2 steps to find the target. We get $x_2 \geq D + (2x_1 + t)w + x_2 w$, from which $D \in [\lambda, (1-w)x_2 - (2x_1 + t)w] = I_1$. When the searcher finds the target in two steps, it actually travels at distance $x_2' = D + (2x_1 + t)w + x_2'w$ during the second step, from which we have $x_2' = \frac{D+(2x_1+t)w}{1-w}$. Therefore,

$$\frac{2x_1 + t + \frac{\lambda+(2x_1+t)w}{1-w}}{\lambda} = \sup_{D\in I_1} \frac{2x_1 + t + x_2'}{D} \leq \gamma. \tag{7}$$

In general, if the searcher does not find the target at step n but finds it at step $n+2$, we get $x_n < D + \left(\sum_{i=1}^{n-1}(2x_i + t)\right)w + x_n w$ and $x_{n+2} \geq D + \left(\sum_{i=1}^{n+1}(2x_i + t)\right)w + x_{n+2}w$, from which $D \in I_{n+1}$, where

$$I_{n+1} = \left](1-w)x_n - \left(\sum_{i=1}^{n-1}(2x_i + t)\right)w, (1-w)x_{n+2} - \left(\sum_{i=1}^{n+1}(2x_i + t)\right)w\right].$$

When the searcher finds the target in $n+2$ steps, it actually travels at distance $x_{n+2}' = D + \left(\sum_{i=1}^{n+1}(2x_i + t)\right)w + x_{n+2}'w$ during the last step, from which we find $x_{n+2}' = \frac{D+\left(\sum_{i=1}^{n+1}(2x_i+t)\right)w}{1-w}$. Therefore,

$$\sum_{i=1}^{n+1}(2x_i + t) + \frac{\left((1-w)x_n - \left(\sum_{i=1}^{n-1}(2x_i+t)\right)w\right) + \left(\sum_{i=1}^{n+1}(2x_i+t)\right)w}{1-w}$$
$$(1-w)x_n - \left(\sum_{i=1}^{n-1} 2x_i\right)w \tag{8}$$

$$= \sup_{D \in I_{n+1}} \frac{\sum_{i=1}^{n+1}(2x_i + t) + x'_{n+2}}{D} \tag{9}$$

$$\leq \gamma. \tag{10}$$

Gal [10, Sect 7.5] proved that the optimal strategy for searching a moving target on a line has a competitive cost of $\frac{(3+w)^2}{(1-w)^3}$ in the worst case. We show that if $\frac{t}{2\lambda} \leq \frac{1+3w}{(1-w)^2}$, a competitive cost of $\frac{(3+w)^2}{(1-w)^3}$ is still achievable in the worst case.

Lemma 3. *If $\frac{t}{2\lambda} \leq \frac{1+3w}{(1-w)^2}$, the strategy*

$$x_i = \left(\frac{1-w}{1+w}\left(\left(\frac{1+3w}{(1-w)^2} - \frac{t}{2\lambda}\right)i + \left(\frac{1}{1-w} + \frac{t}{2\lambda}\right)\right)\left(2\frac{1+w}{1-w}\right)^i - \frac{t}{2\lambda}\right)\lambda$$

is optimal and has a competitive cost of $\frac{(3+w)^2}{(1-w)^3}$.

Proof. Using (7) and (10), we can show that the strategy x_i has the prescribed competitive cost in the worst case. The strategy x_i is optimal since it cannot do better than the optimal strategy for searching a moving target on a line.   □

When $t = 0$, we find $x_i = \left(\frac{1+3w}{(1-w)(1+w)} i + \frac{1}{1+w}\right)\left(2\frac{1+w}{1-w}\right)^i \lambda$, which has a competitive cost of $\frac{(3+w)^2}{(1-w)^3}$ in the worst case. Before presenting the result for the case where $\frac{t}{2\lambda} > \frac{1+3w}{(1-w)^2}$, we need the following technical lemma whose proof is omitted due to lack of space.

Lemma 4. *If $\frac{t}{2\lambda} > \frac{1+3w}{(1-w)^2}$, then $\gamma > \frac{(3+w)^2}{(1-w)^3}$.*

Theorem 2. *If $\frac{t}{2\lambda} \geq \frac{1+3w}{(1-w)^2}$, the strategy*

$$x_i = \left(\left(\frac{1}{1+w} + \frac{1-w}{1+w}\frac{t}{2\lambda}\right)\left(\frac{(1+w)\left((1-w)\frac{t}{2\lambda}+1\right)}{(1-w)2\frac{t}{2\lambda} - 2w}\right)^i - \frac{t}{2\lambda}\right)\lambda.$$

is optimal and has a competitive cost of $\frac{(3+w)^2}{(1-w)^3} + \frac{2\left((1-w)^2\frac{t}{2\lambda} - (1+3w)\right)^2}{(1-w)^3\left((1-w)^2\frac{t}{2\lambda} - 2w\right)}$.

Proof. Using (7) and (10), we can show that the strategy x_i has the prescribed competitive cost in the worst case. Moreover, if $\frac{t}{2\lambda} = \frac{1+3w}{(1-w)^2}$, then $\frac{(3+w)^2}{(1-w)^3} + \frac{2\left((1-w)^2\frac{t}{2\lambda}-(1+3w)\right)^2}{(1-w)^3\left((1-w)^2\frac{t}{2\lambda}-2w\right)} = \frac{(3+w)^2}{(1-w)^3}$. For the rest of the proof, we suppose that $\frac{t}{2\lambda} > \frac{1+3w}{(1-w)^2}$ (hence, $\gamma > \frac{(3+w)^2}{(1-w)^3}$ by Lemma 4).

From (7), we find $0 < \lambda \le x_1 \le \frac{((1-w)\gamma-1)\lambda - t}{2}$ and from (10), we find $x_{n+1} \le \tau_0 x_n - \mu_0 - \nu_0 \sum_{i=1}^{n-1}(x_i + \kappa_0)$ for all $n \ge 1$, where $\tau_0 = \frac{(1-w)^2\gamma - 3}{2} + \frac{w}{2}$, $\mu_0 = t$, $\nu_0 = (1-w)(1+\gamma w)$ and $\kappa_0 = \frac{t}{2}$. The rest of the proof is identical to that of Theorem 1. $\qquad\qquad\qquad\qquad\qquad\qquad\qquad\qquad\qquad\qquad\qquad\quad\square$

3.2   Turn Cost Is Fuel — Competitive with Respect to D

Suppose that t is the amount of fuel needed to make an instantaneous turn, thereby not affecting the target's position. Using the same approach as in Sect. 3.1, we get the following results.

Lemma 5. *If $\frac{t}{2\lambda} \le \frac{(1+3w)^2}{(1+w)(1-w)^3}$, the strategy*

$$x_i = \left(\left(\frac{1+3w}{1-w}\left(\frac{1}{1+w} - \frac{t(1-w)^3}{2\lambda(1+3w)^2}\right)i + \left(\frac{1}{1+w} + \frac{t(1-w)^3}{2\lambda(1+3w)^2}\right)\right)\left(2\frac{1+w}{1-w}\right)^i - \frac{t(1-w)^3}{2\lambda(1+3w)^2}\right)\lambda$$

is optimal and has a competitive cost of $\frac{(3+w)^2}{(1-w)^3}$.

Theorem 3. *If $\frac{t}{2\lambda} \ge \frac{(1+3w)^2}{(1+w)(1-w)^3}$, the strategy $x_i = (pq^i - r)\lambda$, where*

$$p = \frac{(1+3w)^2\left(2+(1-w)(1+w)\frac{t}{2\lambda}\right)}{2(1+w)\left((1+3w)^2+2(1-w)(1+w)^2\frac{t}{2\lambda}\right)} + \frac{1+3w}{2\left((1+3w)^2+2(1-w)(1+w)^2\frac{t}{2\lambda}\right)}\Phi,$$

$$q = \frac{-4w+(1-w)^2(1+w)\frac{t}{2\lambda}}{8w^2} - \frac{(1+w)^2}{8w^2}\Phi,$$

$$r = \frac{(1-w)(1+2w+5w^2)\frac{t}{2\lambda}}{2\left((1+3w)^2+2(1-w)(1+w)^2\frac{t}{2\lambda}\right)} + \frac{1+3w}{2\left((1+3w)^2+2(1-w)(1+w)^2\frac{t}{2\lambda}\right)}\Phi,$$

is optimal and has a competitive cost of $\frac{\frac{t}{2\lambda}(1-w)(1+2w+5w^2)-4w(1-w)-(1+3w)\Phi}{4w^2(1-w)}$,

where $\Phi = \sqrt{\frac{t}{2\lambda}(1-w)\left(\frac{t}{2\lambda}(1-w)^3 - 8w\right)}$.

4   The General Framework

The line-searching problems we solved in Sects. 2 and 3 can all be encapsulated in the same framework. For all these problems, we managed to rewrite the constraints on an optimal strategy x_i so that they satisfy the general inequalities (3) and (4). Inequality (3) is obtained by calculating what happens if the searcher finds the target in two steps. Inequality (4) is obtained by calculating what happens if the searcher needs $n+2$ steps to find the target. It depends on the initial parameters τ_0, μ_0, ν_0 and κ_0. These parameters are functions of the optimal cost γ and the parameters of the problem that is being studied (for instance, λ and t for searching on a line with turn cost). Then, (3) and (4) are converted into an optimization problem which consists in finding the smallest value of γ that satisfies (3) and

$$a = \frac{\tau_0(\tau_0 - r_2) - \nu_0}{r_1(r_1 - r_2)}x_1 - \frac{(\tau_0+1)\mu_0 + (\nu_0-1)\kappa_0 + (\kappa_0 - \mu_0)r_2}{r_1(r_1 - r_2)} \ge 0 \quad (11)$$

(refer to (6)), where

$$r_1 = \frac{\tau_0 + 1 + \sqrt{(\tau_0 - 1)^2 - 4\nu_0}}{2} \quad \text{and} \quad r_2 = \frac{\tau_0 + 1 - \sqrt{(\tau_0 - 1)^2 - 4\nu_0}}{2} \quad (12)$$

are the roots of the polynomial $X^2 - (\tau_0 + 1)X + (\tau_0 + \nu_0)$. In this section, we describe how to solve any line-searching problem for which the constraints on an optimal strategy x_i can be written as (3) and (4). In Sect. 4.1, we solve a general class of line-searching problems using our framework.

Let Θ be a finite set of parameters which contains λ. For instance, in Sect. 2, $\Theta = \{\lambda, t\}$ and in Sect. 3, $\Theta = \{\lambda, t, w\}$. Suppose that, for a given line-searching problem, the cost of travelling at distance x depends on x and parameters from Θ. Suppose that the constraints on an optimal strategy x_i can be written as (3) and (4), where these inequalities depend on γ and parameters from Θ. Then τ_0, μ_0, ν_0 and κ_0 are functions of γ and parameters from θ. We can solve the problem optimally in the following way.

1. Write Inequality (3) as $0 < \lambda \le x_1 \le \rho(\gamma, \Theta)$, where ρ is a function of γ and parameters from Θ.
2. Write Inequality (4) as $x_{n+1} \le \tau_0(\gamma, \Theta)x_n - \mu_0(\gamma, \Theta) - \nu_0(\gamma, \Theta) \sum_{i=1}^{n-1} (x_i + \kappa_0(\gamma, \Theta))$,
   where τ_0, μ_0, ν_0 and κ_0 are functions of γ and parameters from Θ.
3. Find the smallest γ such that $0 < \lambda \le x_1 \le \rho(\gamma, \Theta)$ and $a(\gamma, \Theta, x_1) \ge 0$, where a is a function of x_1, γ and parameters from Θ, defined as in (11). Let $\gamma^*(\Theta)$, which depends on parameters from Θ, be the solution to this optimization problem.
4. Let x_i^* be the solution to the following linear recurrence where $x_1 = \rho(\gamma^*(\Theta), \Theta)$:

$$x_{n+1} = \tau_0(\gamma^*(\Theta), \Theta)x_n - \mu_0(\gamma^*(\Theta), \Theta) - \nu_0(\gamma^*(\Theta), \Theta) \sum_{i=1}^{n-1} (x_i + \kappa_0(\gamma^*(\Theta), \Theta)).$$

5. Notice that x_i^* is defined for as long as $r_1(\gamma^*(\Theta), \Theta)$ and $r_2(\gamma^*(\Theta), \Theta)$ are real numbers, where r_1 and r_2 are functions of γ and parameters from Θ, defined as in (12). Let $\overline{\gamma}(\Theta)$, which depends on parameters from Θ, be the value of γ for which $(\tau_0(\gamma, \Theta) - 1)^2 - 4\nu_0(\gamma, \Theta) = 0$. The optimal solution x_i^* is defined for any values of parameters from Θ for which $\gamma^*(\Theta) \ge \overline{\gamma}(\Theta)$. It has a competitive cost of $\gamma^*(\Theta)$ in the worst case.
6. Let $\overline{x}_i$ be the solution to the following linear recurrence where $x_1 = \rho(\overline{\gamma}(\Theta), \Theta)$:

$$x_{n+1} = \tau_0(\overline{\gamma}(\Theta), \Theta) x_n - \mu_0(\overline{\gamma}(\Theta), \Theta) - \nu_0(\overline{\gamma}(\Theta), \Theta) \sum_{i=1}^{n-1} (x_i + \kappa_0(\overline{\gamma}(\Theta), \Theta)).$$

The solution $\overline{x}_i$ is optimal for all values of parameters from Θ for which $\gamma^*(\Theta) \le \overline{\gamma}(\Theta)$. It has a competitive cost of $\overline{\gamma}(\Theta)$ in the worst case.

4.1   One More Application of the General Framework

Consider an infinite family of line-searching problems. Let $\mathrm{cost}_1(x) = \alpha_1 x + \beta_1$ be the cost of walking distance x away from the origin and $\mathrm{cost}_2(y) = \alpha_2 y + \beta_2$ be the cost of walking distance y back to the origin. For instance, we have $\mathrm{cost}_1(x) = \mathrm{cost}_2(x) = x$ for the problem of searching on a line, and we have

$\text{cost}_1(x) = x$ and $\text{cost}_2(x) = x + t$ for the problem of searching on a line with turn cost. In this section, we suppose that $\alpha_1 \geq 0$, $\alpha_2 \geq 0$, $\alpha_1 + \alpha_2 > 0$, $\beta_1 \geq 0$ and $\beta_2 \geq 0$. Also, we suppose that the target is immobile. Using our general framework with $\Theta = \{\lambda, \alpha_1, \alpha_2, \beta_1, \beta_2\}$, we can prove the following result.

Theorem 4. *If $\frac{3\beta_1 + 2\beta_2}{2(\alpha_1 + \alpha_2)\lambda} \leq 1$, the strategy*

$$x_i = \left(\left(\left(1 - \frac{3\beta_1 + 2\beta_2}{2(\alpha_1 + \alpha_2)\lambda}\right)i + \left(1 + \frac{\beta_1 + \beta_2}{(\alpha_1 + \alpha_2)\lambda}\right)\right)2^i - \frac{\beta_1 + \beta_2}{(\alpha_1 + \alpha_2)\lambda}\right)\lambda$$

is optimal and has a competitive cost of $5\alpha_1 + 4\alpha_2$.

If $\frac{3\beta_1 + 2\beta_2}{2(\alpha_1 + \alpha_2)\lambda} \geq 1$, the strategy $x_i = \left(\left(1 + \frac{\beta_1 + \beta_2}{(\alpha_1 + \alpha_2)\lambda}\right)\Phi^i - \frac{\beta_1 + \beta_2}{(\alpha_1 + \alpha_2)\lambda}\right)\lambda$, where

$$\Phi = 1 + \left(\frac{2\beta_1 + \beta_2 - (\alpha_1 + \alpha_2)\lambda + \sqrt{(2\beta_1 + \beta_2)^2 - \beta_2^2 + (\beta_2 + (\alpha_1 + \alpha_2)\lambda)^2}}{2(\alpha_1 + \alpha_2)\lambda}\right)^{-1}, \text{ is optimal and}$$

has a competitive cost of $\frac{(\alpha_1 + \alpha_2)x_1 + (\beta_1 + \beta_2) + (\alpha_1\lambda + \beta_1)}{\lambda}$.

We can see where the "9" comes from in the original problem of searching on a line by setting $\beta_1 = \beta_2 = 0$.

References

1. Alpern, S., Fokkink, R., Gasieniec, L., Lindelauf, R., Subrahmanian, V.S.: Search Theory: A Game Theoretic Perspective. Springer, New York (2013)
2. Alpern, S., Gal, S.: The Theory of Search Games and Rendezvous. International Series in Operations Research and Management Science, vol. 55. Springer, New York (2003)
3. Baeza-Yates, R.A., Culberson, J.C., Rawlins, G.J.E.: Searching in the plane. Inf. Comp. **106**(2), 234–252 (1993)
4. Bose, P., Morin, P.: Online routing in triangulations. SIAM J. Comput. **33**(4), 937–951 (2004)
5. Bose, P., Morin, P., Stojmenović, I., Urrutia, J.: Routing with guaranteed delivery in ad hoc wireless networks. Wireless Netw. **7**(6), 609–616 (2001)
6. Chrobak, M., Kenyon-Mathieu, C.: Sigact news online algorithms column 10: competitiveness via doubling. SIGACT News **37**(4), 115–126 (2006)
7. Demaine, E.D., Fekete, S.P., Gal, S.: Online searching with turn cost. Theor. Comput. Sci. **361**(2–3), 342–355 (2006)
8. Dudek, G., Jenkin, M.: Computational principles of mobile robotics. Cambridge University Press, Cambridge (2010)
9. Gal, S.: A general search game. Israel J. Math. **12**(1), 32–45 (1972)
10. Gal, S.: Search Games. Mathematics in Science and Engineering, vol. 149. Academic Press, New York (1980)
11. LaValle, S.M.: Planning algorithms. Cambridge University Press, Cambridge (2006)
12. O'Kane, J.M., LaValle, S.M.: Comparing the power of robots. Int. J. Robot. Res. **27**(1), 5–23 (2008)
13. Pruhs, K., Sgall, J., Torng, E.: Handbook of Scheduling: Algorithms, Models, and Performance Analysis. Chapter Online Scheduling. CRC Press, Boca Raton (2004)
14. Zilberstein, S., Charpillet, F., Chassaing, P.: Optimal sequencing of contract algorithms. Annals Math. Artif. Intell. **39**(1–2), 1–18 (2003)

An Optimal Algorithm for Computing the Integer Closure of UTVPI Constraints

K. Subramani and Piotr Wojciechowski[(✉)]

LCSEE, West Virginia University, Morgantown, WV, USA
k.subramani@mail.wvu.edu, pwojciec@mix.wvu.edu

Abstract. In this paper, we study the problem of computing the lattice point closure of a conjunction of Unit Two Variable Per Inequality (UTVPI) constraints. We accomplish this by adapting Johnson's all pairs shortest path algorithm to UTVPI constraints. This problem has been extremely well-studied in the literature, inasmuch as it arises in a number of applications, including but not limited to, program verification and operations research. The complexity of solving this problem has steadily improved over the past several decades with the fastest algorithm for this problem running in time $O(n^3)$ on a UTVPI constraint system with n variables and m constraints. For the same input parameters, we detail an algorithm that runs in time $O(m \cdot n + n^2 \cdot \log n)$. It is clear that our algorithm is superior to the state of the art when the constraint system is sparse ($m \in O(n)$), and no worse than the state of the art when the constraint system is dense ($m \in \Theta(n^2)$). It is worth noting that our algorithm is time optimal in the following sense: The best known running time for computing the closure of a conjunction of difference constraints (m constraints, n variables) is $O(m \cdot n + n^2 \cdot \log n)$, and UTVPI constraints subsume difference constraints.

Keywords: UTVPI constraints · Relaxation · Dijkstra steps · Optimal · Closure

1 Introduction

This paper is concerned with the design of a new algorithm for computing the integer closure of a conjunction of UTVPI constraints. This is accomplished by adapting Johnson's all pairs shortest path algorithm to UTVPI constraints. UTVPI constraints occur in a number of application domains, including but not limited to, constraint solving [LM05], abstract interpretation [Min06], spatial databases, and theorem proving. There are two principal problems associated with UTVPI constraints, viz., the integer feasibility problem and the integer closure problem. Both of these problems are closely related, and many algorithms for the latter depend upon efficient strategies for the former. The class of UTVPI

P. Wojciechowski—This research is supported in part by the National Science Foundation through Award CCF-1305054.

M. Kaykobad and R. Petreschi (Eds.): WALCOM 2016, LNCS 9627, pp. 154–165, 2016.
DOI: 10.1007/978-3-319-30139-6_13

constraints clearly subsumes the class of difference constraints. Furthermore, unlike difference constraint systems, it is possible for a UTVPI system to be feasible, yet lack an integer solution. Likewise, for difference constraints, the linear closure and integer closure are equivalent and can be computed by finding the all pairs shortest path. However, for UTVPI constraints the linear closure is not always equivalent to the integer closure. We formally define the integer closure problem in Sect. 2.

We propose a new algorithm in this paper for the integer closure problem in UTVPI constraints (**IC**). The **IC** problem has received a fair amount of attention in the literature [LM05], with algorithms for this problem steadily improving in time complexity. To date, the fastest algorithm for the **IC** problem is the one proposed in [BHZ09]. Their algorithm runs in time $O(n^3)$ on a UTVPI system having n variables and m constraints. In contrast, our algorithm runs in time $O(m \cdot n + n^2 \cdot \log n)$ on a UTVPI system having the same input parameters. Inasmuch as the best known closure algorithm for difference constraints also runs in time $O(m \cdot n + n^2 \cdot \log n)$, our algorithm is optimal [CLRS01].

The principal contribution of this paper is the design and analysis of a new algorithm for computing the integer closure of a conjunction of UTVPI constraints.

The rest of this paper is organized as follows: In Sect. 2, we formally describe the problem under consideration. The motivation for our work as well as related work in the literature are outlined in Sect. 3. A fast algorithm for a special class of UTVPI constraints is described in Sect. 4. This algorithm forms the basis for the new integer closure algorithm detailed in Sect. 5. The correctness of the integer closure algorithm is discussed in Sect. 6. We conclude in Sect. 7 by summarizing our contributions and identifying avenues for future research.

2   Statement of Problem

In this section, we formally define the integer closure problem in UTVPI constraints.

Definition 1. *A constraint of the form $a_i \cdot x_i + a_j \cdot x_j \leq c_{ij}$ is said to be a Unit Two Variable Per Inequality (UTVPI) constraint if $a_i, a_j \in \{-1, 0, +1\}$ and $c_{ij} \in \mathbb{Z}$.*

A constraint for which $a_i = 0$ or $a_j = 0$ is called an absolute constraint. Such a constraint can be converted into constraints of the form: $a_i \cdot x_i + a_j \cdot x_j \leq c_{ij}$, where both a_i and a_j are non-zero (see Sect. 2.1).

A constraint for which $a_i = -a_j$ is called a difference constraint.

Definition 2. *A conjunction of UTVPI constraints is called a UTVPI constraint system and can be represented in matrix form as $\mathbf{A} \cdot \mathbf{x} \leq \mathbf{b}$.*

If the constraint system has m constraints over n variables, then $\mathbf{A}$ has dimensions $m \times n$.

Observe that a UTVPI system defines a polyhedron in n-dimensional space.

The Integer Feasibility problem (**IF**) in UTVPI systems is defined as follows: Does the defined polyhedron $\mathbf{A} \cdot \mathbf{x} \leq \mathbf{b}$ enclose a lattice point?

Assuming that the answer to the **IF** problem is affirmative, the integer closure problem (**IC**) is defined as follows: For each pair (x_i, x_j), find the following four values: (**1**) $\max_{\mathbf{A} \cdot \mathbf{x} \leq \mathbf{b}, \mathbf{x} \in \mathbb{Z}^n}(x_i - x_j)$, (**2**) $\max_{\mathbf{A} \cdot \mathbf{x} \leq \mathbf{b}, \mathbf{x} \in \mathbb{Z}^n}(x_i + x_j)$, (**3**) $\min_{\mathbf{A} \cdot \mathbf{x} \leq \mathbf{b}, \mathbf{x} \in \mathbb{Z}^n}(x_i - x_j)$, and (**4**) $\min_{\mathbf{A} \cdot \mathbf{x} \leq \mathbf{b}, \mathbf{x} \in \mathbb{Z}^n}(x_i + x_j)$.

For each variable x_i we also find the following two values:
(**1**) $\max_{\mathbf{A} \cdot \mathbf{x} \leq \mathbf{b}, \mathbf{x} \in \mathbb{Z}^n}(x_i)$ and (**2**) $\min_{\mathbf{A} \cdot \mathbf{x} \leq \mathbf{b}, \mathbf{x} \in \mathbb{Z}^n}(x_i)$.

It should be clear that if the **IF** problem does not have a solution, then neither does the **IC** problem. From this point onward, we use the terms max() and min() as shorthand for $\max_{\mathbf{A} \cdot \mathbf{x} \leq \mathbf{b}, \mathbf{x} \in \mathbb{Z}^n}()$ and $\min_{\mathbf{A} \cdot \mathbf{x} \leq \mathbf{b}, \mathbf{x} \in \mathbb{Z}^n}()$. Note that if $\mathbf{x}$ is restricted to $\mathbb{R}^n$ instead of $\mathbb{Z}^n$ then the above bounds constitute the linear closure of $\mathbf{U}$.

2.1 Constraint Network Presentation

Let $\mathbf{U} : \mathbf{A} \cdot \mathbf{x} \leq \mathbf{b}$ denote the UTVPI constraint system, and let $\mathbf{X}$ denote the set of all (fractional and integral) solutions to $\mathbf{U}$. Corresponding to this constraint system, we construct the constraint network $\mathbf{G} = \langle V, E, \mathbf{c} \rangle$ by utilizing the network construction from [SW15].

The input UTVPI system is transformed into a constraint network as follows:

For each variable, one node is added to the constraint network. Each constraint corresponds to a single edge as follows:

1. The constraint $x_i - x_j \leq c$ corresponds to the edge $x_j \overset{c}{\blacksquare} x_i$. We refer to this as a grey edge. This edge is also denoted by $x_i \overset{c}{\blacksquare} x_j$.

2. The constraint $x_i + x_j \leq c$ corresponds to the edge $x_j \overset{c}{\square} x_i$. We refer to this as a white edge.

3. The constraint $-x_i - x_j \leq c$ corresponds to the edge $x_j \overset{c}{\blacksquare} x_i$. We refer to this as a black edge.

To handle absolute constraints we add the vertex x_0. Each absolute constraint corresponds to a pair of edges as follows:

1. The constraint $x_i \leq c$ corresponds to the edges $x_i \overset{c}{\square} x_0$ and $x_i \overset{c}{\blacksquare} x_0$.

2. The constraint $-x_i \leq c$ corresponds to the edges $x_i \overset{c}{\blacksquare} x_0$ and $x_i \overset{c}{\blacksquare} x_0$.

If $\mathbf{U}$ has n variables and m constraints, then $\mathbf{G}$ has $(n+1)$ vertices and up to $(m + 4 \cdot n)$ edges.

We now introduce the notion of edge reductions.

Definition 3. *An edge reduction is an operation which determines a single edge equivalent to a two-edge path and represents the addition of the two UTVPI constraints which correspond to the edges in question. If this addition results in a UTVPI constraint, the reduction is said to be* valid.

We use Definition 3 to define paths in the constraint network.

Definition 4. *We say that a path has type t, if it can be reduced to a single edge of type t, where $t \in \{ \square, \blacksquare, \blacksquare, \blacksquare \}$.*

Let $\mathbf{U} : \mathbf{A} \cdot \mathbf{x} \leq \mathbf{b}$ denote a system of UTVPI constraints. To compute the integer closure of $\mathbf{U}$, we use the following inference rules [LM05].
The transitive rule is

$$\frac{a \cdot x_i + b \cdot x_j \leq c_{ij} \qquad -b \cdot x_j + b' \cdot x_k \leq c_{jk}}{a \cdot x_i + b' \cdot x_k \leq c_{ij} + c_{jk}},$$

and the tightening rule is

$$\frac{a \cdot x_i + b \cdot x_j \leq c_{ij} \qquad a \cdot x_i - b \cdot x_j \leq c'_{ij}}{a \cdot x_i \leq \lfloor \frac{c_{ij} + c'_{ij}}{2} \rfloor}.$$

The integer closure of $\mathbf{U}$ is the closure of $\mathbf{U}$ under the transitive and tightening inference rules [JMSY94]. This means that once the integer closure is computed, additional applications of the transitive and tightening inference rules do not create any additional constraints.

Note that the transitive inference rule corresponds to the addition of two UTVPI constraints. Thus, the transitive inference rule preserves linear solutions in addition to preserving integer solutions. In fact the closure of $\mathbf{U}$ under the transitive inference rule is the linear closure of $\mathbf{U}$ [JMSY94]. This is also true for systems of difference constraints.

3   Motivation and Related Work

In this section, we briefly motivate the study of UTVPI constraints and discuss related work in the literature.

UTVPI constraints occur in a number of problem domains, including but not limited to, program verification [LM05], abstract interpretation [Min06, PC77], real-time scheduling [GPS95], and operations research. Indeed, many software and hardware verification queries are naturally expressed using this fragment of integer linear arithmetic (i.e., the case in which the solutions of a UTVPI system are required to be integral.) We note that when the goal is to model indices of an array or queues in hardware or software, rational solutions are not usable [LM05]. Other application areas include spatial databases [SS00] and theorem proving.

The first known procedure for finding the integer closure of a system of UTVPI constraints is detailed in [JMSY94]. Their algorithm runs in $O(m \cdot n^2)$

time and uses $O(n^2)$ space. [HS97] improves on the approach in [JMSY94] from an ease-of-implementation standpoint. However, this does not improve on the asymptotic running time of the algorithm.

A $O(n^3)$ algorithm for the **IC** problem was mentioned in [LM05], however few details were provided. [BHZ09] implemented a $O(n^3)$ algorithm for this problem and alluded to a $O(m \cdot n + n^2 \cdot \log n)$ algorithm for finding the integer closure of a system of UTVPI constraints. In this paper, we provide a fully justified $O(m \cdot n + n^2 \cdot \log n)$ algorithm for the **IC** problem in UTVPI constraints.

4   A Fast Integer Closure Algorithm for a Special Subclass of UTVPI Constraints

In this section, we describe an algorithm for finding the integer closure of certain systems of UTVPI constraints.

We first describe a modified version of Dijkstra's shortest path algorithm, adapted to handle the network construction used in this paper. Dijkstra's shortest path algorithm is a well-known method for solving the single source shortest path problem in graphs with non-negative edge weights. It runs in $O(m+n \cdot \log n)$ time [CLRS01].

In Algorithm 4.1 we utilize the weight function $c(e)$ defined in Sect. 2.1.

Algorithm 4.1 operates on the same basic principles as Dijkstra's shortest path algorithm. However, it has been adapted to utilize the graph construction described in Sect. 2.1. We use it to find the shortest white, black, and gray paths from x_i to every vertex x_j in **G**. This gives us $\max(x_i + x_j)$, $\max(-x_i - x_j)$, $\max(x_i - x_j)$, $\max(-x_i + x_j)$, $\max(2 \cdot x_i)$, and $\max(-2 \cdot x_i)$. See Sect. 6 for a proof of this statement.

If a finite upper bound cannot be established on one of these values, then it is given the default value of ∞. Just like Dijkstra's shortest path algorithm, Algorithm 4.1 assumes that all edge weights are non-negative.

If we are given a system of pure difference constraints, then we cannot derive $\max(x_i + x_j)$, $\max(-x_i - x_j)$, $\max(x_i)$, or $\max(-x_i)$ for any x_i, x_j. Thus, for these bounds, Algorithm 4.1 returns a value of ∞.

We find the integer closure of a system of UTVPI constraints by converting the system into an equivalent one in each constraint has a non-negative right hand side. We also require that each constraint derivable from the tightening inference rule is included. We run Algorithm 4.2 on the modified graph to obtain the integer closure of the original system. This graph conversion is described in Sect. 5.

The constraints derived by the tightening rule need to be included. Otherwise, the bounds generated only apply to linear solutions, with integer solutions possibly requiring tighter bounds.

Example 1. Let us consider the system, $x_1 + x_2 \leq 1$, $x_1 - x_2 \leq 0$, and $x_2 - x_1 \leq 0$. Without tightening constraints, the best upper bound on $(x_1 + x_2)$ that we can obtain by simply adding constraints is $x_1 + x_2 \leq 1$. However, this bound is only satisfied with equality when $x_1 = x_2 = \frac{1}{2}$. Since no integer solution satisfies this bound with equality, this is not the tightest bound we can obtain.

Instead, if we add the constraints derived by the tightening inference rule, then we see that the two constraints $x_1 \leq 0$ and $x_2 \leq 0$ would be added. This would make the new best upper bound on $(x_1 + x_2)$, $x_1 + x_2 \leq 0$. This bound can be satisfied with equality when $x_1 = x_2 = 0$, which is an integer solution to the original system.

Function UTVPI-DIJKSTRA (network **G**, start vertex x_i)

1: Create array $D[x_j, t]$ of distance labels for $t \in \{\,\square\,,\,\blacksquare\,,\,◧\,,\,◨\,\}$. {Note that $D[x_j, t]$ is the length of a shortest path of type t between x_i and x_j.}
2: Create queue of unvisited vertex label pairs Q.
3: **for** (each x_j) **do**
4:     $D[x_j, t] \leftarrow \infty$ for $t \in \{\,\square\,,\,\blacksquare\,,\,◧\,,\,◨\,\}$.
5:     Add (x_j, t) to Q for $t \in \{\,\square\,,\,\blacksquare\,,\,◧\,,\,◨\,\}$.
6: **end for**
7: $D[x_i, t] \leftarrow 0$ for $t \in \{\,◧\,,\,◨\,\}$.
8: **while** $(Q \neq \emptyset)$ **do**
9:     $(y, t) \leftarrow \arg\min_{(x,t) \in Q}(D[x, t])$.
10:     **for** (each neighbor x of y) **do**
11:        **if** $(t = \square)$ **then**
12:           {Perform all *valid* edge reductions that start with a white edge.}
13:           {A table of these reductions can be found in [SW15].}
14:           **if** $(D[x, \square] > D[y, \square] + c(y\,◨\,x))$ **then**
15:              {Reduce $x_i \,\square\, y\,◨\,x$ to $x_i\,\square\,x$.}
16:              $D[x, \square] \leftarrow D[y, \square] + c(y\,◨\,x)$.
17:           **end if**
18:           **if** $(D[x, ◧] > D[y, \square] + c(y\,\blacksquare\,x))$ **then**
19:              {Reduce $x_i\,\square\,y\,\blacksquare\,x$ to $x_i\,◧\,x$.}
20:              $D[x, ◧] \leftarrow D[y, \square] + c(y\,\blacksquare\,x)$.
21:           **end if**
22:        **end if**
23:        $t \in \{\,\blacksquare\,,\,◧\,,\,◨\,\}$ are handled according to the rules in [SW15].
24:     **end for**
25:     Remove (y, t) from Q.
26: **end while**
27: **return** Distance labels D.

Once we get the system into this form we run Algorithm 4.2. From this, we obtain the tightest bounds on each possible UTVPI constraint.

5   The New Algorithm

Algorithm 4.2 provides a method for obtaining the integer closure when all constraints have non-negative constants, and all results of the tightening inference rule have been found.

To get all constraints to have non-negative constants, it suffices to find an integer solution **d** to the system, and to adjust the constraints accordingly.

That is if the values for x_i and x_j are d_i and d_j, then the adjusted version of the constraint $x_i + x_j \leq c_{ij}$ would be $x_i + x_j \leq c_{ij} - (d_i + d_j)$. Since $d_i + d_j \leq c_{ij}$ we have that $c_{ij} - (d_i + d_j) \geq 0$.

Function UTVPI-JOHNSON (set **U** of UTVPI constraints)

1: Construct constraint network **G** from **U** according to the rules in Section 2.
2: Create array $B[x_i, x_j, t]$ of bounds. {Note that $B[x_i, x_j, 0]$ represents $\max(x_i + x_j)$, $B[x_i, x_j, 1]$ represents $\max(x_i - x_j)$, $B[x_i, x_j, 2]$ represents $\min(x_i + x_j)$, and $B[x_i, x_j, 3]$ represents $\min(x_i - x_j)$.}
3: Create array $D[x_i]$ of distance labels.
4: **for** (each x_j) **do**
5:    $D \leftarrow$ UTVPI-DIJKSTRA(G, x_j). {Run Algorithm 4.1 from every vertex.}
6:    **for** (each x_i) **do**
7:       {Store the results in B.}
8:       $B[x_i, x_j, 0] \leftarrow D[x_i, \square\,]$.
9:       $B[x_i, x_j, 1] \leftarrow D[x_i, \blacksquare\,]$.
10:      $B[x_i, x_j, 2] \leftarrow -D[x_i, \blacksquare\,]$. {Convert $\max(-x_i - x_j)$ to $\min(x_i + x_j)$.}
11:      $B[x_i, x_j, 3] \leftarrow -D[x_i, \square\,]$. {Convert $\max(-x_i + x_j)$ to $\min(x_i - x_j)$.}
12:    **end for**
13: **end for**
14: **return** Array of bounds B.

Similarly, $x_i - x_j \leq c_{ij}$ would become $x_i - x_j \leq c_{ij} - (d_i - d_j)$, $-x_i + x_j \leq c_{ij}$ would become $-x_i + x_j \leq c_{ij} - (-d_i + d_j)$, and $-x_i - x_j \leq c_{ij}$ would become $-x_i - x_j \leq c_{ij} - (-d_i - d_j)$.

Example 2. Consider the UTVPI constraint $x_1 + x_2 \leq -5$. We have that the vector $(x_1, x_2) = (-3, -3)$ satisfies this constraint. If this is used as the initial valid solution, then this constraint would become $x_1 + x_2 \leq 1$ in the new system.

To account for all constraints generated by the tightening inference rule, we need to find the tightest bounds on $(x_i + x_i)$ and $(-x_i - x_i)$ for each x_i.

Once these are found, if the tightest bound on $(x_i + x_i)$ corresponds to a constraint $x_i + x_i \leq 2 \cdot a + 1$ for some integer a, then the constraint $x_i \leq a$ can be added to the system as a result of the tightening inference rule.

To make this process easier, finding these bounds can be performed on the adjusted graph where each constraint has non-negative constant. The proof of this is in Sect. 6.

We need to reduce an arbitrary system of UTVPI constraints to a system in which each constraint has non-negative constant, and all constraints added by the tightening rule are included. This is done in Algorithm 5.1.

5.1 Analysis of Running Time

Algorithm 5.1 consists of two portions. The first converts a general system of UTVPI constraints into one that can be accepted by Algorithm 4.2. Then Algorithm 4.2 is run on this modified graph. We shall analyze these two portions separately.

Function UTVPI-CLOSURE (set **U** of UTVPI constraints)

1: Find an integer solution **d** to **U**.
2: Create system of UTVPI constraints **U'**.
3: Create array $B[x_i, x_j, t]$ of bounds. {Note that $B[x_i, x_j, 0]$ represents $\max(x_i + x_j)$, $B[x_i, x_j, 1]$ represents $\max(x_i - x_j)$, $B[x_i, x_j, 2]$ represents $\min(x_i + x_j)$, and $B[x_i, x_j, 3]$ represents $\min(x_i - x_j)$.}
4: Create array $D[x_i]$ of distance labels.
5: **for** (each constraint e in **U**) **do**
6:     **if** (e is of the form $(x_i + x_j) \leq c_{ij}$) **then**
7:         Add the constraint $(x_i + x_j) \leq (c_{ij} - d_i - d_j)$ to **U'**.
8:     **end if**
9:     **if** (e is of the form $(x_i - x_j) \leq c_{ij}$) **then**
10:         Add the constraint $(x_i - x_j) \leq (c_{ij} - d_i + d_j)$ to **U'**.
11:     **end if**
12:     **if** (e is of the form $(-x_i + x_j) \leq c_{ij}$) **then**
13:         Add the constraint $(-x_i + x_j) \leq (c_{ij} + d_i - d_j)$ to **U'**.
14:     **end if**
15:     **if** (e is of the form $(-x_i - x_j) \leq c_{ij}$) **then**
16:         Add the constraint $(-x_i - x_j) \leq (c_{ij} + d_i + d_j)$ to **U'**.
17:     **end if**
18: **end for**
19: Construct constraint network **G** from **U'** according to the rules in Section 2.
20: **for** (each x_i) **do**
21:     $D \leftarrow$ UTVPI-DIJKSTRA(G, x_i).
22:     **if** ($D[x_i, \square]$ is odd) **then**
23:         {$\max(2 \cdot x_i)$ is odd.}
24:         Add the constraint $x_i \leq \lfloor \frac{D[x_i, \square]}{2} \rfloor$ to **U'**.
25:     **end if**
26:     **if** ($D[x_i, \blacksquare]$ is odd) **then**
27:         {$\max(-2 \cdot x_i)$ is odd.}
28:         Add the constraint $-x_i \leq \lfloor \frac{D[x_i, \blacksquare]}{2} \rfloor$ to **U'**.
29:     **end if**
30: **end for**
31: $B \leftarrow$ UTVPI-JOHNSON$($**U'**$)$.
32: **for** (each x_i) **do**
33:     **for** (each x_j) **do**
34:         $B[x_i, x_j, 0] \leftarrow B[x_i, x_j, 0] + d_i + d_j$.
35:         $B[x_i, x_j, 1] \leftarrow B[x_i, x_j, 1] + d_i - d_j$.
36:         $B[x_i, x_j, 2] \leftarrow B[x_i, x_j, 2] + d_i + d_j$.
37:         $B[x_i, x_j, 3] \leftarrow B[x_i, x_j, 3] + d_i - d_j$.
38:     **end for**
39: **end for**
40: **return** Array of bounds B.

Converting the graph into the form required by our fast integer closure algorithm consists of

- Finding an integer solution: Using the algorithm described in [LM05] this can be done in $O(m \cdot n + n^2 \cdot \log n)$ time.

- Re-weighting the graph: since every edge needs to be re-weighted this portion runs in $O(m)$ time.
- Determining which edges need to be tightened: This consists of n runs of UTVPI-DIJKSTRA(), and so takes $O(m \cdot n + n^2 \cdot \log n)$ time.
- Adding tightened edges: for each x_i we add at most 2 new edges, and so this portion runs in $O(n)$ time.

Thus, the graph conversion procedure runs in $O(m \cdot n + n^2 \cdot \log n)$ time.

Algorithm 4.2 which computes the integer closure for certain types of UTVPI systems consists of

- Creating the graph: since each edge and vertex of the graph need to be created, this process takes $O(m + n)$ time.
- Computing the transitive closure of the tightened graph: This consists of n runs of Algorithm 4.1, and so takes $O(m \cdot n + n^2 \cdot \log n)$ time.
- Determining $\min(x_i - x_j)$, $\max(x_i - x_j)$, $\max(x_i + x_j)$, $\min(x_i + x_j)$, $\max(x_i)$, and $\min(x_i)$ for each x_i and x_j: There are $O(n^2)$ bounds that need to be determined, so this portion of the algorithm runs in $O(n^2)$ time.

Thus, Algorithm 4.2 and the entire integer closure procedure run in $O(m \cdot n + n^2 \cdot \log n)$ time.

6   Correctness

We first show that Algorithm 4.2 generates the correct bounds on integer solutions to $\mathbf{U}$. This is done by showing that each of the desired bounds can be obtained from applying the transitive and tightening inference rules.

Lemma 1. *The bound $x_i - x_j \geq \min(x_i - x_j)$ can be obtained through repeated applications of the transitive and tightening inference rules.*

Proof. Let $c_{ij} = \min(x_i - x_j)$. Thus, there exists a valid solution to $\mathbf{U}$, say $\mathbf{x}'$, such that $x_i' - x_j' \leq c_{ij}$. However, for any $c < c_{ij}$ there is no such value of $\mathbf{x}$. This means that $\mathbf{U} \cup \{x_i - x_j \leq c_{ij}\}$ is feasible but $\mathbf{U} \cup \{x_i - x_j \leq c_{ij} - 1\}$ is infeasible.

Since this second system is infeasible, repeated applications of the transitive and tightening inference rules are able to produce the constraint $0 \leq b < 0$ [JMSY94].

Note that this proof of infeasibility must use the constraint $x_i - x_j \leq c_{ij} - 1$. Thus, removing $x_i - x_j \leq c_{ij} - 1$ from the proof of infeasibility, results in a derivation of the constraint $x_i - x_j \geq c_{ij}'$ such that $c_{ij} - 1 < c_{ij}' \leq c_{ij}$. Since c_{ij}' and c_{ij} are both integers, $c_{ij}' = c_{ij}$.

Thus, through repeated applications of the transitive and tightening inference rules we are able to derive the constraint $x_i - x_j \geq \min(x_i - x_j)$.   □

The ability to construct the remaining bounds is shown without proof since they are analogous to the proof of Lemma 1.

The distance labels, $D[x_j, t]$, in Algorithm 4.1 are computed by adding edge weights. This corresponds to applications of the transitive inference rule. Thus, we have that Algorithm 4.1 always generates valid bounds.

Lemma 2. *After running Algorithm 4.1 from x_i, we have the following:*

(1) $\max(x_i + x_j) = D[x_j, \square]$.
(2) $\max(x_i - x_j) = D[x_j, \blacksquare]$.
(3) $\min(x_i + x_j) = -D[x_j, \blacksquare]$.
(4) $\min(x_i - x_j) = -D[x_j, \square]$.

(5) $\max(x_i) = \lfloor \frac{D[x_i, \square]}{2} \rfloor$.
(6) $\min(x_i) = -\lfloor \frac{D[x_i, \blacksquare]}{2} \rfloor$.

The proof of Lemma 2 will be included in the journal version of this paper. As a direct result of Lemma 2, we have the following.

Corollary 1. *After running Algorithm 4.2, we have the following:*

(1) $B[x_i, x_j, 0] = \max(x_i + x_j)$.
(2) $B[x_i, x_j, 1] = \max(x_i - x_j)$.

(3) $B[x_i, x_j, 2] = \min(x_i + x_j)$.
(4) $B[x_i, x_j, 3] = \min(x_i - x_j)$.

Lemma 3. *All bounds generated by Algorithm 4.2 can be satisfied with equality by valid integer assignments to the variables.*

Proof. Suppose otherwise, thus there exists a system of constraints $\mathbf{U}$ such that the algorithm generates a bound, say $x_i + x_j \le c$, that can only be satisfied with equality by a non-integer solution to the original system of equations. Because the relaxation steps of DIJKSTRA correspond to the addition of constraints, the constraint $x_i + x_j \le c$ can be derived from the original system.

If no such constraint can be derived, then no upper bound on $(x_i + x_j)$ can be generated. Thus, there is no white path between x_i and x_j in the constraint network. In this case, the algorithm correctly gives the upper bound as ∞.

Consider the system $\mathbf{U} \cup \{-x_i - x_j \le -c\}$. This system is linear feasible, but not integer feasible. Thus, through repeated applications of the transitive and tightening inference rules a contradiction can be generated. However, by construction, $\mathbf{U}$ already contains all constraints generated by the tightening inference rule. This means that the contradiction can be obtained by only applying the transitive inference rule.

This would mean that $\mathbf{U} \cup \{-x_i - x_j \le -c\}$ is also linearly infeasible. Thus, $\mathbf{U} \cup \{-x_i - x_j \le -c\}$ must be integer feasible, and so the constraint $x_i + x_j \le c$ can be satisfied with equality by an integer solution.

A similar proof applies to constraints of type $x_i - x_j \le c$, $-x_i + x_j \le c$, and $-x_i - x_j \le c$. $\qquad\square$

Theorem 1. *Algorithm 4.2 correctly computes the integer closure of $\mathbf{U}$.*

Proof. From Lemma 1, and the corresponding results for the other constraint types, we know that all bounds generated by Algorithm 4.2 are derivable from $\mathbf{U}$. Thus, these bounds are satisfied by all integer solutions to $\mathbf{U}$.

From Lemma 3, every bound generated by Algorithm 4.2 is satisfied with equality by some integer solution of $\mathbf{U}$. Thus, these bounds are tight, and Algorithm 4.2 correctly computes the integer closure of $\mathbf{U}$.

We now show that the graph conversion procedure in Algorithm 5.1 generates all constraints derivable from the tightening inference rule.

Lemma 4. *Re-weighting the graph does not change the parity of the bounds on $(x_i + x_i)$ and $(-x_i - x_i)$.*

Proof. When the constraints are re-weighted, the change to the bound on both $(x_i + x_i)$ and $(-x_i - x_i)$ only depends on the value of d_i. We have that d_i is an integer, so the bound changes by $\pm 2 \cdot d_i$. Consequently, the parity of the weight remains unchanged. Thus, re-weighting the graph does not change where the tightening rule is applied.   □

Now, we show that only one pass of tightening is required. We do this by showing that adding a constraint generated by the tightening rule does not result in additional applications of the tightening rule.

Lemma 5. *Adding a constraint generated by the tightening rule does not necessitate adding any additional constraints as a result of the tightening inference rule.*

Proof. Suppose that adding the new constraint $x_i + x_i \leq 2 \cdot c_i$ results in an odd tightest bound for $(x_j + x_j)$.

The path p responsible for this bound must use the constraint $x_i + x_i \leq 2 \cdot c_i$. Thus, the path consists of a path p_1 from x_j to x_i, the newly added edge, and a path p_2 from x_i to x_j.

We have that p_1 corresponds to a constraint of the form $x_j - x_i \leq w_1$, and that the path p_2 corresponds to a constraint of the form $-x_i + x_j \leq w_2$.

Since p is of odd weight, we have that $(2 \cdot c_i + w_1 + w_2)$ is odd. Thus, $(w_1 + w_2)$ is also odd. This means that, since w_1 and w_2 are both integers, either $w_1 < w_2$ or $w_2 < w_1$.

If $w_1 < w_2$, then we can construct the bound $x_j + x_j = x_i + x_i + 2 \cdot (x_j - x_i) \leq 2 \cdot (c_i + w_1) < 2 \cdot c_i + w_1 + w_2$. This contradicts the assumption that $(2 \cdot c_i + w_1 + w_2)$ is the tightest bound on $(x_j + x_j)$

If $w_2 < w_1$, then we can construct the bound $x_j + x_j = x_i + x_i + 2 \cdot (-x_i + x_j) \leq 2 \cdot (c_i + w_2) < 2 \cdot c_i + w_1 + w_2$. This contradicts the assumption that $(2 \cdot c_i + w_1 + w_2)$ is the tightest bound on $(x_j + x_j)$

A similar proof applies when the edge added is $-x_i - x_i \leq 2 \cdot c_i$, or when the odd bound is created for $(-x_j - x_j)$.   □

Theorem 2. *Algorithm 5.1 computes all constraints derivable by the tightening inference rule.*

Proof. The detailed proof is provided in the journal version of this paper.

7   Conclusion

In this paper, we described an optimal algorithm for computing the integer closure of a conjunction of UTVPI constraints. The proposed algorithm is a generalization of Johnson's algorithm for difference constraints [CLRS01]. Our algorithm runs in $O(m \cdot n + n^2 \cdot \log n)$ time on a UTVPI constraint system with m constraints and n variables, which is superior to the $O(n^3)$ algorithm described in [BHZ09].

From our perspective, there are is one issue worth pursuing, viz. implementation. As discussed previously, there exist several algorithms in the literature for the **IC** problem, with various asymptotic running times. From a practical viewpoint, it is important to establish the empirical efficiency of these algorithms. A comprehensive empirical analysis should go a long way towards establishing the relative efficiency of the various methodologies discussed in this paper.

References

[BHZ09] Bagnara, R., Hill, P.M., Zaffanella, E.: Weakly-relational shapes for numeric abstractions: improved algorithms and proofs of correctness. Formal Meth. Syst. Des. **35**(3), 279–323 (2009)

[CLRS01] Cormen, T.H., Leiserson, C.E., Rivest, R.L., Stein, C.: Introduction to Algorithms. MIT Press, Cambridge (2001)

[PC77] Cousot, P., Cousot, R.: Abstract interpretation: a unified lattice model for static analysis of programs by construction or approximation of fixpoints. In: Proceedings of the 4th ACM Symposium on the Principles of Programming Languages (POPL), pp. 238–252. ACM Press (1977)

[GPS95] Gerber, R., Pugh, W., Saksena, M.: Parametric dispatching of hard real-time tasks. IEEE Trans. Comput. **44**(3), 471–479 (1995)

[HS97] Harvey, W., Stuckey, P.J.: A unit two variable per inequality integer constraint solver for constraint logic programming. In Proceedings of the 20th Australasian Computer Science Conference, pp. 102–111 (1997)

[JMSY94] Jaffar, J., Maher, M.J., Stuckey, P.J., Yap, H.C.: Beyond finite domains. In: Borning, A. (ed.) PPCP 1994. LNCS, vol. 874, pp. 86–94. Springer, Heidelberg (1994)

[LM05] Lahiri, S.K., Musuvathi, M.: An efficient decision procedure for UTVPI constraints. In: Gramlich, B. (ed.) FroCos 2005. LNCS (LNAI), vol. 3717, pp. 168–183. Springer, Heidelberg (2005)

[Min06] Miné, A.: The octagon abstract domain. Higher-Order Symbolic Comput. **19**(1), 31–100 (2006)

[SS00] Sitzmann, I., Stuckey, P.J.: O-trees: a constraint-based index structure. In: Australasian Database Conference, pp. 127–134 (2000)

[SW15] Subramani, K., Wojciechowski, P.: A graphical theorem of the alternative for UTVPI constraints. In: Leucker, M., Rueda, C., Valencia, F.D. (eds.) ICTAC 2015. LNCS, vol. 9399, pp. 328–345. Springer, Heidelberg (2015)

Covering Points with Convex Sets
of Minimum Size

Hwan-Gue Cho[1], William Evans[2], Noushin Saeedi[2], and Chan-Su Shin[3]([✉])

[1] Pusan National University, Busan, South Korea
hgcho@pusan.ac.kr
[2] University of British Columbia, Vancouver, Canada
{will,noushins}@cs.ubc.ca
[3] Hankuk University of Foreign Studies, Yongin, South Korea
cssin@hufs.ac.kr

Abstract. For a set P of n points in the plane and a fixed integer $k \geqslant 2$, we present algorithms for finding k bounded convex sets that cover P such that the total area or perimeter of the convex sets is minimized in time $O(n^{2k(k-1)} \log^2 n)$ and $O(n^{k(k-1)} \log^2 n)$, respectively. The algorithms can be applied to detect road intersections from the GPS traces of moving objects.

1   Introduction

Let P be a set of n points in the plane. We want to find two bounded convex sets whose union contains P and whose size is minimized. We measure the size of a convex set by its area or perimeter. We consider two different optimization versions; one to minimize the sum of their sizes, called *min-sum* optimization, and the other to minimize their maximum size, called *min-max* optimization.

The optimal two bounded convex sets (A, B) covering P are, in fact, two convex hulls on a partition (P_1, P_2) of P, that is, $A = \text{conv}(P_1)$ and $B = \text{conv}(P_2)$, where $\text{conv}(P)$ denotes the convex hull of P. Thus the problem is equivalent to finding a partition (P_1, P_2) of P such that $\mu(\text{conv}(P_1)) + \mu(\text{conv}(P_2))$ or $\max(\mu(\text{conv}(P_1)), \mu(\text{conv}(P_2)))$ is minimized, where $\mu(C)$ is either the area area(C) of C or the perimeter peri(C) of C. For the degenerate case that a convex hull is a line segment, its area is zero, and its perimeter is defined as twice the length of the segment. The problem naturally extends to covering P with $k > 2$ convex hulls. To the best of our knowledge, no previous results on this problem have been reported.

Related Work. Hershberger et al. [15] summarized the shape of streaming point data using limited memory by covering the points approximately with a bounded number of convex hulls, called a "ClusterHull". They used a combined cost function area$(H) + c \cdot$ peri(H) for each convex hull H, and clustered the points

W. Evans and N. Saeedi are supported by NSERC Discovery Grant. C.-S. Shin is supported by Research Grant of Hankuk Univ. of Foreign Studies.

M. Kaykobad and R. Petreschi (Eds.): WALCOM 2016, LNCS 9627, pp. 166–178, 2016.
DOI: 10.1007/978-3-319-30139-6_14

with k convex hulls to minimize total cost, where c is a constant, and k depends on the memory size $m \ll n$. This is an "on-line" algorithm to cover the streaming data approximately, so it is essentially different than our off-line algorithm to cover the static data exactly.

Many researchers have focused on the problem of finding a subset $P' \subseteq P$ of size k whose convex hull has the minimum area or perimeter [1,4,9–11,13]. The minimum-area subset P' of size k can be found in $O(kn^3)$ time [11] and $O((k^3 + \log n)n^2)$ time [10], while the minimum-perimeter subset P' in $O(k^2 n \log n + k^4 n)$ time [9].

Yet another related line of work addresses convex hull construction from imprecise point data [14,19,20]. Löffler and van Kreveld [19] gave algorithms for selecting the points from imprecise points modelled as line segments or squares such that the convex hull of the selected points has a maximum/minimum area/perimeter. They solved the minimum area and minimum perimeter problems for squares in $O(n^2)$ and $O(n^7)$ time. No polynomial time algorithm is known for the minimum area problem on the line segments, nor is the problem known to be NP-hard.

Covering a point set P by geometric objects is a widely investigated topic. Perhaps the most related results consider covering P by two objects such as disks [6], rectangles [5,18], and squares [17,22]. These problems search for two optimal objects whose union covers P and the larger area is minimized. This is a typical min-max optimization. The main difference with our problem is that the two optimal objects are determined by a constant number of points in P, which always guarantees polynomial algorithms.

Our Results. We present two algorithms to cover an n-point set P in the plane with two convex sets (or hulls) of minimum total area and minimum total perimeter, which run in $O(n^4 \log^2 n)$ and $O(n^2 \log^2 n)$ time, respectively. These are the first results to cover P with two convex hulls of the minimum total area or perimeter. They can be easily extended to cover P with k (> 2) convex hulls with minimum total area or perimeter, running in time $O(n^{2k(k-1)} \log^2 n)$ and $O(n^{k(k-1)} \log^2 n)$, respectively.

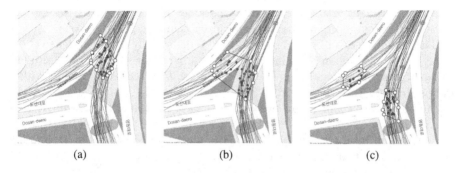

(a)   (b)   (c)

Fig. 1. Snapshots of GPS trajectories and the convex hulls at a 3-way intersection.

Applications. A recent advance in mobile services using GPS information triggered the generation of massive GPS tracking data, which allows at a low cost the generation of new street maps or refinement of existing maps using vehicular GPS trajectories [2,3,7,8,12,23]. One of the issues in such problems is to identify the 3-way or 4-way intersections of the road precisely. If the car trajectories are traced near the intersection within some time interval, then the change of convex hulls that cover a point set (car locations at a certain time) can be used to decide where or when the intersection exists. As in Fig. 1, the cars start to move in the same group along the road, but they separate at a 3-way intersection, and take different roads afterwards. This change can be captured by the convex hull coverage. In other words, the point set was initially covered with minimum cost (possibly with minimum perimeter) by one convex hull, but after the intersection, the cover by two convex hulls would cost less than that by one convex hull. One can use this as a criterion in recognizing the intersection.

We first explain algorithms for finding two convex hulls to cover P with minimum total area in Sect. 3 and the minimum total perimeter in Sect. 4, i.e., $k = 2$. We extend the algorithms to finding the optimal k convex hulls to cover P for general $k > 2$ in Sect. 5.

2   Preliminary

Let P be a set of n points in the plane. We assume for the sake of the simple description that points of P are in general position. This assumption can be easily discarded by dealing with collinear cases carefully.

A partition (P_1, P_2) of P is said to be *linearly separable* if there is a line ℓ such that P_1 lies on one side of ℓ and P_2 lies on the other side. Note that one of P_1 and P_2 may be an empty subset. If there is an optimal partition (P_1, P_2) which is linearly separable, then we can find its separation line by looking at all lines defined by two points from P and their associated partitions. Since we have $O(n^2)$ partitions, and compute two convex hulls for each partition in $O(n \log n)$ time, we can compute an optimal partition in $O(n^3 \log n)$ time, which can be improved by the data structure for the dynamic convex hull.

As we show later, an area-optimal partition is not linearly separable, but a perimeter-optimal partition is linearly separable. This makes the area-optimal partition problem more interesting, but harder than the perimeter one.

The boundary of a bounded convex set C is denoted by ∂C. We refer to C (or a polygon P) as a union of its interior and boundary. For simplicity, we also use the notations area(P) and peri(P) for a point set P to represent area(conv(P)) and peri(conv(P)), respectively.

Suppose that A and B are two convex sets of minimum total size covering P, equivalently, $A = \text{conv}(P_1)$ and $B = \text{conv}(P_2)$ for a partition (P_1, P_2) of P with minimum total size. To handle the degenerate case when A or B is a line segment, we simply assume that its boundary consists of two edges doubling the segment, thus its perimeter is twice the length of the segment. We now observe that:

Observation 1. *There is an optimal partition* (P_1, P_2) *of* P *such that no point of* P_1 *on* ∂A *is contained in* B, *and no point of* P_2 *on* ∂B *is contained in* A.

Proof. It is trivially true that $\mu(P') \leqslant \mu(P)$ for any $P' \subseteq P$, where μ is either area or peri. Suppose that $p \in P_1$ lying on ∂A is contained in B. Setting $A' = \mathrm{conv}(P_1')$ where $P_1' := P_1 \setminus \{p\}$ and $B' = \mathrm{conv}(P_2')$ where $P_2' := P_2 \cup \{p\}$ gives $\mu(A') \leqslant \mu(A)$ since $P_1' \subset P_1$, and $\mu(B') = \mu(B)$ since p is in B. This implies that (P_1', P_2') has total size no larger than that of (P_1, P_2).   □

Observation 2. *If* ∂A *and* ∂B *intersect, then they intersect at least even* $t \geqslant 4$.

Proof. By Observation 1, no points of P_1 on ∂A (and of P_2 on ∂B) are contained in B (and in A). So all the intersections are proper ones, i.e., defined between the interiors of two edges of ∂A and ∂B. This furthermore guarantees at least four even number of intersections. Note here that if at least one of them is a degenerate set, i.e., a line segment, then the segment is counted twice as edges of the hull, thus the boundaries of the hulls intersect exactly four times.   □

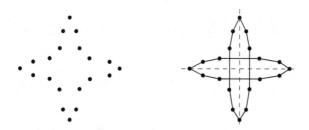

Fig. 2. Example that the area-optimal partition is not linearly separable.

3   Area-Optimal Covering with Two Convex Hulls

We first explain that there is a point set P whose area-optimal partition is not linearly separable. Let us consider P as in Fig. 2. It is not hard to see that its optimal hulls A and B are longish, and they intersect at their middle parts. Their area sum can be made sufficiently small by putting the points closer to the axes. For any arbitrary line separating P into two subsets P_1' and P_2', $\mathrm{area}(P_1') + \mathrm{area}(P_2')$ is at least a quarter of $\mathrm{area}(P)$, which is much larger than $\mathrm{area}(A) + \mathrm{area}(B)$. However, we can prove a crucial property that ∂A and ∂B intersect at most four times. This allows us to find an area-optimal partition in polynomial time.

Lemma 1. *There exist two convex sets of minimum total area whose union contains a given set* P *of points such that their boundaries intersect at most four times.*

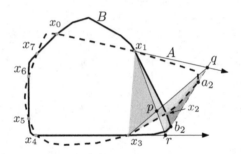

Fig. 3. Two convex hulls that intersect. Rays $\overrightarrow{x_{i-2}x_{i-1}}$ and $\overrightarrow{x_{i+2}x_{i+1}}$ do not diverge ($i = 2$).

Proof. Suppose A and B are two convex sets of minimum total area whose union contains P and whose boundaries intersect a minimum number of times. If either A or B is a line segment, then their boundaries intersect at most four times, so we assume from now on that neither A nor B is a line segment.

For the sake of contradiction, suppose the boundaries of A and B intersect $t > 4$ times. By Observation 2, their boundaries intersect an even number of times. Let $x_0, x_1, \ldots, x_{t-1}$ be these intersection points in clockwise order around $\partial(A \cup B)$. These intersection points are not on the boundary of the convex hull of $A \cup B$. Let $a_i b_i$ be the segment on this hull boundary, with $a_i \in A$ and $b_i \in B$, that "covers" x_i; that is, the ray from a point in $A \cap B$ through x_i intersects $a_i b_i$.

Let x_i be such that the rays $\overrightarrow{x_{i-2}x_{i-1}}$ and $\overrightarrow{x_{i+2}x_{i+1}}$ do not diverge, where index arithmetic is modulo t. In Fig. 3, the intersection point x_2 satisfies this property. Assume without loss of generality, that a_i precedes b_i in clockwise order around the convex hull of $A \cup B$. Let δ_i be the area of the "pocket" formed by the segment $a_i b_i$ and the boundaries of A and B between a_i and x_i and x_i and b_i (the pseudo-triangular-shaped shaded region in Fig. 3). We first show:

Claim. δ_i is at most the area of triangle $x_{i-1}x_i x_{i+1}$.

Proof. Refer to Fig. 3 for notation.

$$\delta_i \leqslant \text{area}(qrx_i) \leqslant \text{area}(pqr) \leqslant \text{area}(x_{i-1}px_{i+1}) \leqslant \text{area}(x_{i-1}x_i x_{i+1}).$$

These inequalities all follow from the convexity of A and B except for the third one, which follows from the fact that since $\overrightarrow{x_{i-2}x_{i-1}}$ and $\overrightarrow{x_{i+2}x_{i+1}}$ do not diverge, $\text{area}(qrx_{i+1}) \leqslant \text{area}(x_{i-1}rx_{i+1})$. $\qquad\square$

We now use Claim to prove the lemma.

Consider the lines $\overleftrightarrow{x_i x_{i-1}}$ and $\overleftrightarrow{x_{i+2}x_{i+3}}$. First, suppose that $\overrightarrow{x_i x_{i-1}}$ and $\overrightarrow{x_{i+2}x_{i+3}}$ do not diverge, as in Fig. 4. Since these two rays do not diverge and A is convex, the rays $\overrightarrow{x_i x_{i-1}}$ and $\overrightarrow{x_{i-4}x_{i-3}}$ do not diverge. In this case, we create convex sets A' and B', whose union contains the union of A and B, and whose total area is no bigger. The boundary of the convex set A' is the same as that of

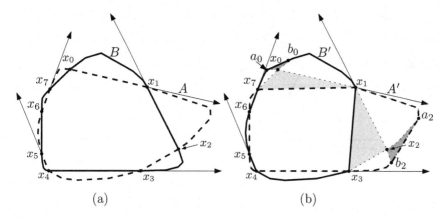

Fig. 4. (a) Two convex hulls that intersect. Rays $\overrightarrow{x_{i-2}x_{i-1}}$ and $\overrightarrow{x_{i+2}x_{i+1}}$ do not diverge and rays $\overrightarrow{x_ix_{i-1}}$ and $\overrightarrow{x_{i+2}x_{i+3}}$ do not diverge ($i = 2$). (b) Convex sets A' (dashed) and B' (solid) cover $A \cup B$ and have smaller total area.

A from x_{i-1} to a_i, then it consists of the segment a_ib_i followed by the boundary of B from b_i to x_{i-3} followed by the segment $x_{i-3}x_{i-1}$. The boundary of the convex set B' is the same as that of B from b_{i-2} to x_{i-1}, then it consists of the segment $x_{i-1}x_{i+1}$ followed by the boundary of A from x_{i+1} to a_{i-2} followed by the segment $a_{i-2}b_{i-2}$. See Fig. 4. Notice that the union of A' and B' includes $A \cup B$.

The convex set A' includes the pocket at x_i and B' includes the pocket at x_{i-2}, while neither pocket is in $A \cup B$. On the other hand, *both* A and B include the triangles $x_{i-1}x_ix_{i+1}$ and $x_{i-3}x_{i-2}x_{i-1}$, while only A' includes one and only B' includes the other. Since by Claim the area δ_i of the pocket at x_i is at most area$(x_{i-1}x_ix_{i+1})$ and the area δ_{i-2} of the pocket at x_{i-2} is at most area$(x_{i-3}x_{i-2}x_{i-1})$, we conclude that area(A') + area$(B') \leqslant$ area(A) + area(B). Notice that the boundaries of A' and B' intersect two fewer times than those of A and B.

The other possibility is that $\overrightarrow{x_{i-1}x_i}$ and $\overrightarrow{x_{i+3}x_{i+2}}$ converge, as in Fig. 5. Again we create convex sets A' and B' whose union contains $A \cup B$ and whose total area is no more than area(A) + area(B). In this case, the boundary of the convex set A' is the same as that of A from a_{i+1} to a_i, then it consists of the segment a_ib_i followed by the boundary of B from b_i to b_{i+1} followed by the segment $b_{i+1}a_{i+1}$. The boundary of the convex set B' is the same as that of B from x_{i+2} to x_{i-1}, then it consists of the segment $x_{i-1}x_{i+2}$. See Fig. 5. Notice that the union of A' and B' includes $A \cup B$.

The convex set A' includes the pockets at x_i and x_{i+1}, while neither pocket is in $A \cup B$. On the other hand, both A and B include the quadrilateral $x_{i-1}x_ix_{i+1}x_{i+2}$, while only A' includes it. As shown above, the area δ_i of the pocket at x_i is at most area$(x_{i-1}x_ix_{i+1})$ and the area δ_{i+1} of the pocket at x_{i+1} is at most area$(x_ix_{i+1}x_{i+2})$. Together these areas are at most area$(x_{i-1}x_ix_{i+1}x_{i+2})$, since area$(px_ix_{i+1}) <$ area$(px_{i-1}x_{i+2})$ because $\overrightarrow{x_{i-1}x_i}$

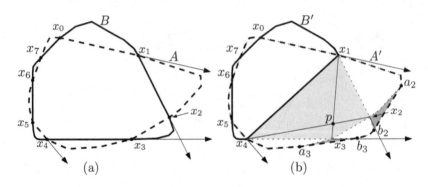

Fig. 5. (a) Two convex sets that intersect. Rays $\overrightarrow{x_{i-2}x_{i-1}}$ and $\overrightarrow{x_{i+2}x_{i+1}}$ do not diverge. Rays $\overrightarrow{x_{i-1}x_i}$ and $\overrightarrow{x_{i+3}x_{i+2}}$ converge ($i = 2$). (b) Convex sets A' (dashed) and B' (solid) cover $A \cup B$ and have smaller total area.

and $\overrightarrow{x_{i+3}x_{i+2}}$ converge and hence the sides of the quadrilateral $\overrightarrow{x_{i-1}x_i}$ and $\overrightarrow{x_{i+2}x_{i+1}}$ converge. We conclude that $\mathrm{area}(A') + \mathrm{area}(B') \leqslant \mathrm{area}(A) + \mathrm{area}(B)$.

Theorem 1. *For a set P of n points in $\mathbb{R}^2$, an area-optimal partition of P can be computed in $O(n^4 \log^2 n)$ time.*

Proof. Lemma 1 directly gives an algorithm to find an optimal partition in $O(n^5 \log n)$ time as follows: There are two situations whether ∂A and ∂B intersect or not. When they do not intersect, they are linearly separable, thus we can compute an optimal partition in $O(n^3 \log n)$ time. When they intersect, they do exactly four times by Observation 2 and Lemma 1. We first specify two edges e and e' of ∂A which properly intersect the two edges f and f' of ∂B. Then the points of P in the slab (or wedge) defined by the two lines extending e and e' are covered by A, and the others by B. Since there are $O(n^4)$ candidates of such edge pairs, it takes $O(n^5 \log n)$ time.

The time-consuming case is when ∂A and ∂B intersect four times. Using dynamic convex hull algorithms, we can handle this case in $O(n^4 \log^2 n)$ time. See Fig. 6(a). Assume that e is horizontal and above e'. P_1 is specified as a set of points below the line $\ell(e)$ extending e and above the line $\ell(e')$ extending e'. P_2 is the union of P' and P'', where P' is the set of points above $\ell(e)$, and P'' is the set of points below $\ell(e')$. Then $A = \mathrm{conv}(P_1)$, $B = \mathrm{conv}(P_2) = \mathrm{conv}(P' \cup P'')$, and the two edges f and f' of ∂B are defined as the two "outer" tangent segments between $\mathrm{conv}(P')$ and $\mathrm{conv}(P'')$.

Let p and q be the endpoints of e, and let p' and q' be the endpoints of e'. Assume that p is to the left of q, and p' is to the left of q'. We fix a pair (e, p'). Then P' is determined, so is $\mathrm{conv}(P')$. Once a point among the ones below $\ell(e)$ is further fixed as q', P_1 and P'' are completely determined. We maintain $\mathrm{conv}(P_1)$ and $\mathrm{conv}(P' \cup P'')$ dynamically for such a point q' in angular sorted order with respect to p'. We check if q' is *valid*, that is, it satisfies both (1) e and e' appear on the boundary of $\mathrm{conv}(P_1)$, and (2) the two tangent edges f and f' between

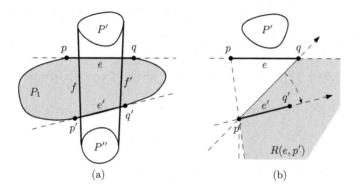

Fig. 6. The situation that two hulls intersect four times.

$\text{conv}(P')$ and $\text{conv}(P'')$ intersect e and e' properly. If q' is valid, we compute the total area of $\text{conv}(P_1)$ and $\text{conv}(P_2)$.

It is easy to see that e and e' appear on the boundary of $\text{conv}(P_1)$ if and only if (1) q' is below $\ell(e)$, (2) q' is on the same side of the line $\ell(p, p')$ as q, and (3) q' is on the opposite side of $\ell(p', q)$ as p. Denote such a region by $R(e, p')$. Note that $R(e, p')$ is determined only by e and p', and is an unbounded convex region with a constant complexity; for instance, refer to Fig. 6(b).

More precisely, we angular-sort the points below $\ell(e)$ with respect to p'. Initially, we build well-known dynamic data structures, presented by Overmars and van Leeuwen [21], on $\text{conv}(P_1)$ and $\text{conv}(P' \cup P'')$ for the first point q' in the sorted list. These data structures can support two update operations, insert and delete, in $O(\log^2 n)$ time, and several queries in $O(\log n)$ time, such as intersecting the hull with a line or a line segment or finding the tangent from an exterior point. As q' varies through the sorted list, we maintain $\text{conv}(P_1)$ and $\text{conv}(P' \cup P'')$ dynamically in $O(\log^2 n)$ time. After each update, we check if q' is valid. The first condition can be verified in constant time by checking whether q' is in $R(e, p')$ or not. If so, we check the second one. For this, we identify the edges of the boundary of $\text{conv}(P' \cup P'')$ which e and e' intersect. This type of query is supported by the data structure in $O(\log n)$ time. If both e and e' intersect the same two edges properly, then these edges are f and f', and the second condition holds. If q' is valid, we record the area of $\text{conv}(P_1)$ and $\text{conv}(P' \cup P'')$ obtained from the data structure in constant time. This is possible because the area can be maintained for each update operation. As a result, for a fixed pair (e, p'), we can find the minimum total area by considering all q' in $O(n \log^2 n)$ time. Since there are $O(n^3)$ such pairs, the total time is $O(n^4 \log^2 n)$ time.   □

Remark 1. We actually know when each q' is inserted and then deleted. This allows us to use the semi-dynamic convex hull data structure that supports the update operations and queries in $O(\log n)$ time [16]. But, since it stores the hull in an implicit way, it cannot answer the queries related with the whole hull structure in logarithmic time; for example, reporting the number of the points on the hull, and the area/perimeter of the hull. That is why we do not use this data structure.

4   Perimeter-Optimal Covering with Two Convex Hulls

Unlike the area-optimal partition, the perimeter-optimal partition is linearly separable.

Lemma 2. *For a point set P, there is a linearly separable partition (P_1, P_2) of P with minimum total perimeter.*

Proof. Suppose (P_1, P_2) is a minimum total perimeter partition of P, with $|P_1| \leqslant |P_2|$, chosen so that the boundaries of $A = \mathrm{conv}(P_1)$ and $B = \mathrm{conv}(P_2)$ intersect the minimum number of times. We assume that P_1 and P_2 are not degenerate sets, that is, none of A and B are line segments; for the degenerate case, the same argument can be easily applied. Let t be the number of times they intersect and suppose, for the sake of contradiction, that $t \geqslant 4$ by Observation 2. We further assume that $|P_2| \geqslant |P_1| \geqslant 2$ because they are linearly separable otherwise.

Like the area-optimal partition, let $x_0, x_1, \ldots, x_{t-1}$ be the intersection points of A and B in clockwise order. They are not on the boundary of the convex hull of $A \cup B$. Let $a_i b_i$ be the segment on the boundary of $\mathrm{conv}(A \cup B)$, with $a_i \in \partial A$ and $b_i \in \partial B$, that "covers" x_i.

Consider four consecutive intersections, x_{i-1}, x_i, x_{i+1}, and x_{i+2} as in Fig. 7(a). Let $p \in P_2$ be the last point on ∂B that appears before x_i, and let $p' \in P_2$ be the first point on ∂B that appears after x_{i+2}. Let ℓ be the line through x_{i-1} and x_{i+2}. Assume that the four intersections lie above ℓ. Let P' be the points of P_2 above or on ℓ.

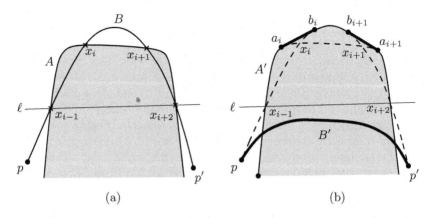

(a)   (b)

Fig. 7. (a) ∂A and ∂B intersect four or more times. A is shaded. (b) Set new convex sets A' and B'. A' is shaded. The dashed chains are deleted from ∂A and ∂B, and the thickest chains are added to $\partial A'$ and $\partial B'$.

We now consider new convex sets $A' = \text{conv}(P_1 \cup P')$ and $B' = \text{conv}(P_2 \setminus P')$ as in Fig. 7(b). The boundary of B' has a new chain as its boundary from p to p' that lies below ℓ. Since the new chain is upward convex, its length is less than $|C_1| + |C_2| + |C_3|$, where C_1 is a chain between p and x_i along ∂B, C_2 is a chain between x_i and x_{i+1} along ∂A, and C_3 is a chain between x_{i+1} and p' along ∂B. The boundary of A' gains two bridge segments $a_i b_i$ and $b_{i+1} a_{i+1}$. The length of $a_i b_i$ is less than the sum of the lengths of two convex chains C_4 and C_5; C_4 from a_i to x_i along ∂A, and C_5 from x_i to b_i along ∂B. For the other bridge $b_{i+1} a_{i+1}$, its length is less than the sum of the lengths of two convex chains C_6 and C_7; C_6 from b_{i+1} to x_{i+1} along ∂B, and C_7 from x_{i+1} to a_{i+1} along ∂A. Note that the chains $C_1, \ldots, C_7$ of $(\partial A \cup \partial B)$ are all disappeared in $(\partial A' \cup \partial B')$. As a result, the new segments added to $\partial A'$ and $\partial B'$ have smaller total length than the old segments deleted from ∂A and ∂B. In other words, $\text{peri}(A') + \text{peri}(B') < \text{peri}(A) + \text{peri}(B)$, which is a contradiction. □

Theorem 2. *For a set P of n points in the plane, a perimeter-optimal partition of P can be computed in $O(n^2 \log^2 n)$ time.*

Proof. Lemma 2 gives a straightforward way to find a perimeter-optimal partition for P in $O(n^3 \log n)$ time. This can be easily improved as we did in the area-optimal partition. By Lemma 2, there must be an optimal line ℓ separating P into P_1 and P_2, and passing through $p \in P_1$ and $q \in P_2$ such that $A = \text{conv}(P_1)$ and $B = \text{conv}(P_2)$. In fact, ℓ is an inner tangent supporting A and B. Fix a point of P as one end $p \in P_1$ of the inner tangent, and sort the other points angularly with respect to p. Then we can maintain two convex hulls of the partition incrementally with q from the sorted list [21]. Thus we can find a perimeter-optimal partition for a fixed point p in $O(n \log^2 n)$ time. This results in an $O(n^2 \log^2 n)$-time algorithm. □

5   Extensions

We can generalize this optimization problem in several directions as follows.
Min-sum partition for general $k \geqslant 2$. Let $A_1, \ldots, A_k$ be k convex hulls such that $A_i = \text{conv}(P_i)$ where $(P_1, \ldots, P_k)$ is a perimeter-optimal partition of P. Any pair of A_i and A_j is linearly separable by Lemma 2. Consider P_1. There are $(k-1)$ lines H_j separating A_1 and A_j for any $j > 1$ such that they pass through two points, each of P_1 and P_j. Then A_1 can be either in the half-plane above H_j, denoting H_j^+, or in the half-plane below H_j, denoting H_j^-. We denote by $\text{sign}(j) \in \{+, -\}$ the sign of the half-plane of H_j in which A_1 (and P_1) lies. There are $2^{(k-1)}$ combinations of the signs, so we simply check all the sign combinations. For a fixed sign combination, define $\mathcal{H} = \bigcap_{j=2}^{k} H_j^{\text{sign}(j)}$, then we can say that $P_1 = P \cap \mathcal{H}$. Each H_j is determined by a pair of points, so there are $O(|P|^{2(k-1)})$ point pairs. Therefore, there are total $O(2^{(k-1)} |P|^{2(k-1)})$ "configurations" to determine P_1. For a fixed configuration, we identify the points of P_1 and compute A_1 in $O(|P| \log |P|)$ time. What remains is to cover the points in $P \setminus P_1$ optimally by $(k-1)$ convex hulls. Let $S_P(k)$ denote the time complexity

to find k convex hulls to cover P with minimum total perimeter. By Theorem 2, $S_P(2) = |P|^2 \log^2 |P|$. Then

$$S_P(k) = (S_{P \setminus P_1}(k-1) + O(|P| \log |P|)) \times O(2^{(k-1)} |P|^{2(k-1)})$$
$$= O(2^{\frac{k(k-1)}{2}} |P|^{k(k-1)} \log^2 |P|).$$

This gives an algorithm to find a perimeter-optimal k convex hulls to cover P for a fixed $k \geqslant 2$, running in $O(n^{k(k-1)} \log^2 n)$ time.

For the area-optimal cover for $k \geqslant 2$, ∂A_i and ∂A_j for any pair $i \neq j$ can intersect at most four times by Lemma 1. Assume that A_1 is the convex hull with the maximum intersections, and further assume that ∂A_1 intersects m different ∂A_j's for some $0 \leqslant m < k$. For the case that $m = 0$, i.e., A_i's are pairwise distinct, we apply the same argument used for the perimeter-optimal problem. This can be done in $O(n^{k(k-1)+2} \log^2 n)$ time. For other cases that $1 \leqslant m < k$, ∂A_1 intersect m ∂A_j's (at most) four times at two edges e_j and e'_j of ∂A_1. Let W_j be the wedge bounded by $\ell(e_j)$ and $\ell(e'_j)$ that includes e_j an e'_j. Define $\mathcal{W} = \bigcap_j W_j$. Since $P_1 \subseteq W_j$, we know that $P_1 = P \cap \mathcal{W}$. Once the total $2m$ edges of ∂A_1 that intersect m A_j's are fixed, P_1 and A_1 are determined. The other points $P \setminus P_1$ are optimally covered by $(k-1)$ convex hulls. Let $T_P(k)$ denotes the time complexity to find k convex hulls to cover P with minimum total area. By Theorem 1, $T_P(2) = |P|^4 \log^2 |P|$. Since there are $O(|P|^{4m})$ edge combinations, we have that,

$$T_P(k) = (T_{P \setminus P_1}(k-1) + O(|P| \log |P|)) \times O(|P|^{4m}) = O(|P|^{2k(k-1)} \log^2 |P|).$$

Summing up over all $0 \leqslant m < k$, $T_P(k) = O(kn^{2k(k-1)} \log^2 n)$.

Theorem 3. *For any fixed $k \geqslant 2$ and a set P of n points, we can find k convex hulls with minimum area and minimum perimeter in $O(n^{2k(k-1)} \log^2 n)$ time and $O(n^{k(k-1)} \log^2 n)$ time, respectively.*

Min-Max Optimization Problems. We have mentioned the min-sum optimization problems so far, but we can also consider k convex hulls to cover P such that the maximum area or perimeter of them is minimized. We found an interesting fact that the min-max perimeter partition is not linearly separable unlike the min-sum perimeter partition. This makes the perimeter version much harder. Of course, the min-max area partition is not linearly separable as in Fig. 2. But we have no clue yet on how many times two optimal hulls for both problems can intersect each other, e.g., at most four times as in Lemma 1.

Let us consider an example on the min-max perimeter problem when $k = 2$. See n points of P as in Fig. 8(a), in which most points are collinear, but one can easily make a similar configuration of the points in general position. For simplicity, we assume here that n is odd. Such points in P are aligned along two unit-length orthogonal segments so that the distance between two adjacent points on a segment is $\frac{2}{n}$.

We claim that the optimal convex hulls are indeed two segments which cross at the origin o, thus the min-max perimeter becomes two. Consider an arbitrary

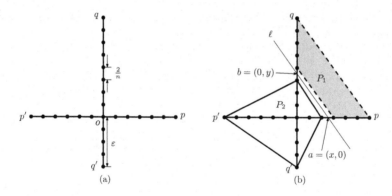

Fig. 8. A non-separable example for the min-max perimeter problem.

line ℓ separating P into two subsets P_1 and P_2, where P_1 is above (or in the right of) ℓ and P_2 is below (or in the left of) ℓ, as in Fig. 8(b). It passes two points $a = (x, 0)$ and $b = (0, y)$. For the special cases that $x = 0$ or $y = 0$, we can easily verify that the maximum of $\text{peri}(P_1)$ and $\text{peri}(P_2)$ is larger than 2, so it is not optimal. For another case that $y < 0$, $\text{peri}(P_1) > \text{peri}(\Delta opq) > 2$ for any $\varepsilon < 0.25$, so not optimal either. The remaining case is that $y > 0$ as in Fig. 8(b). Let Q_1 be a quadrangle $apqb$, and let Q_2 be a quadrangle $bp'q'a$. Then we have that $\text{peri}(P_1) \geqslant \text{peri}(Q_1) - \frac{4}{n}$ and $\text{peri}(P_2) \geqslant \text{peri}(Q_2) - \frac{8}{n}$. Set $\varepsilon = \frac{8}{n}$, then $\varepsilon \leqslant 0.1$ for any $n \geqslant 80$. We can check by a simple calculation that $\max(\text{peri}(P_1), \text{peri}(P_2)) \geqslant \max(\text{peri}(Q_1), \text{peri}(Q_2)) - \varepsilon > 2$ for $\varepsilon \leqslant 0.1$. As a result, there is an example whose optimal two convex hulls to cover P of any large n intersect each other.

References

1. Aggarwal, A., Imai, H., Katoh, N., Suri, S.: Finding k points with minimum diameter and related problems. J. Algorithms **12**(1), 38–56 (1991)
2. Ahmed, M., Karagiorgou, S., Pfoser, D., Wenk, C.: A comparison and evaluation of map construction algorithms using vehicle tracking data. GeoInformatica **19**(3), 601–632 (2015)
3. Ahmed, M., Wenk, C.: Constructing street networks from GPS trajectories. In: Algorithms - European Symposium on Algorithms, Ljubljana, Slovenia, September 10–12, pp. 60–71 (2012)
4. Atanassov, R., Morin, P., Wuhrer, S.: Removing outliers to minimize area and perimeter. In: Proceedings of Canadian Conference on Computational Geometry, August 14–16, Ontario, Canada (2006)
5. Bespamyatnikh, S., Segal, M.: Covering a set of points by two axis-parallel boxes. Inf. Process. Lett. **75**(3), 95–100 (2000)
6. Chan, T.M.: More planar two-center algorithms. Comput. Geom. **13**(3), 189–198 (1999)
7. Chen, D., Guibas, L.J., Hershberger, J., Sun, J.: Road network reconstruction for organizing paths. In: Proceedings of ACM-SIAM Symposium on Discrete Algorithms, Texas, USA, January 17–19, pp. 1309–1320 (2010)

8. Davies, J.J., Beresford, A.R., Hopper, A.: Scalable, distributed, real-time map generation. Pervasive Comput. **5**(4), 47–54 (2006)
9. Dobkin, D., Drysdale, R., Guibas, L.: Finding smallest polygons. Comput. Geom. Theor. Appl. **1**, 181–214 (1983)
10. Eppstein, D.: New algorithms for minimum area k-gons. In: Proceedings of ACM-SIAM Symposium on Discrete Algorithms, Orlando, Florida, January 27–29, pp. 83–88 (1992)
11. Eppstein, D., Overmars, M.H., Rote, G., Woeginger, G.J.: Finding minimum area k-gons. Discrete Comput. Geom. **7**, 45–58 (1992)
12. Fathi, A., Krumm, J.: Detecting road intersections from GPS traces. In: Fabrikant, S.I., Reichenbacher, T., van Kreveld, M., Schlieder, C. (eds.) GIScience 2010. LNCS, vol. 6292, pp. 56–69. Springer, Heidelberg (2010)
13. Guibas, L.J., Overmars, M.H., Robert, J.: The exact fitting problem in higher dimensions. Comput. Geom. **6**, 215–230 (1996)
14. Hassanzadeh, F.: Minimum perimeter convex hull of a set of line segments: An approximation. Master's thesis, Queen's University (2008)
15. Hershberger, J., Shrivastava, N., Suri, S.: Summarizing spatial data streams using clusterhulls. ACM J. Exp. Algorithmics **13** (2009)
16. Hershberger, J., Suri, S.: Off-line maintenance of planar configurations. J. Algorithms **21**(3), 453–475 (1996)
17. Jaromczyk, J., Kowaluk, M.: Orientation independent covering of point sets in $\mathbb{R}^2$ with pairs of rectangles or optimal squares. In: European Workshop on Computational Geometry, pp. 71–78 (1996)
18. Kim, S., Bae, S.W., Ahn, H.: Covering a point set by two disjoint rectangles. Int. J. Comput. Geometry Appl. **21**(3), 313–330 (2011)
19. Löffler, M., van Kreveld, M.J.: Largest and smallest convex hulls for imprecise points. Algorithmica **56**(2), 235–269 (2010)
20. Mukhopadhyay, A., Kumar, C., Greene, E., Bhattacharya, B.K.: On intersecting a set of parallel line segments with a convex polygon of minimum area. Inf. Process. Lett. **105**(2), 58–64 (2008)
21. Overmars, M.H., van Leeuwen, J.: Maintenance of configurations in the plane. J. Comput. Syst. Sci. **23**(2), 166–204 (1981)
22. Saha, C., Das, S.: Covering a set of points in a plane using two parallel rectangles. Inf. Process. Lett. **109**(16), 907–912 (2009)
23. Wu, J., Zhu, Y., Ku, T., Wang, L.: Detecting road intersections from coarse-gained GPS traces based on clustering. J. Comput. **8**(11), 2959–2965 (2013)

Data Structures

Efficient Generation of Top-k Procurements in a Multi-item Auction

Biswajit Sanyal[1]([✉]), Subhashis Majumder[2], and Wing-Kai Hon[3]

[1] Department of Information Technology, Government College of Engineering
and Textile Technology, Serampore, West Bengal, India
Biswajit_sanyal@yahoo.co.in
[2] Department of Computer Science and Engineering,
Heritage Institute of Technology, Kolkata, West Bengal, India
subhashis.majumder@heritageit.edu
[3] Department of Computer Science, National Tsing Hua University, Hsinchu, Taiwan
wkhon@cs.nthu.edu.tw

Abstract. We consider the top-k procurement decision problem in a multi-item auction, where there exists only one prospective buyer and the input consists of (i) a set of items, where each item is partitioned into equal number of shares, (ii) a set of suppliers, and (iii) for each supplier, her bids of selling different shares of each item; our target is to find k procurements with the least total costs. Kelly and Byde [SIGEcom 2006] studied the case where for each item, we may buy from each supplier at most once; this setting is suitable in most of the cases where total cost of a procurement is the primary concern.

In this paper, we assume a slightly different version where for any item, we may buy from a supplier multiple times (but each time with a different number of shares). We propose an *anytime* algorithm which can report successive best procurements until a buyer is satisfied (or terminates the algorithm). Our solution is based on preprocessing a metadata structure (without knowing the actual bids) to speed up the reporting steps. The metadata structure consists of (i) a DAG M_{local} for the valid combination of shares for each item, and (ii) a DAG M_{global} that coordinates the information from all M_{local}. We further show that the metadata structure can be generated on the fly, thereby saving a considerable amount of storage space.

1 Introduction

The *auction winner determination problem* [1,4] is a procurement decision problem in a multi-item auction, where multi-sourcing of the items are allowed (i.e., fractions of an item may be procured from different suppliers). In a typical multi-item auction, several suppliers can submit their bids for supplying different items. In some cases, a buyer may request the same item to be purchased on a large scale, making it difficult for a small-scale supplier to participate in the auction for supplying all amount of the item for the request. Multi-sourcing of an item should then be allowed so that the same item can be purchased from different

© Springer International Publishing Switzerland 2016
M. Kaykobad and R. Petreschi (Eds.): WALCOM 2016, LNCS 9627, pp. 181–193, 2016.
DOI: 10.1007/978-3-319-30139-6_15

suppliers in different units. This will be beneficial not only for the small-scale suppliers, but also for the buyers (as they will get more competitive prices for their items). In such situations, buyers become interested in identifying the best winning bids, so that all the items can be procured entirely (from a combination of suppliers) at the overall lowest cost. In general, it is better for the buyers to have the k procurements with the least total costs, so that from which they can choose the best procurement, had they got other intangible or strategic concerns.

We consider the top-k procurement decision problem in a multi-item auction, where the input consists of the following parts:

1. A set $\mathcal{I}$ of n items, $\{I_1, I_2, \ldots, I_n\}$, where each item is partitioned into equal number Q of shares.
2. A set $\mathcal{S}$ of m suppliers, $\{S_1, S_2, \ldots, S_m\}$.
3. Supplier S_j's bids for selling q shares of item I_i, for each $i \in [1, n]$, $j \in [1, m]$, and integer $q \in [1, Q]$.

Our target is to find k procurements with the least total costs. Kelly and Byde [4] studied the case where for each item, we may buy from each supplier at most once; this setting is suitable in most of the cases where total cost of a procurement is the primary concern. They proposed an algorithm that runs in $O(n\,R\,m \log m)$ time, where $R = (Q + m - 1)!/(Q!(m - 1)!)$ denotes the number of ways to partition Q into m nonnegative integral parts. Later, Byde et al. [1] improved the solution that runs in $O(nm(Q^2 + k))$ time. Both results are based on reducing the procurement problem into finding k shortest paths in a directed acyclic graph (DAG).

In this paper, we assume a slightly different version of the problem, where for any item we may buy from a supplier multiple times (but each time with a different number of shares). We call this variant *the basic problem*. As opposed to the original variant proposed by Kelly and Byde, this variant finds applications in the cases where cost is a primary concern, but not the *only* concern. As an example, consider a restaurant which wants to buy $100\,L$ (L) of cooking oil; suppliers may be selling oil with containers of size $1\,L$, $2\,L$, $10\,L$, and $20\,L$. For storage and convenience concerns, some combinations (say, 10 containers with $10\,L$ each, or 3 containers of $20\,L$ + 4 containers of $10\,L$) may be more preferable than other combinations (say, 5 containers of $20\,L$ each, or 100 containers with $1\,L$ each), even if the former combinations cost more than the latter combinations. In such a case, it is better to retain *all possible combinations* for consideration, rather than *over-simplifying* the problem to treat costs as the only concern. By applying the same reduction technique as used by Byde et al., we show that for the basic problem, we can obtain the k cheapest procurements, in unsorted order, in $O(mnQ(k + Q))$ time; an additional $O(k \log k)$ time can further be applied to sort the procurements by their total costs.

We also consider designing an *anytime* algorithm which can report successive best procurements until a buyer is satisfied (or terminates the algorithm). This algorithm will be useful when there is no prior knowledge how large (or how small) k should be set to obtain meaningful procurements. Our solution is based on preprocessing a metadata structure (without knowing the actual bids) to speed up the reporting steps. The metadata structure consists of (i) a DAG M_{local} for the valid combination of shares for each item, and (ii) a DAG M_{global} that coordinates the information from all M_{local}. We further show that the metadata structure can be generated on the fly, thereby saving a considerable amount of storage space. Precisely, we show with $O(nQ(m \log m + \texttt{partition}(Q)))$ preprocessing time, the tth cheapest procurement can be iteratively extracted from the metadata structure in $O(n(Q + \log t))$ time; here, $\texttt{partition}(Q)$ denotes the number of integer partition of Q, which is bounded by $e^{O(\sqrt{Q})}$. Note that the preprocessing time is now exponential in $\sqrt{Q}$, but as Kelly and Byde [4] have pointed out, Q would be a small constant in practice, so that we may consider $Q = O(1)$, and thus $\texttt{partition}(Q) = O(1)$,[1] so that the proposed algorithm can still work well in practice.

The paper is organised as follows. Section 2 shows how to reduce our problem into finding k shortest paths in a DAG. In Sect. 3, we introduce the metadata structure M_{local} for organising suppliers' bids for a single item, while in Sect. 4 we continue to show how to use M_{local} structures to give an anytime algorithm for our problem, and discuss how the framework can be adapted for other variants of the problem. We conclude the paper in Sect. 5.

2   Method I: Reduction to k Shortest Paths Problem

In this section, we use an analogous strategy by Byde et al. [1] to reduce a top-k procurement decision problem into finding k shortest paths in an edge-weighted DAG, where the latter problem can then be directly solved by Eppstein's linear-time algorithm [3]. For simplicity, we first discuss the special case where $n = |\mathcal{I}| = 1$, that is, there is only one item I to consider.

Initially, create a tuple (i, q) for each supplier S_i and quantile q, and fix a total ordering $\mathcal{O}$ among these tuples (any ordering will work). For instance, we may order the tuples by the reverse lexicographical ordering as follows: $(1, 1), (2, 1), (3, 1), \ldots, (m - 1, Q), (m, Q)$. Next, we create a directed graph that contains $mQ + 2$ columns of nodes as follows (See Fig. 1 for an example):

1. Column 0 contains a single node s, the source node.
2. Column $mQ + 1$ contains a single node t, the sink node.
3. For each r with $1 \leq r \leq mQ$, Column r contains $Q + 1$ nodes, labeled as $(i, q, 0), (i, q, 1), \ldots, (i, q, Q)$, where (i, q) is the rth tuple in the ordering $\mathcal{O}$.
4. Let (i, q) and (i', q') be the rth and the $r + 1$th tuples in $\mathcal{O}$. For any r and any j, (i) create a weighted edge from (i, q, j) to (i', q', j) with cost 0, and (ii) a weighted edge from (i, q, j) to $(i', q', j + q')$ with cost $b_{i', q'}$, where $b_{i', q'}$ is the bid of supplier $S_{i'}$ for q' quantiles of item I.

---

[1] We quote some specific values of $\texttt{partition}(Q)$ below: $\texttt{partition}(10) = 42$, $\texttt{partition}(20) = 627$, $\texttt{partition}(30) = 5604$, $\texttt{partition}(40) = 37338$. See *Online Encyclopedia of Integer Sequences* (http://oeis.org/A000041) for more details.

5. Let $(\hat{\imath}, \hat{q})$ be the first tuple in the ordering $\mathcal{O}$. Create an edge from s (the source) to $(\hat{\imath}, \hat{q}, 0)$ with cost 0, and an edge from s to $(\hat{\imath}, \hat{q}, \hat{q})$ with cost $b_{\hat{\imath}, \hat{q}}$.
6. Let $(\tilde{\imath}, \tilde{q})$ be the last tuple in the ordering $\mathcal{O}$. Create an edge from to $(\tilde{\imath}, \tilde{q}, Q)$ to t (the sink) with cost 0.

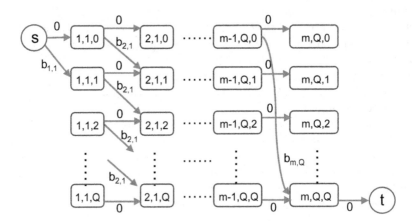

Fig. 1. Reduction to k shortest paths. We assume that the corresponding total ordering $\mathcal{O}$ orders the tuples (i, q) in reverse lexicographical ordering.

It is easy to verify that the graph is a DAG, and there is a one-to-one correspondence between a path from s to t and a valid combination of suppliers' bids for the procurement of item I. Thus, the k shortest paths in the DAG will correspond to the top-k procurements for the item I. Also, the DAG contains $O(mQ^2)$ nodes and $O(mQ^2)$ edges, and each path from s to t contains $O(mQ)$ edges. Using Eppstein's algorithm [3], the k shortest paths can be obtained, explicitly, in a total of $O(kmQ)$ time (but these paths are not ordered by their costs). This gives the following lemma.

Lemma 1. *We can obtain the k cheapest procurements of an item I (in the basic problem) in $O(mQ(k + Q))$ time. An extra $O(k \log k)$ time can be applied to order these procurements according to their costs.*

To solve the general case with two or more items, we simply construct a separate DAG for each item, and then for each $i \in [1, n - 1]$, link the sink of the DAG for item i to the source of the DAG for item $i + 1$ with a cost 0 edge. This gives the following theorem.

Theorem 1. *Let n denote the number of items and m denote the number of suppliers. Given a value k, the top-k procurement decision problem can be solved in $O(mnQ(k + Q) + k \log k)$ time.*

3   The Metadata Structure $M_{\texttt{local}}$

In this section, we describe the metadata structure $M_{\texttt{local}}$ that is maintained for a fixed item with Q quantiles. We first introduce its key component, which is a metadata structure to organise all the combinations of suppliers' bids, with those bids corresponding to the *same* partition of Q. After that, we show how to combine the different key components (one for each possible way to partition Q) into the desired $M_{\texttt{local}}$ structure.

3.1   Metadata Structure Per Partition

Let $\pi = \pi_1 + \pi_2 + \cdots + \pi_j$ be an integer partition of Q, such that each element π_r is an integer in $[1, Q]$, and $\sum_{r=1}^{j} \pi_r = Q$. Alternatively, we may represent π by a vector $(f_1, f_2, \ldots, f_Q)$ where f_r is the frequency of r as a tuple in π. Given a particular item in $\mathcal{I}$, we define a *valid combination* of suppliers' bids *with respect to* π to be a combination that has exactly f_r suppliers' biddings for r quantiles (thus, the total number of quantiles is Q). In the following, we define a metadata structure that organises all the possible valid combinations with respect to π, so that it supports efficient way to extract, iteratively, the next combination with the least total cost.

Suppose that for each quantile r/Q, the m suppliers' bids are sorted in nondecreasing order. The cheapest combination with respect to π must be to select the first f_r bids for each quantile r/Q. Such a combination may be represented by a bitvector with $Q \times m$ bits, logically divided into Q blocks of m bits, such that the first f_r bits in block r is marked 1, and all other bits 0. For instance, suppose $S = 4$, $Q = 4$, and $\pi = 1 + 1 + 2$ which is represented by $(2, 1, 0, 0)$. Then, the bitvector corresponding to the cheapest combination with respect to π is: 1100 1000 0000 0000. In fact, each valid combination has a one-to-one correspondence with a bitvector that contains for each r exactly f_r 1s in block r.

Using a similar concept from an earlier work [5], we define a bitvector v as a *one-shift* of another vector u, if we can obtain v by swapping a 1 in u with a neighboring 0 placed immediately on its right, without crossing block boundaries. For instance, 1100 1000 0000 0000 has two possible one-shifts, which are 1010 1000 0000 0000 and 1100 0100 0000 0000; in contrast, 1001 0001 0000 0000 has only one possible one-shift 0101 0001 0000 0000, while 0011 0001 0000 0000 does not have any one-shift. Also, a vector may be a one-shift of two or more distinct vectors. For instance, 0110 0100 0000 0000 is a one-shift of both 1010 0100 0000 0000 and 0110 1000 0000 0000.

Each vector has at most j one-shifts, where j is the number of parts in the partition π (or equivalently, the number of 1s in the vector); in the worst case, $j = Q$. If v is a one-shift of u, the corresponding combination of v cannot have total cost smaller than that of u (or, u is at least *better* than v). This gives a natural way to organise all possible combinations as a DAG (each node corresponds to a valid combination, and we will abuse the notation to refer the node and the corresponding combination by the same label), with the cheapest

combination as the root, and the children of a node u are all the one-shifts of u.[2]
See Fig. 2 for an example.

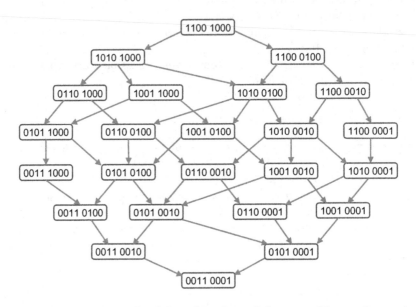

Fig. 2. A DAG containing all valid combinations of the case with $n = 4, m = 4, \pi = 1 + 1 + 2 = 4$. A node u is linked to another node v if v is a one-shift of u. For any node in the DAG, the last 8 bits of its vector are 0000 0000, so that we omit them for brevity.

Given a rooted DAG, we say a node u *covers* another node w, if $w = u$, or there is a directed path from u to w; thus, the root of the DAG covers all the nodes in the DAG. Similarly, we say a set U of nodes *covers* w if some node in U covers w. Suppose that we want to find the cheapest combination among a set W of combinations in the DAG; then, it is sufficient to consider a subset $Z \in W$ that covers all nodes in W, and find the cheapest combination in Z. This gives the following algorithm for reporting iteratively the next cheapest combination (when needed):

1. Maintain a set Z, which is initialized to contain the cheapest combination of all combinations (i.e., the root of the DAG).
2. When reporting the next cheapest combination, find the cheapest combination u among Z.
3. After reporting u, extract u from Z, and insert to Z all the children (i.e., one-shifts) of u that is not currently in Z.

Note that the DAG can be constructed in advance, but the evaluation of the cheapest combination u among Z (Step 2) can be done when the suppliers' bids

---

[2] We defer the formal proof to the full paper.

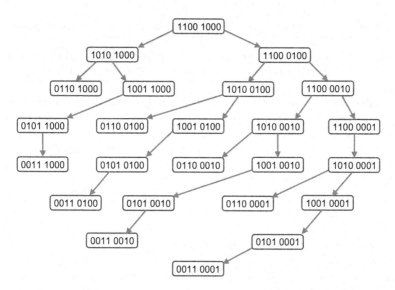

Fig. 3. The implicit DAG corresponding to the DAG in Fig. 2. A node u is linked to another node v if v is a mandatory-one-shift of u. For any node in the DAG, the last 8 bits of its vector are 0000 0000, so that we omit them for brevity.

are known. By maintaining a binomial heap on Z [2], and the DAG as well (to speed up the checking for repeated insertion in Step 3), we obtain the following lemma.

Lemma 2. *Let W be a set of combinations with respect to a partition π of Q, and Z be a subset in W such that Z covers all the combinations in W. Suppose that the suppliers' bids are all known, and for each quantile, the bids are sorted in non-decreasing order. Let t_{eval} be the time to evaluate the cost of a combination. Then, the cheapest combination in W can be extracted in $O(\log|Z|)$ time, and the set Z can be updated in $O(Q \times t_{\mathrm{eval}})$ time.*

In the above algorithm, storing the DAG seems to be unavoidable, or else there may be multiple insertions of the same combination into Z. Interestingly, the DAG can be avoided, if we slightly *strengthen* the definition of one-shift as follows: We say v is a *mandatory* one-shift of u, if (i) v is a one-shift of u, and (ii) among all the vectors of which v is its one-shift, u is lexicographically the largest. For instance, $v = 0110\ 0100\ 0000\ 0000$ is a mandatory one-shift of $1010\ 0100\ 0000\ 0000$, but v is *not* a mandatory one-shift of $0110\ 1000\ 0000\ 0000$.

We may define a DAG based on the mandatory one-shifts (instead of one-shifts) as before. See Fig. 3 for an example. There are some interesting observations about this DAG:

1. **Each valid combination can be reached from the root.** Note that to obtain a particular vector from the root vector, we use a series of mandatory one-shifts to move the rightmost 1 to its correct position, and then a series of

mandatory one-shifts to move the second rightmost 1 to its correct position, and so on.

2. **The DAG is effectively a rooted tree, where each node (except the root) has exactly one parent.** It follows directly from the definition of mandatory one-shift.

3. **All children of a node can be deduced from the combination corresponding to the node.** Precisely, consider only those blocks that contain 1s. Let ℓ be the number of contiguous blocks, from left to right, with all 1s appearing before all 0s. Then, there are ℓ children corresponding to swapping the last 1 appearing in each of these blocks. Moreover, there may be 2 more children corresponding to the swapping in the $\ell + 1$th block: If this block begins with 1, then one child corresponds to swapping the last 1 appearing in the first contiguous sequence of 1s. The other child corresponds to swapping the 1 that immediately appears after the first sequence of 0; moreover, the right neighbour of this 1 must be a 0 for swapping to be valid. For instance, 1100 1010 0000 0000 has exactly three mandatory one-shifts: 1010 1010 0000 0000, 1100 0110 0000 0000, and 1100 1001 0000 0000. In contrast, 1100 0110 0000 0000 has only one mandatory one-shift 1010 0110 0000 0000; this is because its other one-shift, 1100 0101 0000 0000, would be a mandatory one-shift of 1100 1001 0000 0000 instead.

4. **The above algorithm works for this DAG, with a minor change in Step 3, where we always insert all the children (i.e., mandatory one-shifts) of u after u is extracted, without the need to check Z for repeated insertion.**

Thus, the DAG need not be stored, and we call this an *implicit* DAG. Furthermore, although we have described the algorithm based on bitvectors, this is simply for the ease of illustration. The actual representation of a combination may always be its j parts (i.e., the positions of 1s inside the vector). Alternatively, we may refer a combination v by a rank ρ, such that v is the ρth child of u in the implicit DAG; thus, apart from the root of the DAG, each node is represented in $O(1)$ space. Since we evaluate a combination v only if its parent u in the implicit DAG was extracted before, and there are only $O(1)$ differences between u and v, evaluating a combination can be bounded by $O(1)$ time, if the combination of u is known. Thus, based on a particular combination u, we can examine u and generate the costs of all its mandatory one-shifts in a total of $O(Q)$ time.This gives the following lemma.

Lemma 3. *Let W be a set of combinations with respect to a partition π of Q, and Z be a subset in W such that Z covers all the combinations in W. Suppose that the suppliers' bids are all known. Then, the cheapest combination in W can be extracted from Z in $O(\log |Z|)$ time, and the set Z can be updated in $O(Q)$ time.*

3.2 Metadata Structure Per Item

On the basis of the metadata structures for each partition, the metadata structure $M_{\texttt{local}}$, for each item in $\mathcal{I}$, consists of $\texttt{partition}(Q)$ implicit DAGs, one for each partition of Q. To find the next cheapest combination iteratively, we use an analogous idea as follows:

1. Maintain a set Y, which is initialized to contain the cheapest combination in each of the implicit DAGs. *At anytime, Y covers all the remaining combinations that have not been reported.*
2. When reporting the next cheapest combination, find the cheapest combination u among Y.
3. After reporting u, extract u from Y, and from the implicit DAG D that contains u, insert the children of u (i.e., mandatory one-shifts) to Y, and update D.

Yamanaka et al. [6] showed that all the partitions of a positive integer Q can be generated (implicitly) in $O(\texttt{partition}(Q))$ time, so that the cheapest combinations in all implicit DAGs can be computed and evaluated in $O(Q \times \texttt{partition}(Q))$ times. Consequently, we have the following lemma.

Lemma 4. *Suppose that each item in $\mathcal{I}$ is divided into Q quantiles. Furthermore, suppose that the suppliers' bids are all known, and for each item in $\mathcal{I}$ the bids for each of the Q quantiles are sorted in non-decreasing order. Then, we can initialize an index for storing a set Y in $O(Q \times \texttt{partition}(Q))$ time, such that the next cheapest combination can be iteratively extracted from Y in $O(Q + \log |Y|)$ time, and the set Y can be updated in $O(Q)$ time. The size of Y is bounded by $O(\texttt{partition}(Q) + tQ)$, where t is the total number of combinations extracted, and the index space is bounded by $O(Q(\texttt{partition}(Q) + t))$.*

4   Method II: Using $M_{\texttt{local}}$ to Find Top-k Procurements

In this section, we first show how to use the $M_{\texttt{local}}$ structure as a building block to construct an index for solving the basic problem. After that, we discuss how our method can be adapted to solve other variants of the procurement decision problem.

4.1   The Basic Problem

In the previous section, we first designed a metadata structure to organise all combinations with respect to a particular partioning of Q, and then designed $M_{\texttt{local}}$ to organise the metadata structure for all different partitionings of Q. Since each $M_{\texttt{local}}$ is designed for a particular item in $\mathcal{I}$, to solve the basic problem, we will apply a similar trick, where we design a further metadata structure $M_{\texttt{global}}$ to organise all different $M_{\texttt{local}}$ structures that correspond to all items in $\mathcal{I}$. The details are as follows.

Firstly, each procurement of the $n = |\mathcal{I}|$ items can be described as a choice vector $c = (c_1, c_2, \ldots, c_n)$, which indicates that the procurement takes, for each item I_r, the corresponding c_rth cheapest combination of suppliers' bids. Using the same concept as before, we organise all possible procurements as an implicit

DAG (each node corresponds to a valid procurement), with the cheapest procurement being the root. Each procurement with a choice vector c may be represented by a bitvector with n blocks, such that in the rth block, the c_rth bit is 1 while all other bits are 0. The children of a node u in the implicit DAG are all the mandatory one-shifts of u, and each node has at most n children. However, to improve the time complexity we can also represent the choice vector c as a string d of n digits, $d = (d_1 d_2 \ldots d_n)$, where $d_r \geq 1$ represents that we use the d_rth cheapest combination of suppliers' bids for item I_r. Then, we obtain the following lemma.

Lemma 5. *If the suffix of the digit string d of a node is a block of b consecutive 1's, where $0 \leq b < n$, then the node will have $b + 1$ children whose digit strings can be generated by increasing each of the last $b + 1$ digits one at a time.*

Proof. We use the notation $node(\delta)$ to denote the unique node whose digit string is δ. Firstly, if d' is obtained by incrementing a digit in d at position p among the last $b+1$ digits, then $node(d')$ must be a child of $node(d)$, since any string d'' that can construct d' by incrementing exactly one of its digits, d is the one that is lexicographically the largest (so that its corresponding choice vector is lexicographically the largest). On the other hand, if d' is obtained by incrementing a digit in d at position p *not* among the last $b + 1$ digits, then $node(d')$ cannot be a child of $node(d)$. It is because in the digit string d, there exists a digit at some position p' on the right of p with value greater than 1. Then, a digit string d'' obtained by *decrementing* the digit at position p' in d' will be lexicographically larger than d, which indicates that $node(d')$ cannot be the child of $node(d)$. $\square$

For illustration of the above lemma, note that a node with digit string 21134112 will have only one child node with digit string 21134113, whereas a node with digit string 12113111 will have 4 children with digit strings 12113112, 12113121, 12113211, and 12114111, respectively. Hence every node will have at least one child, as long as the last digit can be increased till it reaches k (as the search is for top-k elements).

We maintain a cover X of all the procurements that have not been reported, and use the following algorithm for reporting iteratively the next cheapest procurement (when needed):

1. Maintain a set X, which is initialized to contain the cheapest procurement of all items (i.e., the roots of all the DAGs). That is, X contains the procurement which selects, for each item, the corresponding cheapest combination of suppliers' bids.
2. When reporting the next cheapest combination, find the cheapest procurement u among X.
3. After reporting u, extract u from X, and insert to X all the children (i.e., mandatory one-shifts) of u.

This gives the main theorem of the paper.

Theorem 2. *Let $\mathcal{I}$ be a set of n items, and $\mathcal{S}$ be a set of m suppliers. Suppose that each item is divided into Q quantiles, and the suppliers' bids for each quantile of each item are given. Then, we can initialize an index for storing a set X in $O(nQ(m \log m + \texttt{partition}(Q)))$ time, such that the next cheapest procurement can be iteratively extracted from X in $O(\log |X| + nQ)$ time, and the set X can be updated in $O(n(Q + \log t))$ time, where t is the total number of procurements extracted. The size of X is bounded by $O(nt + 1)$, and the index space is bounded by $O(nQ(\texttt{partition}(Q) + t))$.*

Proof. For the initialization, it takes (i) $O(nQ \times m \log m)$ time to sort the m suppliers' bids into non-decreasing order, for all of the nQ combinations of items and quantiles; (ii) $O(n(Q \times \texttt{partition}(Q)))$ time to initialize n metadata structure $M_{\texttt{local}}$; and (iii) $O(n)$ time to construct the cheapest procurement, and insert it to X.

Once X is maintained by an implicit DAG, the next cheapest procurement can be extracted from X in $O(nQ + \log |X|)$ time, where the term nQ comes from recovering, for each item and quantile, the corresponding suppliers and bids in the procurement. As for the update after an extraction of u in X, it involves the calculation of n procurements (corresponding to the mandatory one-shifts of u), which in turn involves a total of at most n extractions and updates in the $M_{\texttt{local}}$ structures. The total time for the latter part is $O(n \log(\texttt{partition}(Q) + tQ) + nQ) = O(n(Q + \log t))$, and then we can perform the former part in $O(n)$ time. The update time thus follows.

The size of X is trivially $O(nt + 1)$, as we insert at most n procurements for each extraction. The index space includes that of the n $M_{\texttt{local}}$ structures (one for each item), and one $M_{\texttt{global}}$ structure for X. The space of the former one is bounded by $O(nQ(\texttt{partition}(Q) + t))$ in total, while that of the latter one is bounded by $O(nt)$. Moreover, there is an additional $O(tnQ)$ space for the outputs, which are needed so that each node in the implicit DAGs can be represented in $O(1)$ space, which include a pointer to its parent which appears in some output, and how they differ. □

4.2 Other Variants

In the following, we list out some of the variants which can be solved by slightly adapting our method.

1. **When items are divided into different number of quantiles:** For each item I, we simply define its corresponding $M_{\texttt{local}}$ structure, such that each structure is defined on the basis of the number of quantiles I is divided into. The algorithm remains correct.
2. **When the same bid can be selected multiple times:** A supplier may be willing to offer the same bid (of a certain quantile for a certain item) multiple times. In general, for a particular item I, a supplier S may want to offer her bid for q quantiles x_q times. In such a case, we simply create a total of $\max\{x_1, x_2, \ldots, x_Q\}$ copies of the supplier S, so that the same bid

for q quantiles is offered by exactly x_q copies of S. The algorithm remains the same.

3. **Top-k version:** The naive way is to apply Theorem 2 directly to report the k cheapest procurements. However, since the number of procurements, k, is already determined, it is easy to see that there will be no point to maintain more than k procurements in the implicit DAG of M_{global}, and similarly, there will be no point to maintain more than k combinations in the implicit DAG of each M_{local} structures. Thus, if we actively control the size of each of these data structures by deleting redundant combinations or procurements, the index space can be reduced by $O(nk)$, though there will be an increase in update time due to the deletion of those redundant objects from the binomial heaps. The latter method will be particularly useful when index space or working space is a major concern.

4. **Kelly-Byde's problem:** We solve the problem by a post-filtering approach, where we apply Theorem 2 to report, iteratively, the next cheapest procurement, and each time verify if the reported procurement is valid (i.e., for each item, no distinct bids from the same supplier); we stop until we obtain k valid procurements. The post-filtering approach can be used when there are other constraints, such as we may want to have at least a certain number of suppliers, or no supplier is dominating the supply for a certain item (these variants are suggested in Byde et al.'s paper [1]). Unfortunately, the post-filtering approach is a heuristic method, and there is no worst case guarantee about the running time.

5   Conclusions and Future Research

We have proposed a variant of the procurement decision problem, where we may buy from a supplier multiple times (but each time with a different number of shares), and designed an *anytime* algorithm which can report successive best procurements until a buyer is satisfied. Our solution is based on preprocessing a metadata structure M_{global} (without knowing the actual bids) to speed up the reporting steps. We further show that M_{global} can be generated on the fly, thereby saving a considerable amount of storage space.

We have also shown how to adapt our method to solve other variants of the problem, including the existing variant proposed by Kelly and Byde [4]. Adding different constraints to the problem, and designing tailormade algorithm with a similar approach remains an interesting direction for future research.

References

1. Byde, A., Kelly, T., Zhou, Y., Tarjan, R.: Efficiently generating k-best solutions to procurement auctions. In: Goldberg, A.V., Zhou, Y. (eds.) AAIM 2009. LNCS, vol. 5564, pp. 68–84. Springer, Heidelberg (2009)
2. Cormen, T.H., Leiserson, C.E., Rivest, R.L., Stein, C.: Introduction to Algorithms, 3rd edn. MIT press, Cambridge (2009)

3. Eppstein, D.: Finding the k shortest paths. SIAM J. Comput. (SICOMP) **28**(2), 652–673 (1998)
4. Kelly, T., Byde, A.: Generating k-best solutions to auction winner determination problems. ACM SIGecom Exch. **6**(1), 23–34 (2006)
5. Majumder, S., Sanyal, B., Gupta, P., Sinha, S., Pande, S., Hon, W.-K.: Top-K query retrieval of combinations with sum-of-subsets ranking. In: Zhang, Z., Wu, L., Xu, W., Du, D.-Z. (eds.) COCOA 2014. LNCS, vol. 8881, pp. 490–505. Springer, Heidelberg (2014)
6. Yamanaka, K., Kawano, S., Kikuchi, Y., Nakano, S.: Constant time generation of integer partitions. IEICE Trans. Fundam. Electron. Commun. Comput. Sci. **E90–A**(5), 888–895 (2007)

Counting Subgraphs in Relational Event Graphs

Farah Chanchary[1(✉)] and Anil Maheshwari[1]

School of Computer Science, Carleton University, Ottawa, ON K1S 5B6, Canada
farah.chanchary@carleton.ca, anil@scs.carleton.ca

Abstract. Analysis of the structural properties of social networks has gained much interest nowadays due to its diverse range of applications. When communications between entities (i.e., edges) of a social network (graph) are stamped with time, we want to analyze all subgraphs within an arbitrary query time slice so that the number of a specific subgraph can be counted and reported quickly. We present data structures to answer such queries for triangles, quadrangles, complete subgraphs, and maximal complete subgraphs.

Keywords: Dominance counting · Relational event graph · Subgraph counting · Timestamp

1 Introduction

Problem Definition: A *relational event graph* G is an undirected graph with a fixed set of vertices V and a sequence of edges E between pairs of vertices, where each edge has a unique *timestamp*. Suppose the entire relational event graph G is given. We want to construct data structures to efficiently count the number of subgraphs (e.g., triangles, quadrangles, complete subgraphs of order l, maximal cliques) in G having their timespans lying within a query pair of indices $[i, j]$. To be precise, let $G = (V, E = \{e_1, e_2, \cdots, e_m\})$ be a relational event graph and without loss of generality we assume that $t(e_1) < t(e_2) < \cdots < t(e_m)$, where $t(e_i)$ is the timestamp of the edge e_i. For a given query pair of indices $[i, j]$, where $i \leq j$, we want to report the total number of a specific subgraph, for example triangle, in the graph slice $G' = (V, E' = \{e_i \cup e_{i+1} \cup \cdots \cup e_j\})$. A relational event (RE) graph generally represents communication events between pair of entities, where each event carries a timestamp.

The RE graph model was first proposed by Bannister et al. [2]. The authors presented data structures to find the number of connected components, number of components containing cycles, number of vertices with some predetermined degree and number of reachable vertices on a time-increasing path.

In this paper, we present data structures to solve subgraph counting problems in RE graphs. The triangle counting problem is fundamental to many graph applications. This problem has been studied in various contexts, for example as a base case for counting complete subgraphs of given orders [14], in minimum cycle detection problem [12], and as a special case of counting given length cycles [1].

© Springer International Publishing Switzerland 2016
M. Kaykobad and R. Petreschi (Eds.): WALCOM 2016, LNCS 9627, pp. 194–206, 2016.
DOI: 10.1007/978-3-319-30139-6_16

For analyzing large networks such as WWW or social networks, triangle counting problem is also crucial as it can be used to compute important network structures such as clustering coefficients [21] and transitivity coefficients [11]. However, when analyzing bipartite graphs or two-mode networks, these coefficients require total number of quadrangles [15,22]. Both complete subgraphs and maximal complete subgraphs counting problems have applications in combinatorics (e.g., [19,20]) and network analysis (e.g., [9,17]).

In this paper, we use an edge searching technique, originally proposed by Chiba and Nishizeki [3], to count the total number of subgraphs, i.e., triangles, quadrangles, complete subgraphs and maximal complete subgraphs, in a relational event graph. We maintain geometric structures such as, dominance counting structure, interval tree and range tree to store all subgraphs so that the structure can efficiently answer queries for arbitrary query time slices.

Previous Work: Some known results, i.e., by Itai and Rodeh [12] and Chiba and Nishizeki [3], show that an n-vertex graph with m edges may have $\Theta(m^{3/2})$ triangles and quadrangles, and all triangles and quadrangles can be found in $O(m^{3/2})$ time. It is also known that all cycles of length up to seven can be counted in $O(n^w)$ time [1] where w is the exponent from the asymptotically fastest known matrix multiplication algorithms. Other general problems of finding and counting small cliques in graphs and hypergraphs have also utilized the fast matrix multiplication technique [5,16].

Eppstein [6,7] shows that in planar graphs, or more generally graphs of bounded local treewidth, the number of copies of any fixed subgraph may be found in linear time. Papadimitriou and Yannakakis also show that all complete subgraphs in planar graphs ($K_l, l \leq 4$) can be found in linear time [18].

Our work presented in this paper utilizes an edge searching technique to count all instances of subgraphs from a relational event graph. This technique was first presented by Chiba and Nishizeki [3] for listing copies of all complete subgraphs ($K_l, l \geq 3$) in $O(la(G)^{l-2}m)$ time and an implicit representation of quadrangles (C_4) in $O(a(G)m)$ time, where $a(G)$ is the graph arboricity. Arboricity $a(G)$ of a graph $G = (V, E)$ having $m = |E|$ edges and $n = |V|$ vertices is defined as the minimum number of edge-disjoint spanning forests into which G can be partitioned [10]. This graph variant has also been used in other studies as parameters to bound the running time of algorithms. Eppstein [8] showed a linear time algorithm to list all maximal complete bipartite subgraphs $K_{k,l}$ for any constant k in graphs of bounded arboricity.

New Results: Let G be a relational event graph consisting of m edges and let $a(G)$ be the arboricity of G. Let $\mathcal{K}$ be the total number of subgraphs in G and w be the total number of subgraphs in $G_{i,j}$. The main contributions of this paper are,

1. The problem of determining the total number of triangles in the query time slice $[i, j]$ can be reduced to dominance counting in $O(a(G)m)$ time. The dominance counting takes $O(\log w / \log \log \mathcal{K})$ time and linear space to report the total number of triangles in $G_{i,j}$ (Theorem 2).

2. We can construct a data structure in $O(a(G)m + min\{\binom{|V|}{2}, a(G)m\} \log |V|)$ time using $O(a(G)m \log |V|)$ space so that the number of quadrangles in the query time slice $[i, j]$ can be reported in $O(\gamma \log |V| + w)$ time, where $\gamma \leq min\{\binom{|V|}{2}, a(G)m\}$ (Theorem 1).

3. The problem of determining the total number of complete subgraphs of order $l(\geq 2)$ in the query time slice $[i, j]$ can be reduced to dominance counting in $O(a(G)^{l-2}m)$ time. The dominance counting takes $O(\log w / \log \log \mathcal{K})$ time and $O(\mathcal{K})$ space to report the total number of K_l within $G_{i,j}$ (Theorem 3).

4. The problem of determining the total number of complete subgraphs K_k of order $3 \leq k \leq l$ in the query time slice $[i, j]$ can be reduced to dominance counting in $O(a(G)^{l-1}m)$ time. The dominance counting takes $O((\log \mathcal{K} / \log \log \mathcal{K})^2)$ time and $O(\log \mathcal{K} / \log \log \mathcal{K})$ space to report the total number of K_k within $G_{i,j}$ (Theorem 4).

5. The problem of determining the total number of maximal complete subgraphs in the query time slice $[i, j]$ can be reduced to dominance counting in $O(a(G)m\mathcal{K})$ time and using $O(m\mathcal{K})$ space so that the total number of maximal complete subgraphs in $G_{i,j}$ can be reported in $O(\log w / \log \log \mathcal{K})$ time and $O(\mathcal{K})$ space (Theorem 5).

Our results for subgraph counting are summarized in Table 1.

Table 1. Summarized results for subgraph counting.

Problems	Preprocessing time	Query time	
Triangle	$O(a(G)m)$	$O(\log w / \log \log \mathcal{K})$	(Theorem 2)
Quadrangle	$O(a(G)m +$ $min\{\binom{|V|}{2}, a(G)m\} \log |V|)$	$O(\gamma \log |V| + w)$	(Theorem 1)
CS ($l \geq 3$)	$O(a(G)^{l-2}m)$	$O(\log w / \log \log \mathcal{K})$	(Theorem 3)
CS ($3 \leq k \leq l$)	$O(a(G)^{l-1}m)$	$O((\log \mathcal{K} / \log \log \mathcal{K})^2)$	(Theorem 4)
MCS	$O(a(G)m\mathcal{K})$	$O(\log w / \log \log \mathcal{K})$	(Theorem 5)

2   Preliminaries

Definition 1. *A relational event graph G is defined to be a graph with set of vertices V and a sequence of edges (or relational events) $E = \{e_k | 0 \leq k < m\}$ between pairs of vertices.*

We assume that the graph is undirected so the pairs are unordered. We also assume that each edge (or relational event) has a timestamp. We denote timestamp of edge $e = (u, v)$ by $t(u, v)$. All timestamps are unique.

Definition 2. *Given a relational event graph G, we define the slice graph $G_{i,j}$ to be the graph with vertices V and edges $\{e_k | i \leq k \leq j\}$.*

Figure 1(a) illustrates a relational event graph with 19 edges $e_0, e_1, .., e_{18}$. Figure 1(b) represents edges and vertices of slice graph $G_{5,15}$ for range $[5, 15]$.

In a given slice $G_{i,j}$, a subgraph (i.e., triangle, clique, quadrangle) forms only if all its participating edges appear in that slice. So, to count these subgraphs we define the *high event* (also mentioned as *triad closure event* in [2]) and the *low event* as follows.

Definition 3. *A high (low) event in an undirected relational event graph G is defined to be an edge e_k within a given slice $G_{i,j}$ such that e_k is the edge with the highest (lowest) timestamp among all edges that form a particular subgraph.*

We refer back to Fig. 1(a) and this time we are interested in counting how many quadrangles (C_4) are there in query slice $[i, j]$. For a given slice $G_{0,11}$, edges $\{e_0, .., e_{11}\}$ do not form any C_4, thus no *high* or *low event* occurs in this slice. However, edges e_8, e_9, e_{11} and e_{12} create a $C_4 = (e_8, e_9, e_{11}, e_{12})$ in $G_{0,12}$. Therefore, edges e_8 and e_{12} become the *low* and the *high event* respectively for this C_4. It is possible that some edges participate in multiple subgraphs and thus become *high* or *low events* for more than one subgraphs in the same slice. There are three C_4's in $G_{4,18}$, $(e_4, e_5, e_{17}, e_{18})$, $(e_8, e_{10}, e_{12}, e_{18})$ and $(e_9, e_{10}, e_{11}, e_{18})$. For all of these subgraphs, e_{18} is the *high event*, though each of them have different *low events*.

Chiba and Nishizeki [3] gave an upper bound on $a(G)$ for a general graph G as $a(G) \leq \lceil (2m + n)^{1/2}/2 \rceil$. Thus, for a connected graph G, $a(G) \leq O(m^{1/2})$. We state the following lemma from their paper.

Lemma 1. *(Chiba and Nishizeki [3] Lemma 2.1) If graph $G = (V, E)$ has n vertices and m edges, then $\sum_{(u,v)\in E} \min\{d(u), d(v)\} \leq 2a(G)m$, where $d(x)$ denotes the degree of vertex x in G.*

Proof. Let G be partitioned into edge-disjoint forests F_i, where $1 \leq i \leq a(G)$, such that $E(G) = \cup_{1 \leq i \leq a(G)} E(F_i)$. For each tree $T \in F_i$, choose an arbitrary vertex u as the root of T and consider all edges are now directed from the root to descendants in T. Each edge e of T is now associated with its head vertex $h(e)$. Thus, except the roots, every vertex of F_i is associated with exactly one edge of F_i. So, we have $\sum_{(u,v)\in E} \min\{d(u), d(v)\} \leq \sum_{1 \leq i \leq a(G)} \sum_{e \in E(F_i)} d(h(e)) \leq \sum_{1 \leq i \leq a(G)} \sum_{v \in V} d(v) = 2a(G)m$. □

We represent each subgraph as a 2-dimensional point in R^2 and apply dominance counting [2] to determine the total number of subgraphs that exist within a given query slice (Lemma 2). A point $q = (q_x, q_y)$ dominates a point $p = (p_x, p_y)$ if $p_x \leq q_x$ and $q_y \leq p_y$. Given a point set P with $\mathcal{K}$ points in plane, where $\forall p \in P$: $0 \leq p_y < p_x < m$, the dominance counting query is to determine the total number of points of P dominated by a query point q.

Lemma 2. *The subgraphs of G can be mapped to points in R^2 such that the total number of subgraphs in slice $G_{i,j}$ can be determined by a dominance counting query.*

Proof. For each instance of a specific type of subgraph in G, we continuously update the *high* and *low* events by scanning the timestamp of each participating edge. Once a subgraph is identified, we plot its corresponding $(high, low)$ in R^2. For a given query slice $[i, j]$, we translate the slice indices to a query point $q = (j, i)$ to count all points $(high, low)$ dominated by q. For all cases we have $high > low$, so the dominance counting query only have to look for points between the range $(high \leq j)$ and $(i \leq low)$.   □

3   Counting Quadrangles

Throughout the paper we use adjacency linked lists to represent a relational event graph G. There will be two copies of each edge (u, v) for each endpoints u and v. Each node of the linked list for u stores its neighbour v and the timestamp $t(u, v)$. We follow the representation of quadrangles presented by Chiba and Nishizeki [3]. A set of quadrangles is represented by a tuple $(y, z, \{a, b, c, ...\})$, where y and z are vertices on two opposite sides of all quadrangles of this set and each vertex $v \in \{a, b, c..\}$ shares edges with both y and z. Within this setting, any two vertices from $\{a, b, c..\}$ together with y and z represent a quadrangle.

Figure 1(c) illustrates a relational event graph with eight quadrangles. The search algorithm mentioned above represents these quadrangles using four tuples, $(a, b, \{e, f, g\}), (a, d, \{c, e, h\}), (b, n, \{m, l\})$ and $(e, i, \{j, k\})$. We can see that the first tuple $(a, b, \{e, f, g\})$ contains three quadrangles $(a, e, b, f), (a, f, b, g)$ and (a, e, b, g).

To process G we need the following data structures.

- For each tuple $(y, z, \{a, b, c, ...\})$, we use two linked lists E and E' to store edges adjacent to y and z respectively. i.e., $E = \{(y, a), (y, b), (y, c), ..\}$ and $E' = \{(z, a), (z, b), (z, c), ..\}$. Each node in E points to a node in E' that contains edge having one common endpoint, i.e., (y, a) points to (z, a), (y, b) points to (z, b) and so on.
- For each tuple, we use a two-dimensional range tree to store points $\{(t(e_1), t(e_1')), (t(e_2), t(e_2')), ..., (t(e_p), t(e_p'))\}$, where $t(e_i)$ is the timestamp of edge e_i, $e_i \in E$ and $e_i' \in E'$.
- We store the timespan of each tuple using an interval $[x, x']$. For the given graph G, we use an interval tree to store the set of horizontal intervals $I = \{[x_1, x_1'], [x_2, x_2'], ..., [x_p, x_p']\}$, where each interval represents the timespan of a tuple.

Quadrangle counting algorithm contains a *Preprocessing* step and a *Query* step.

Preprocessing Step: The Preprocessing step takes a RE graph G consisting of n vertices and m edges as input. G is processed by staring with the vertex having the highest degree. So, the vertices $v_1, v_2, \ldots, v_n$ are first sorted in non-increasing

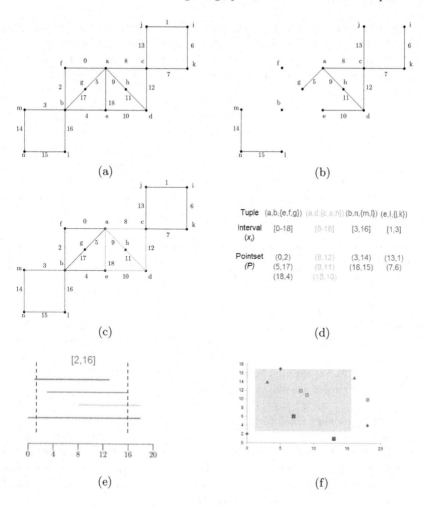

Fig. 1. Counting quadrangles in a RE graph.

order of their degrees and without loss of generality, let $d(v_1) \geq d(v_2) \geq \cdots \geq d(v_n)$. Once all quadrangles containing v_1 are identified correctly, v_1 is deleted from G to avoid duplication and the loop continues with the next vertex in the sequence. We first describe below two major components of this *Preprocessing* step.

(a) *Finding Quadrangles:* For each vertex $y \in V$ of G, we apply the following technique to find all quadrangles containing y: for each vertex $z \in V$ within distance 2 from y, we find all vertices which are adjacent to both y and z. We store these vertices in a set $U[z]$. Thus, the tuple $(y, z, U[z])$ represents a set of quadrangles, where every quadrangle has vertices y and z as two opposite corner points.

(b) *Finding Intervals:* Recall that we want to count the number of quadrangles within a given time slice $[i, j]$, but all we have is the representation of a set of quadrangles or tuples. So, for each such tuple we mark its timespan and maintain some geometric data structures so that we can answer the query.

First, we store each tuple $(y, z, U[z] = \{a, b, c, ...\})$ using two linked lists $E = \{(y, a), (y, b), (y, c), ..\}$ and $E' = \{(z, a), (z, b), (z, c), ..\}$ according to the order of the vertices in $\{a, b, c, ..\}$. We also represent each tuple with a point set $P = \{(t(y, a), t(z, a)), (t(y, b), t(z, b)), ..\}$, where $t(y, a)$ is the timestamp of the edge (y, a) and so on. We store P using a 2-dimensional range tree.

Next, we compute the timespan of each tuple and store it as an interval segment using an interval tree. We sort linked lists E and E' according to the timestamps of their edges in non-decreasing order, such as, $E = (e_1, e_2, ...e_p)$, and $E' = (e'_1, e'_2, ...e'_p)$, where $t(e_1) \leq t(e_2) \leq ..t(e_p)$ and $t(e'_1) \leq t(e'_2) \leq ..t(e'_p)$. Now we can create an interval segment $x = [min(e_1, e'_1), max(e_p, e'_p)]$ for each tuple to mark its timespan. Similarly, we compute segments for all tuples and store each segment x_i in an interval tree.

The example presented in Fig. 1(c) shows a RE graph G with four tuples, each tuple is highlighted with different colors. We first store the tuple $(a, b, \{e, f, g\})$ using linked lists $E = (18, 0, 5)$ and $E' = (4, 2, 17)$. Then we create a point set $P = \{(18, 4), (0, 2), (5, 17)\}$ and store in a two-dimensional range tree. Next we sort timestamps of edges in $E = (0, 5, 18)$ and $E' = (2, 4, 17)$. Thus we store an interval segment $[0, 18]$ for this tuple in the interval tree. Interval segments and point sets for all tuples are shown in Fig. 1(d).

Query Step: We query the interval tree T with a horizontal query segment $[i, j]$ and obtain a set of valid interval segments $I' \subseteq I$. We consider each segment $x_i \in I'$ valid, if it intersects with the query slice indicating that x_i might contain quadrangles that exist within the query slice. As the next step to get the exact number of quadrangles ($\#Quad$), we need to know which of these valid segments contains at least two sets of paired edges $\{(y, v'), (z, v')\}$ and $\{(y, v''), (z, v'')\}$ such that it makes a quadrangle (y, v', z, v''). Recall that for each segment we already stored a set of points P that contains information of these paired edges. So for each valid segment, we perform a 2-D rectangular range query on P using a query rectangle with four corner points $(i, i), (i, j), (j, j)$ and (j, i) that returns a set of edges S. If $|S| \geq 2$ we add $\binom{|S|}{2}$ to $\#Quad$.

Figure 1(e) shows all valid segments after an interval tree query with query point $q = [2, 16]$. Finally a rectangular query shows that only tuple $(a, d, \{c, e, h\})$ has 2 sets of edges that fall within the query range (Fig. 1(f)). Therefore, we obtain $\binom{2}{2} = 1$ quadrangle ($\#Quad$) within the query slice [2,16].

Preprocessing Analysis. Sorting vertices and identifying all quadrangles containing v_i by traversing all vertices take at most $O(m + n) + \sum_{v_i \in V} O(d(v_i) + \sum_{u \in N(v_i)} d(u))$ time, where $N(v_i)$ is the set of neighbours of v_i. This time can be bounded as $O(a(G)m)$ by Lemma 1.

The interval tree stores at most $a(G)m$ intervals, which is a weak upper bound. For complete graphs $a(G) = \lceil n/2 \rceil$, whereas for planar graphs $a(G) = O(1)$. Each interval considers unique pair of vertices on opposite sides of a quadrangle, so there can be at most $\binom{|V|}{2}$ quadrangles in G. Thus the number of intervals can be bounded as $min\{a(G)m, \binom{|V|}{2}\}$. Hence, the interval tree can be created in time $O(min\{a(G)m, \binom{|V|}{2}\} \log\{min\{a(G)m, \binom{|V|}{2}\}\}) \leq O(min\{a(G)m, \binom{|V|}{2}\} \log |V|)$ [4].

Query Analysis. Suppose I' is the set of all valid interval segments and $|I'| = \gamma$, where $\gamma \leq min\{\binom{|V|}{2}, a(G)m\}$. Reporting all γ segments requires $O(\log min\{\binom{|V|}{2}, a(G)m\} + \gamma) = O(\log |V| + \gamma)$ time [4]. Rectangular range queries on γ range trees takes time $O(\sum_{i=1}^{\gamma} \log |V| + k_i) = O(\gamma \log |V| + w)$, where k_i is the number of reported quadrangles in segment i and w is $\#Quad$ in $G_{i,j}$. Total number of quadrangles in our problem is bounded by $a(G)m$. Each segment that represents a quadrangle is stored using a range tree. So, total space required is $O(a(G)m \log a(G)m) = O(a(G)m \log |V|)$. The following theorem summarizes the results for quadrangle counting in a relational event graph.

Theorem 1. *Given a RE graph G with m edges the number of quadrangles in the query time slice $[i, j]$ can be determined in $O(\gamma \log |V| + w)$ time, where $\gamma \leq min\{\binom{|V|}{2}, a(G)m\}$. Preprocessing takes $O(a(G)m + min\{\binom{|V|}{2}, a(G)m\} \log |V|)$ time and $O(a(G)m \log |V|)$ space.*

4   Counting Triangles and Other Subgraphs

We first present the general overview of the techniques to count triangles, complete subgraphs and maximal complete subgraphs. Similar to the previous algorithm, all algorithms presented in this section contain a *Preprocessing* and a *Query* step.

Preprocessing Step: Again the relational event graph G consisting of n vertices and m edges is processed by staring with the vertex having the highest degree. So, we sort the vertices $v_1, v_2, \ldots, v_n$ in non-increasing order of their degrees and without loss of generality, let $d(v_1) \geq d(v_2) \geq \cdots \geq d(v_n)$. Then the search process starts with vertex v_1 by marking the subgraph induced by the neighbours $N(v_1)$ of v_1, and finding all instances of a specified subgraph containing v_1. For each of these instances, the *high* and *low* values are computed by comparing the timestamps of participating edges. Each subgraph is then represented by a point $(high, low)$ in R^2, where $[low - high]$ shows the timespan of that subgraph in G. After timespans of all subgraphs containing v_1 are plotted in R^2, v_1 is deleted from G to avoid duplication and the loop continues with the next vertex in the sequence. This process continues until all subgraphs are identified, represented by a point set $P = \{p_1, p_2, ..\}$, and plotted accordingly.

Query Step: The *Query* step reports the number of specified subgraphs within a given query time slice $[i,j]$. As input, it takes the point set P created in the preprocessing step, and a time slice $[i,j]$. The given slice $[i,j]$ is considered as the query point $q = (j,i)$. By Lemma 2, the dominance counting query with respect to q on P returns the number of subgraphs within the query time slice.

4.1   Counting Triangles

Description of the Algorithm: Procedure *PreprocessTriangle(G)* pre-processes G and identifies all instances of triangles in G (see Algorithm 1). The triangle search process starts with the highest degree vertex v_i by marking the subgraph induced by the neighbours $N(v_i)$ of v_i. Then it finds all marked neighbours $z \in u(v_i)$ such that $z \neq v_i$. Each z forms a triangle (v_i, u, z). The *high* and *low* values for each of these triangles are then respectively the maximum and the minimum timestamps of edges (v_i, u), (u, z) and (z, v_i). Each triangle is now represented by a point $(high, low)$ in R^2. For a given query time slice $[i,j]$, we report the number of points in P dominated by the point (j, i).

Algorithm 1

Input. Relational event graph G.
Output. A point set P plotted in R^2.

1: **procedure** PREPROCESSTRIANGLE(G)
2:     sort vertices $v_1, v_2, \ldots, v_n$ in non-increasing order of their degrees. without loss
       of generality, let $d(v_1) \geq d(v_2) \geq \cdots \geq d(v_n)$
3:     **for** each vertex v_i from $i = 1$ to $n - 2$ **do**
4:         mark all neighbours of v_i
5:         **for** each marked vertex u **do**
6:             **for** each vertex z adjacent to u **do**
7:                 **if** z is marked **then**
8:                     $high \leftarrow \max\{t(v_i, u), t(u, z), t(z, v_i)\}$
9:                     $low \leftarrow \min\{t(v_i, u), t(u, z), t(z, v_i)\}$
10:                    plot $(high, low)$ in R^2
11:                **end if**
12:                unmark u
13:            **end for**
14:        **end for**
15:        delete v_i from G and without loss of generality, let G be the resulting graph.
16:    **end for**
17: **end procedure**

Analysis: Lines 5–14 of Algorithm 1 require at most $O(\sum_{u \in N(v_i)} d(u))$ time, where $d(u)$ denotes the degree of vertex u in G and $N(v_i)$ denotes the set of neighbours of v_i in the current graph. So, total time required for the outer **for** loop is $\sum_{v_i \in V} O(d(v_i) + \sum_{u \in N(v_i)} O(d(u)))$. In the current graph, v_i is the vertex with the highest degree and for all other neighbouring vertices $u \in N(v_i)$, $d(u) \leq d(v_i)$. We explore all neighbouring vertices of v_i and at the end of the **for** loop, v_i is deleted (line 15). Since $\sum_{v_i \in V} O(d(v_i) + \sum_{u \in N(v_i)} O(d(u)))$ is same as $O(\sum_{(u,v) \in E} min\{d(u), d(v)\})$ by [3], total time needed for the outer **for** loop can be reduced to $O(\sum_{(u,v) \in E} min\{d(u), d(v)\})$. Thus the total time required for preprocessing can be bounded by $O(m + n) + O(\sum_{(u,v) \in E} min\{d(u), d(v)\})$, and by Lemma 1 it is $O(a(G)m)$. By using dominance counting algorithm in [2], we can determine the total number of triangles in slice $G_{i,j}$. The following theorem summarizes our results.

Theorem 2. *Given a RE graph G with m edges and $\mathcal{K}$ triangles, the problem of determining the number of triangles in the query time slice $[i, j]$ can be reduced to dominance counting in $O(a(G)m)$ time. The dominance counting takes $O(\log w / \log \log \mathcal{K})$ time and $O(\mathcal{K})$ space to count the total number of triangles in $G_{i,j}$, where w is the total number of triangles in $G_{i,j}$.*

4.2   Counting Complete and Maximal Complete Subgraphs

Counting Complete Subgraphs of Order l. We apply the basic strategy of triangle counting to find all complete subgraphs of any fixed order l. That is, we can find a complete subgraph K_l containing a vertex v by detecting a complete subgraph K_{l-1} in a subgraph induced by the neighbours of v. The algorithm plots all points $P = \{p_1, p_2, ..p_\mathcal{K}\}$ in R^2, where $\mathcal{K}$ is the number of complete subgraphs of order l and each point $p_i = (a, b)$ represents a complete subgraph K_l, and a and b are the *high* and *low* values of K_l, respectively. We can find the number of complete subgraphs K_l within a given query range $[i, j]$ by using the dominance counting data structure. We obtain the following result and detailed proof of this theorem can be found in the full version of the paper.

Theorem 3. *Given a RE graph G with m edges and $\mathcal{K}$ complete subgraphs of order $l(\geq 2)$, the problem of determining the number of complete subgraphs of order l in the query time slice $[i, j]$ can be reduced to dominance counting in $O(a(G)^{l-2} m)$ time. The dominance counting takes $O(\log w / \log \log \mathcal{K})$ time to count the total number of complete subgraphs of order l in $G_{i,j}$, where w is the total number of complete subgraphs of order l in $G_{i,j}$.*

Proof. (Sketch) The main function recursively calls a procedure with parameters $k = l$ and $G_k = G$. Let $T(k, m, n)$ be the time required by procedure to find all K_k's in G_k. When $k = 2$: for each edge in G_2 we update *low* and *high* values of K_k by comparing timestamps of edges connecting at most l vertices. This time is upperbounded by $O(m + n)$. So, we can state that $T(2, m, n) = O(m + n)$.

Next for $k \geq 3$: we find the subgraph G_{k-1} induced by the neighbours of the vertex with the current highest degree v_i in $O(d_k(v_i) + \Sigma_{u \in N(v_i)} d_k(u))$ time. The recursive call to find and store all complete subgraphs containing v_i requires $T(k - 1, (\Sigma_{u \in N(v_i)} d_k(u))/2, d_k(v_i))$ time. Thus, total time requires for each v_i is, $O(d_k(v_i) + \Sigma_{u \in N(v_i)} d_k(u)) + O(1) + T(k - 1, (\Sigma_{u \in N(v_i)} d_k(u))/2, d_k(v_i))$.

Therefore, to complete searching for all the complete subgraphs containing all vertices of G requires,
$\Sigma_{v_i \in V_k}(O(d_k(v_i) \quad + \quad \Sigma_{u \in N(v_i)} d_k(u)) \quad + \quad O(1)) \quad + \quad \Sigma_{v_i \in V_k} T(k - 1, (\Sigma_{u \in N(v_i)} d_k(u))/2, d_k(v_i))$ time. Each vertex v_i of G_k satisfies $d_k(v_i) \geq d_k(u)$ for every neighbour vertex u of v_i. Therefore Lemma 1 implies that,
$T(k,m,n) = O(a(G_k)m + n) + \Sigma_{v_i \in V_k} T(k - 1, \Sigma_{u \in N(v_i)} d_k(u)/2, d_k(v_i))$.
Since $a(G_{k-1}) \leq a(G_k)$ for all k, we state that, $T(k,m,n) = O(a(G_k)^{k-2}m+n)$. Recall that main algorithm calls the recursive procedure with $k = l$ and $G_k = G$. Therefore, finding all K_l's in G requires at most $O(a(G)^{l-2}m)$ time.   □

Counting Complete Subgraphs of Orders $l \geq 3$. We extend our algorithm for computing all complete subgraphs of orders 3 to l. That is, if we are given an order l, we can find all complete subgraphs K_k, where $3 \leq k \leq l$, and their corresponding timespans $[low - high]$. To accommodate the problem of multiple orders, we modify our algorithm so that, for each complete subgraph of order k, we plot a point $p = (a, b, k)$ in R^3 where $a = high$, $b = low$ and $k = order$. Now we can report total number of complete subgraphs of all orders within $G_{i,j}$ by using dominance counting in 3-dimensions [13]. Thus,

Theorem 4. *Given a RE graph G with m edges, the problem of determining the total number of complete subgraphs K_k of orders $3 \leq k \leq l$ in the query time slice $[i, j]$ can be reduced to dominance counting in $O(a(G)^{l-1}m)$ time. The dominance counting query takes $O((\log \mathcal{K}/ \log \log \mathcal{K})^2)$ time and $O(\mathcal{K} \log \mathcal{K}/ \log \log \mathcal{K})$ space to report the total number of complete subgraphs in $G_{i,j}$, where $\mathcal{K}$ is the total number of complete subgraphs in G.*

Maximal Complete Subgraphs. For counting maximal complete subgraphs, we utilize algorithm CLIQUE [3] and assume that CLIQUE returns C, a set of vertices representing a maximal complete subgraph, to the main calling program. Our algorithm maintains an adjacency list MCS where each record represents a maximal complete subgraph and stores all its vertices using a linked list. We summarize our findings with the following theorem.

Theorem 5. *Given a RE graph G with m edges and $\mathcal{K}$ maximal complete subgraphs, the problem of determining the number of maximal complete subgraphs in the query time slice $[i, j]$ can be can be reduced to dominance counting in $O(a(G)m\mathcal{K})$ time and $O(m\mathcal{K})$ space. The dominance counting query takes $O(\log w/ \log \log \mathcal{K})$ time to report the total number of complete subgraphs in $G_{i,j}$.*

5   Conclusion

In this paper, we have developed data structures for relational event graphs to efficiently count and report total number of subgraphs in an arbitrary query time slice. We have applied an edge searching technique to identify and count total number of subgraphs in a relational event graph. We have stored the timespans of these subgraphs using geometric data structures. We have shown that these structures can efficiently report the number of specified subgraphs within a query time slice.

The techniques we presented here require the type of subgraphs to be specified at the preprocessing step. An interesting variation of this problem would be to develop data structures that can count total number of subgraphs of any pattern or a fixed graph on a small number of vertices within a given query time slice.

References

1. Alon, N., Yuster, R., Zwick, U.: Finding and counting given length cycles. Algorithmica **17**(3), 209–223 (1997)
2. Bannister, M.J., DuBois, C., Eppstein, D., Smyth, P.: Windows into relational event: data structures for contiguous subsequences of edges. In: Proceedings of the 24th Annual ACM-SIAM Symposium on Discrete Algorithms. SIAM (2013)
3. Chiba, N., Nishizeki, T.: Arboricity and subgraph listing algorithms. SIAM J. Comput. **14**(1), 210–223 (1985)
4. de Berg, M., Van Kreveld, M., Overmars, M., Schwarzkopf, O.: Computational Geometry. Algorithms and Applications, 3rd edn. Springer, New York (1998)
5. Eisenbrand, F., Grandoni, F.: On the complexity of fixed parameter clique and dominating set. Theoret. Comput. Sci. **326**(1–3), 57–67 (2004)
6. Eppstein, D.: Subgraph isomorphism in planar graphs and related problems. J. Graph Algorithms Appl. **3**(3), 1–27 (1999)
7. Eppstein, D.: Diameter and treewidth in minor-closed graph families. Algorithmica **27**, 275–291 (2000)
8. Eppstein, D.: Arboricity and bipartite subgraph listing algorithms. Inf. Process. Lett. **51**, 207–211 (1994)
9. Fortunato, S.: Community detection in graphs. Phys. Rep. **486**(3), 75–174 (2010)
10. Harary, F.: Graph Theory, revised, Addision-Wesley Publishing Company, Reading (1969)
11. Harary, F., Kommel, H.J.: Matrix measures for transitivity and balance. J. Math. Sociol. **6**, 199–210 (1979)
12. Itai, A., Rodeh, M.: Finding a minimum circuit in a graph. SIAM J. Comput. **7**(4), 413–423 (1978)
13. JáJá, J., Mortensen, C.W., Shi, Q.: Space-efficient and fast algorithms for multidimensional dominance reporting and counting. In: Fleischer, R., Trippen, G. (eds.) ISAAC 2004. LNCS, vol. 3341, pp. 558–568. Springer, Heidelberg (2004)
14. Kloks, T., Kratsch, D., Muller, H.: Finding and counting small induced subgraphs efficiently. Inf. Process. Lett. **74**(3–4), 115–121 (2000)
15. Opsahl, T.: Triadic closure in two-mode networks: redefining the global and local clustering coefficients. Soc. Netw. **35**(2), 159–167 (2013)
16. Nešetřil, J., Poljak, S.: On the complexity of the subgraph problem. Commentationes Mathematicae Universitatis Carolinae **26**(2), 415–419 (1985)
17. Newman, M.E.: Detecting community structure in networks. Eur. Phys. J. B-Condensed Matter Complex Syst. **38**(2), 321–330 (2004)
18. Papadimitriou, C., Yannakakis, M.: The clique problem for planar graphs. Inf. Process. Lett. **13**, 131–133 (1981)

19. Ullmann, J.R.: An algorithm for subgraph isomorphism. J. ACM (JACM) **23**(1), 31–42 (1976)
20. Valiant, L.G.: The complexity of enumeration and reliability problems. SIAM J. Comput. **8**(3), 410–421 (1979)
21. Watts, D.J., Strogatz, S.H.: Collective dynamics of "smallworld" networks. Nature **393**, 440–442 (1998)
22. Zhang, P., Wang, J., Li, X., Li, M., Di, Z., Fan, Y.: Clustering coefficient and community structure of bipartite networks. Phys. A **387**(27), 6869–6875 (2008)

Computational Complexity

Large Independent Sets
in Subquartic Planar Graphs

Matthias Mnich[1]([⊠])

Universität Bonn, Bonn, Germany
mmnich@uni-bonn.de

Abstract. By the famous Four Color Theorem, every planar graph admits an independent set that contains at least one quarter of its vertices. This lower bound is tight for infinitely many planar graphs, and finding maximum independent sets in planar graphs is NP-hard. A well-known open question in the field of Parameterized Complexity asks whether the problem of finding a maximum independent set in a given planar graph is fixed-parameter tractable, for parameter the "gain" over this tight lower bound. This open problem has been posed many times [4,8,10,13,17,20,31,32,35,38].

We show fixed-parameter tractability of the independent set problem parameterized above tight lower bound in planar graphs with maximum degree at most 4, in subexponential time.

1 Introduction

In this paper we deal with independent sets in planar graphs G; let $\alpha(G)$ be the size of a maximum independent set in G. Let n denote the number of vertices in G, and call $i(G) := \alpha(G)/n$ the *independence ratio* of G. In the 1960s, Erdős proposed what is known nowadays as the Erdős-Vizing Conjecture [3, p. 280]: any planar graph has independence ratio at least $1/4$. The lower bound of $1/4$ on the independence ratio is tight for infinitely many planar graphs, such as a set of disjoint K_4's connected by edges in a planar way.

The Erdős-Vizing Conjecture was proved in 1976 when Appel and Haken [2] announced the Four Color Theorem: every planar graph admits a coloring of its vertices with four colors such that adjacent vertices receive distinct color. Their lengthy computer-assisted proof was later simplified by Robertson et al. [36], who gave another computer-assisted proof and an algorithm that in time $O(n^2)$ produces a proper vertex coloring of G with four colors. As of today, these computer-assisted proofs of the Four Color Theorem are the only proofs of the Erdős-Vizing Conjecture.

Our concern is a wide generalization of the Erdős-Vizing Conjecture. We would like to find an algorithm that decides, for a given n-vertex planar graph G and integer $k \in \mathbb{N}$, whether G has an independent set of size at least $(n+k)/4$, in time $f(k) \cdot n^{O(1)}$, for some computable function f. We call this problem PLANAR

This research was supported by ERC Starting Grant 306465 (BeyondWorstCase).

M. Kaykobad and R. Petreschi (Eds.): WALCOM 2016, LNCS 9627, pp. 209–221, 2016.
DOI: 10.1007/978-3-319-30139-6_17

INDEPENDENT SET ABOVE TIGHT LOWER BOUND, or PLANAR INDEPENDENT SET-ATLB for short. As the maximum independent set problem is NP-hard in planar graphs, the problem PLANAR INDEPENDENT SET-ATLB is NP-complete. The question of whether PLANAR INDEPENDENT SET-ATLB for parameter the "gain" k is fixed-parameter tractable has been raised several times: first by Niedermeier [35], later by Bodlaender [4], Mahajan et al. [31], by Sikdar [38], and Crowston et al. [8]. Then the problem was raised as a "tough customer" at WorKer 2012 [17], and asked again by Giannopoulou et al. [20], by Dvořák and Mnich [13] and at the Bedlewo School on Parameterized Algorithms and Complexity [10]. We remark that as of now, there is not even a polynomial-time algorithm known for the case of $k = 1$ of PLANAR INDEPENDENT SET-ATLB, but such an algorithm for $k = 1$ is certainly necessary for a fixed-parameter algorithm (or even an $n^{O(k)}$ time algorithm) for the general problem.

Recently, Dvořák and Mnich [13] (ESA 2014) considered the problem PLANAR INDEPENDENT SET-ATLB for triangle-free planar graphs. There, the lower bound on the independence ratio is $1/3$ instead of $1/4$, as Grötzsch' theorem [22] guarantees a proper vertex-coloring using only 3 colors. Dvořák and Mnich [13] showed that the problem is fixed-parameter tractable, and admits an algorithm with run time $2^{O(\sqrt{k})}n$ which is optimal under the Exponential-Time Hypothesis. Their algorithm is based on combinatorial arguments around the treewidth of triangle-free planar graphs, as well as certain precoloring extension theorems (which extend certain proper 3-colorings of subgraphs to proper 3-colorings of larger subgraphs). Unfortunately, it is unlikely their techniques could be used for PLANAR INDEPENDENT SET-ATLB in general planar graphs. The reason is that almost any similar precoloring extension claim is false for general planar graphs, and furthermore there exist planar graphs on n vertices with independence ratio exactly $1/4$ and arbitrarily large tree-width.

Our Contributions. Our main result is a proof that PLANAR INDEPENDENT SET-ATLB is fixed-parameter tractable in subquartic (maximum degree at most 4) planar graphs. This is the first non-trivial case of the problem with respect to maximum degree, because an easy argument yields a fixed-parameter algorithm (and indeed linear kernel) for subcubic planar graphs (where the maximum independent set problem remains NP-hard [19]): remove any K_4 from the input graph G to obtain a graph G', and answer "yes" if $k \leq |V(G')|/3$, or return (G', k) as a kernel with at most $3k$ vertices otherwise. Correctness follows by Brooks' Theorem [5].

For planar graphs G with maximum degree at most 4, Brooks' theorem [5] only ensures 4-colorability and thus $\alpha(G) \geq |V(G)|/4$. But until now there was no combinatorial characterization or polynomial-time recognition algorithm even to decide the case of $k = 1$; hence even finding an $n^{g(k)}$-time algorithm is open. We show fixed-parameter tractability of PLANAR INDEPENDENT SET-ATLB in subquartic planar graphs, in subexponential time.

Theorem 1. *There is an algorithm that, given any n-vertex subquartic planar graph G and $k \in \mathbb{N}$, decides in time $2^{O(\sqrt{k})}n + O(nk^2 \log k + n^2)$ if $\alpha(G) \geq \frac{n+k}{4}$.*

Thus, Theorem 1 solves the open question posed in [4, 8, 10, 13, 17, 20, 31, 32, 35, 38] for this restricted class of graphs. For $k = O(\log^2 n)$, the algorithm runs in polynomial time, which raises the boundary of polynomial-time tractability of finding maximum independent sets in planar graphs of maximum degree 4.

To prove Theorem 1, we first apply certain reduction rules to the input planar graph. We then resort to *fractional colorings*, which to the best of our knowledge have not been used before in the design of fixed-parameter algorithms. For $t \in \mathbb{R}$, a graph G is fractionally t-colorable, if for every assignment of weights to $V(G)$ there is an independent set that contains at least $(1/t)$-fraction of the total weight. In particular, every fractionally t-colorable graph G contains an independent set of size at least $|V(G)|/t$. However, reduced graphs G' can still contain K_4-subgraphs and so we can only guarantee $t \geq 4$ and hence $\alpha(G') \geq |V(G')|/4$. Fortunately, with more care we can use fractional colorings to argue that if G' has many (in terms of k) vertices outside K_4's then its independence number exceeds $|V(G')|/4$ by a linear function of k. In the other case, when almost all vertices of G' belong to K_4-subgraphs, we can prove a sublinear bound (in terms of k) on the treewidth of G; this case can then be solved by dynamic programming.

Proving upper bounds on the fractional chromatic number $\chi_f(G)$ for classes of planar graphs G often turns out to be quite challenging: when not relying on the Four Color Theorem, only $\chi_f(G) \leq 9/2$ is known for planar graphs G [7].

Regarding complexity lower bounds, we note that the run time of our algorithm is asymptotically optimal in k assuming the Exponential Time Hypothesis (ETH), under which results by Cai and Juedes [6, Corollary 5.1] rule out the existence of an algorithm for PLANAR INDEPENDENT SET-ATLB that runs in time $2^{o(\sqrt{k})} \cdot n^{O(1)}$. If instead of ETH we assume that $\mathsf{P} \neq \mathsf{NP}$, then it follows from results of Fleischner et al. [18, Theorem 5.3] that deciding $\alpha(G) \geq (n+k)/4$ is not polynomial-time solvable for $k = n/48$ even if G is 4-regular and Hamiltonian.

As second result, we give a way to decide $\alpha(G) \geq (n+k)/4$ from a class $\mathcal{C}$ of planar graphs G with maximum degree 5. The step from maximum degree 4 to maximum degree 5 poses major technical difficulties, which forces us to impose severe restrictions on $\mathcal{C}$. A vertex v is a d-vertex (resp. $(\geq d)$-vertex) if its degree is equal to (resp. at least) d.

Theorem 2. *There is an algorithm that, given any n-vertex planar graph G with maximum degree 5 with no (≥ 4)-vertex in a triangle and 5-vertices independent and $k \in \mathbb{N}$, decides in time $2^{O(\sqrt{k})}n + O(nk^2 \log k + n^2)$ if $\alpha(G) \geq (n+k)/4$.*

As third result, we answer a special case of a question by Gutin et al. [23] about the independence ratio of degenerate graphs. Recall that a graph G is d-degenerate for some $d \in \mathbb{N}$ if every subgraph of G contains some vertex of degree at most d. Thus, the greedy algorithm shows that d-degenerate n-vertex graphs are $(d + 1)$-vertex-colorable, and thus admit independent sets of size at least $n/(d + 1)$. Gutin et al. [23] asked whether the following problem is fixed-parameter tractable: given a d-degenerate n-vertex graph G and integer $k \in \mathbb{N}$, decide whether G admits an independent set of size at least $(n + k)/(d + 1)$.

Gutin et al.'s question is still unresolved for every value of d. Here we give a result for $(d, k) = (3, 1)$:

Theorem 3. *There is a polynomial-time algorithm that, given any 3-degenerate planar n-vertex graph, decides if it has an independent set of size at least $\frac{n+1}{4}$.*

Finally, we consider the case of $k = 1$ for PLANAR INDEPENDENT SET-ATLB in general planar graphs; recall that no polynomial-time algorithm is known for this case. Already Erdős [3] in 1968 suggested that proving $i(G) \geq 1/4$ for planar graphs G might be easier than proving the Four Color Theorem. We make partial progress in this direction by providing various families of planar graphs with independence ratio exactly 1/4. To the best of our knowledge, this is the first such description in the literature. Another motivation comes from the fact that there are such graphs of small maximum degree (at most 5) that we need to consider in our algorithms for general k.

Due to space constraints, the proofs of Theorems 2 and 3 are deferred to the full version of this paper.

Related Work. The significance of parameterizing above a polynomial-time guaranteed lower bound on the solution value was first recognized by Mahajan and Raman [30]. Only scattered results were known until a survey of Mahajan et al. [31] stimulated the discovery of new techniques for fixed-parameter algorithms based on algebra [8], combinatorics [9,33], probability theory [24,25] and linear programming [12], to efficiently solve problems parameterized above tight lower bound.

Quite some work is currently done on finding large independent sets in subclasses of planar graphs. Heckman and Thomas [26] show that every subcubic triangle-free planar graph G has $\alpha(G) \geq \frac{3}{8}|V(G)|$. For fullerenes, which are planar 3-regular 3-connected graphs with only pentagonal and hexagonal faces, Faria et al. [16] show that $\alpha(G) \geq |V(G)|/2 - \sqrt{3|V(G)|/5}$, and this bound is sharp. However, all graphs considered in the literature have maximum degree 3.

2 Preliminaries

Throughout, we consider graphs that are finite, undirected and simple. For a graph G, let $V(G)$ denote its vertex set and $E(G)$ its set of edges. For each vertex $v \in V(G)$, let $N_G(v)$ be the set of vertices adjacent to v in G. The *degree* of a vertex $v \in V(G)$ is the number $d_G(v) = |N_G(v)|$ of vertices adjacent to it. A vertex of degree exactly (at most, at least) d is called a d-vertex ($(\leq d)$-vertex, $(\geq d)$-vertex). Let $\Delta(G)$ denote the maximum degree over all vertices in G, and let $\omega(G)$ denote the maximum clique size in G. For $d \in \mathbb{N}$, a graph is d-degenerate if each of its subgraphs has a $(\leq d)$-vertex.

A graph is *planar* if it admits an embedding in the plane such that no two edges cross; a *plane graph* is an embedding of a planar graph without any edge crossings. A graph is *complete* if every two of its vertices are connected by an edge; let K_t denote the complete graph on t vertices. For $t \in \mathbb{N}_{\geq 3}$, the t-*cycle* is the graph C_t with vertex set $V(C_t) = \{v_0, \ldots, v_{t-1}\}$ and edge set $E(C_t) = \{\{v_i, v_{i+1 \ (\mathrm{mod}\ t)}\} \mid i = 0, \ldots, t-1\}$. A graph is *connected* if there is a path between any two of its vertices, and a *connected component* of a graph G is a vertex-maximal connected subgraph of G. For $t \in \mathbb{N}_0$, a connected graph G is *t-connected* if removing any subset of strictly less than t vertices from G leaves a graph that is connected. A *t-separator* for G is a set S of t vertices such that the graph $G - S$ has strictly more connected components than G. A set $I \subseteq V(G)$ of vertices is *independent* for G if the vertices of I are pairwise non-adjacent. The maximum size of an independent set of G is denoted by $\alpha(G)$, and the *independence ratio* of G is defined as $i(G) := \alpha(G)/|V(G)|$. A *separation* of G is a pair $\{A, B\}$ of subsets of G with $A \cup B = V(G)$ such that there are no edges between $A \setminus B$ and $B \setminus A$. The *square* of G is the graph G^2 with vertex set $V(G^2) = V(G)$ and in which two vertices are connected by an edge if their distance in G is at most 2; thus G^2 is a supergraph of G.

Tree Decompositions. For a graph G, a *tree decomposition* is a pair $(T, \mathcal{B})$, where T is a tree and $\mathcal{B} = \{B_t \mid t \in V(T)\}$ is a collection of *bags*, such that

(TW-1) for each edge $\{u, v\} \in E(G)$, there is a bag B_t such that $u, v \in B_t$; and
(TW-2) for each $v \in V(G)$, the set $\{t \mid v \in B_t\}$ induces a non-empty tree of T.

The *width* of a tree decomposition $(T, \mathcal{B})$ is $\max_{B \in \mathcal{B}}\{|B| - 1\}$, and the *treewidth* $\mathsf{tw}(G)$ of G is the minimum width over all tree decompositions of G. A tree decomposition for which T is a path is a *path decomposition*, and the *pathwidth* $\mathsf{pw}(G)$ of G is the minimum width over all path decompositions of G.

Fractional Vertex Colorings. Fractional vertex colorings are a refinement of ordinary vertex colorings. There are several definitions of fractional vertex colorings and fractional chromatic number: as in the introduction, or by linear programming, by Lebesgue measures, combinatorial and by homomorphisms to Kneser graphs; the equivalence of the definitions is established by Molloy and Reed [34, Chap. 21.1], Scheinerman and Ullman [37, Sects. 3.1–3.2], and Dvořák et al. [14, Sect. 2]. Let G be a graph. The *chromatic number of* G is the smallest integer $\chi(G)$ for which there is a function $c : V(G) \rightarrow \{1, \ldots, \chi(G)\}$ such that $c(u) \neq c(v)$ for all edges $\{u, v\} \in E(G)$. An $a : b$-*coloring* of G is a function that assigns a non-empty set of b colors to each vertex such that adjacent vertices receive disjoint sets of colors, in which each color is drawn from a palette of a colors. The *b-fold coloring number* of G is the smallest integer $\chi^b(G)$ such that G admits an $\chi^b(G) : b$-coloring. Note that $\chi^1(G) = \chi(G)$. The *fractional chromatic number of* G is $\chi_f(G) = \inf_{b \to \infty}(\chi^b(G)/b)$. We refer to Molloy and Reed [34, Chap. 21] and Scheinerman and Ullman [37] for background on fractional colorings. By definition, $\chi_f(G) \leq \chi(G)$, and further $\chi_f(G) \geq \frac{|V(G)|}{\alpha(G)}$.

3   Fixed-Parameter Algorithm in Subquartic Planar Graphs

We provide an algorithm that, given any n-vertex planar graph G of maximum degree 4 and $k \in \mathbb{N}$, in time $2^{O(\sqrt{k})}n + O(nk^2 \log k + n^2)$ decides if $\alpha(G) \geq \frac{n+k}{4}$.

First, the algorithm exhaustively applies reduction rules that successively remove certain subgraphs of G until those are no longer present.

Reduction Rule 1. *Remove any induced subgraph H from G that is isomorphic to K_4 and satisfies $d_G(v) = 3$ for some vertex $v \in V(H)$.*

Reduction Rule 2. *Remove any connected component of G isomorphic to C_8^2.*

The reduction rules can run in time $O(n)$ per application, as constant-size subgraphs can be found in linear time in planar graphs, e.g., using an algorithm by Eppstein [15].

The proof of the following lemma is immediate.

Lemma 1. *Let G be an n-vertex planar graph with $\Delta(G) \leq 4$, and let G' be the n'-vertex graph obtained from G by applying Reduction Rule 1 or Reduction Rule 2. Then G' is a planar graph with $\Delta(G') \leq 4$. Further, for any integer $k \geq 0$, it holds that $\alpha(G) \geq (n + k)/4$ if and only if $\alpha(G') \geq (n' + k)/4$.*

Call a subquartic planar graph $\tilde{G}$ *reduced* if applying Reduction Rules 1 and 2 do not change it. Hence, we can obtain a reduced graph in time $O(n^2)$.

Second, we handle clique separators, using a result by Lu and Peng [29].

Proposition 1 [29]. *Let G_1, G_2 be subgraphs of a graph G with $G_1 \cup G_2 = G$ and $G_1 \cap G_2 = K_r$ for some $r \in \mathbb{N}$. Then $\chi_f(G) = \max\{\chi_f(G_1), \chi_f(G_2)\}$.*

We next employ an upper bound on the fractional chromatic number of subquartic $\{K_4, C_8^2\}$-free graphs, due to King et al. [28].

Proposition 2 [28]. *Any subquartic $\{K_4, C_8^2\}$-free graph admits a $532 : 134$-coloring.*

This directly yields a non-trivial upper bound on the fractional chromatic number of reduced graphs $\tilde{G}$ that are K_4-free.

Corollary 1. *Any K_4-free reduced graph $\tilde{G}$ admits a fractional coloring μ that assigns each vertex $134p$ colors from a palette $P = \{1, \ldots, 532p\}$ for some $p \in \mathbb{N}$.*

This strengthens a theorem by Albertson et al. [1], who proved that $\{K_4, C_8^2\}$-free connected 4-regular planar graphs have independence ratio larger than $1/4$ (and so were not counterexamples to the Erdős-Vizing Conjecture, now -Theorem).

The advantage of proving the stronger claim of the upper bound on the fractional chromatic number compared to proving a lower bound on the size of a maximum independent set is a stronger inductive hypothesis when exhibiting a hypothetical minimum counterexample.

However, our reduced graphs are not guaranteed to be K_4-free. Therefore, we need a stronger version of Proposition 2 for reduced graphs. To this end, for a reduced plane graph G, consider a subgraph H of G that is isomorphic to K_4. Let T_H denote the triangle of H such that the fourth vertex $\hat{v}_H \in V(H) \setminus T_H$ is embedded inside T_H. We say that a vertex $v \in V(G) \setminus T_H$ is *owned* by T_H if v is not separated from $\hat{v}_H$ by T_H (that is, v is also embedded inside T_H) and there is no subgraph $H' \neq H$ isomorphic to K_4 such that $T_{H'}$ is embedded inside T_H and v is embedded inside $T_{H'}$. Intuitively, this means that v is owned by the "innermost" K_4 subgraph whose triangle encloses it. Notice that in particular, $\hat{v}_H$ is owned by T_H. We illustrate this notion in Fig. 1.

Fig. 1. A vertex v that is owned by T_H.

We can now show that any subquartic plane graph admits a fractional coloring with few colors in which vertices belonging to K_4 or C_8^2 receive fewer colors than the other vertices.

Lemma 2. *For every subquartic plane graph G there is an integer $q = q(G) \in \mathbb{N}$ such that G admits a fractional coloring φ that assigns $134q$ colors from a palette $Q = \{1, \ldots, 532q\}$ to each vertex outside K_4 and C_8^2, and every other vertex is assigned $133q$ colors from Q.*

Proof. We prove the statement of the lemma by induction on the number of vertices of G. Clearly, we can assume that G is connected.

If $G = K_4$ or $G = C_8^2$, then let $\psi : V(G) \to \{0, 1, 2, 3\}$ be a 4:1-coloring of G, and set $q = 1$ and $\varphi(v) = \{133\psi(v) + 1, \ldots, 133\psi(v) + 133\}$ for each $v \in V(G)$.

Hence, from now on assume that $G \notin \{K_4, C_8^2\}$. Since G is connected and has maximum degree at most 4, it does not contain the 4-regular C_8^2 as subgraph.

So, from now on assume that G is plane, subquartic, connected, and C_8^2-free.

If G is K_4-free then (with the assumptions) G is reduced, and so by Corollary 1 it has a fractional coloring μ with colors from $P = \{1, \ldots, 532p\}$ so that $|\mu(v)| = 134p$ for each $v \in V(G)$. We set $q = p$, and define $\varphi(v) = \mu(v)$ for each $v \in V(G)$.

The final case is that G contains a K_4-subgraph; fix one such K_4 subgraph H. The unique embedding of H inside the plane graph G divides the plane into four regions each of which is bounded by a triangle of H. Since G is not equal to K_4, at least one of these regions contains a vertex outside H; thus, H induces a separating triangle T in G (i.e., a 3-separator that is a clique). Let G_1 and G_2 be the two connected induced subgraphs of G that intersect exactly in T and whose union equals G; notice that both G_1, G_2 are C_8^2-free. Now each of G_1, G_2 is a subquartic plane graph with fewer vertices than G; therefore, we can apply the induction hypothesis to each of G_1, G_2. Thus, for $i = 1, 2$ there is a q_i and a fractional coloring φ_i of G_i where each $v \in V(G_i)$ outside K_4 and C_8^2 in G_i is assigned a set $\varphi_i(v)$ of $134q_i$ colors from a palette $Q_i = \{1, \ldots, 532q_i\}$, and each $v \in V(G_i)$ in some K_4 or C_8^2 is assigned a set $\varphi_i(v)$ of $133q_i$ colors from Q_i. Assume, without loss of generality, that the unique vertex $\hat{v}_H \in V(H) \setminus T$

belongs to G_1. Then H is a K_4-subgraph in G_1, and therefore each $v \in V(T)$ has $|\varphi_1(v)| = 133q_1$. However, some vertices $v \in V(T)$ potentially have $|\varphi_2(v)| = 134q_2$, namely if they are outside of K_4-subgraphs of G_2. Thus, to "normalize", for $i = 1, 2$ let φ_i' be a fractional coloring obtained from φ_i by removing an arbitrary subset of exactly q_i colors from the color set $\varphi_i(v)$ of those $v \in V(T)$ that are not contained in K_4-subgraphs of G_i. At this point, we know that each $v \in V(T)$ has $|\varphi_i'(v)| = 133q_i$ for $i = 1, 2$.

Let $q = q_1 q_2$, and let φ_i'' be the fractional coloring of G_i with the palette $\{1, \ldots, 532q\}$ defined by $\varphi_i''(v) = \{q_{3-i}(c - 1) + 1, \ldots, q_{3-i}c \mid c \in \varphi_i'(v)\}$ for each $v \in V(G_i)$. Then for any pair $\{u, v\} \in V(T)$ the sets $\varphi_i''(u)$ and $\varphi_2''(v)$ are disjoint, and for any $v \in V(T)$ it holds $|\varphi_1''(v)| = |\varphi_2''(v)|$. Hence, there is a bijection $\pi : \{1, \ldots, 532q\} \to \{1, \ldots, 532q\}$ such that $\pi(\varphi_2''(v)) = \varphi_1''(v)$ for each $v \in V(T)$. Then the function φ defined by $\varphi(v) = \varphi_1''(v)$ for $v \in V(G_1)$ and $\varphi(v) = \pi(\varphi_2''(v))$ for $v \in V(G_2)$ is a fractional coloring of G as claimed. $\qquad\square$

Next, from Lemma 2 we derive that G has a large independent set (larger than $|V(G)|/4$) if many vertices of G are outside K_4 and C_8^2.

Lemma 3. *For any $r \in \mathbb{N}$, any reduced subquartic planar graph G with at least r vertices outside K_4 has an independent set of size at least $\frac{|V(G)|}{4} + \frac{r}{532}$.*

Proof. Let $q = q(G)$ be an integer, let $Q = \{1, \ldots, 532q\}$ and let $\varphi : V(G) \to Q$ be a fractional coloring of G according to Lemma 2. For each $i \in Q$ let $\varphi^{-1}(i)$ denote the set of vertices in G that get assigned color i by φ. (Note that φ is a set function as each vertex $v \in V(G)$ gets assigned a list of $133q$ or $134q$ colors from Q; so $\varphi^{-1}(i)$ is the list of vertices $v \in V(G)$ for which $i \in \varphi(v)$.) Then

$$\sum_{i \in Q} |\varphi^{-1}(i)| = \sum_{v \in V(G)} |\varphi(v)| = \sum_{\substack{v \in V(G) \\ v \in K_4}} 133q + \sum_{\substack{v \in V(G) \\ v \notin K_4}} 134q \geq (133|V(G)| + r)q.$$

Note that $\varphi^{-1}(i)$ is an independent set for each $i \in Q$. Thus, G has an independent set of size at least $\frac{(133|V(G)| + r)q}{|Q|} = \frac{|V(G)|}{4} + \frac{r}{532}$. $\qquad\square$

Consequently, for a reduced graph G, the algorithm checks if G has at least $r = 133k$ vertices outside K_4; in this case, it accepts (G, k) as a "yes"-instance.

So henceforth assume that G has less than $133k$ vertices outside K_4. We will show that this implies that G has treewidth $O(\sqrt{k})$.

Lemma 4. *For any $r \in \mathbb{N}_0$, any reduced subquartic planar graph G with at most r vertices outside K_4 has treewidth at most $O(\sqrt{r})$.*

Proof. Note that any K_4 in G induces a separating triangle, unless $G = K_4$.

We first argue that for any separating triangle T of G, both the interior and exterior of T contain some vertex outside any K_4. To this end suppose, for sake of contradiction, that all vertices in the interior T belong to some K_4. Then consider the separating triangle T' (of some K_4 subgraph) in the interior of T whose interior does not contain any other separating triangles; possibly $T' = T$.

Then there must be a single vertex $v_{T'}$ embedded inside T' that is adjacent to all vertices of T' and no other vertices of G; however, such a vertex would have been removed by Reduction Rule 1, contradicting that G is reduced. A symmetric argument shows the existence of a vertex outside K_4 in the exterior of T.

Relabel the initial graph $G = G^0$. Given G^i for some $i \geq 0$, we will recursively divide G^i into smaller graphs G_1^i, G_2^i. Then we proceed with $G^i = G_j^i$ for $j = 1, 2$. Precisely, we apply Proposition 1 to the graph $G = G^i$ and a separation $\{A, B\}$ of $V(G^i)$ such that $A \cap B$ induces a separating triangle of G^i. That is, as long as G^i admits a separating triangle T we divide G^i into induced subgraphs G_1^i, G_2^i with $V(G_1^i) \cup V(G_2^i) = V(G^i)$ and $V(G_1^i) \cap V(G_2^i) = V(T)$. We then repeat to recursively divide $G^i = G_j^i$ for $j = 1, 2$ whenever it has a separating triangle.

Once G^i (obtained by this recursive process) has no more separating triangles, any vertex of G^i either does not belong to any K_4 in the *initial* graph G^0 (type I-vertex) or lies on a separating triangle of G^0 (type II-vertex). From the recursive division process we obtain a family $\mathcal{F}$ of separating triangles; the size of $\mathcal{F}$ is bounded by r (the number of vertices outside K_4 in G^0), since for each such triangle we find a vertex outside K_4 in its interior or exterior by the initial claim.

Let G' be the subquartic planar graph that is reached at the base case of the recursion, that is, G' is the union of subquartic planar graphs without separating triangles. (In particular, G' is either isomorphic to K_4, or is K_4-free.) Hence, it consists of at most r vertices that were outside K_4 in G^0 (type I-vertices), along with at most $3r$ vertices on the boundary of separating triangles of G^0 (type II-vertices). Therefore, G' is a planar graph with $O(r)$ vertices, and planar graphs on $O(r)$ vertices have treewidth $O(\sqrt{r})$ [21, Corollary 24]. Thus, since we only recurse along separating triangles, by Proposition 1 also $\mathrm{tw}(G^0) = O(\sqrt{r})$. □

Since $r \leq 133k$, the graph G has treewidth $O(\sqrt{k})$. We use the linear time constant-factor approximation by Kammer and Tholey [27] to compute a tree decomposition $(T, \mathcal{B})$ of G of width $w = O(\sqrt{k})$. The algorithm by Kammer and Tholey runs in time $O(nw^2 \log w)$ and produces no repeated bags; therefore, $(T, \mathcal{B})$ has $O(nw^2 \log w)$ bags. Given $(T, \mathcal{B})$, we turn it into a tree decomposition $(T', \mathcal{B}')$ of no larger width and $|V(T')| = O(wn)$ bags in time $O(w^2 \cdot \max\{V(T), V(G)\}) = O(n \cdot w^4 \log w)$; this transformation is described by Cygan et al. [11, Lemma 7.4]. Then we use a dynamic programming algorithm [35, Theorem 10.14] over $(T', \mathcal{B}')$ for computing a maximum independent set of G in time $O(2^w \cdot w \cdot |V(T')|)$. Since $w = O(\sqrt{k})$, in the bounded treewidth case we can decide if $\alpha(G) \geq (|V(G)| + k)/4$ in time $2^{O(\sqrt{k})} \cdot O(\sqrt{k}n) + O(n \cdot k^2 \log k) = 2^{O(\sqrt{k})}n + O(nk^2 \log k)$. In summary, INDEPENDENT SET-ATLB in n-vertex subquartic planar graphs can be solved in time $2^{O(\sqrt{k})}n + O(nk^2 \log k + n^2)$, completing the proof of Theorem 1.

Remark 1. The idea of proving a sublinear bound on the treewidth in terms of k was suggested to us by an anonymous reviewer. We had shown instead:

Lemma 5. *For any* $r \in \mathbb{N}_0$, *any reduced subquartic planar graph* G *with at most* r *vertices outside* K_4 *has pathwidth at most* $8r - 1$.

A disadvantage of using pathwidth over treewidth is that it is only linearly bounded in k rather than sublinear. Also, we are not aware of any constant-factor approximation for pathwidth in planar graphs that runs in (near-)linear time; naïvely applying the treewidth approximation algorithm by Kammer and Tholey only guarantees a path decomposition of width $O(k \log n)$ and an algorithm for INDEPENDENT SET-ATLB with run time $2^{O(k \log n)} \cdot n^{O(1)}$.

Remark 2. Notice that when our algorithm finds that G has at least $133k$ vertices outside K_4, it concludes that $\alpha(G) \geq (n+k)/4$—but how can we *efficiently* find an independent set in G of this size? A simple self-reducibility argument does not suffice, as this could blow up of the parameter beyond any function of k.

Remark 3. Is there a polynomial kernel for PLANAR INDEPENDENT SET-ATLB in subquartic planar graphs G? If every K_4 in G has some 3-vertex, then exhaustively apply reduction rules 1 and 2 to obtain a graph G' on n' vertices. Then if $k/4 \leq \frac{n'}{532}$, we accept (G, k) as a "yes"-instance, as $\alpha(G') \geq \frac{134}{532}n' = n'/4 + n'/532$ by Corollary 1. Else, if $k/4 > \frac{n'}{532}$, we return (G', k) as a kernel with $n' \leq 133k$ vertices. However, this argument does not work in general subquartic graphs.

4   Discussion

We showed fixed-parameter tractability for PLANAR INDEPENDENT SET-ATLB in graphs with maximum degree 4, and a certain (very restricted) class with maximum degree 5. This resolves an often-posed question [4,8,10,13,17,31,32,35,38] on these restricted classes. It remains to resolve the parameterized complexity of PLANAR INDEPENDENT SET-ATLB in all planar graphs. As a first step towards this result, one might consider for example planar graphs of average degree at most four (recall that general planar graphs have average degree less than six). Notice, however, that an application of Reduction Rule 1 to a planar graph might yield a graph whose average degree is strictly larger than 4.

We observe that how to solve the PLANAR INDEPENDENT SET-ATLB problem for planar graphs with few triangles. Let G be an n-vertex planar graph with $t \leq n/12$ triangles; we decide if $\alpha(G) \geq (n + k)/4$, for parameter $k \geq 0$. First, if G contains $n \leq 18k/4$ vertices, then (G, k) forms a kernel. Hence, suppose that $k/4 \leq n/18$. Clearly, any (maximum) independent set of G contains at most one vertex of each triangle of G. Therefore, remove a set $V_T \subseteq V(G)$ of at most t vertices from G to obtain a triangle-free planar graph G' with $n' := |V(G')|$ vertices. By Grötzsch' theorem, $\alpha(G') \geq n'/3$. Since any maximum independent set of G' is an independent set of G, it follows that $\alpha(G) \geq \alpha(G') \geq n'/3 \geq (n - t)/3 \geq (n - n/12)/3 = (11/36)n = n/4 + n/18$. Thus, as $k/4 \leq n/18$, we can accept (G, k) as a "yes"-instance. In summary, this solves PLANAR INDEPENDENT SET-ATLB in planar graphs with at most

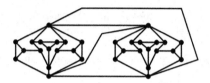

Fig. 2. A non-3-colorable planar graph with triangles at pairwise distance 3.

$n/12$ triangles in fixed-parameter time. Can one raise $n/12$ to $n/4$?—Notice that there are connected non-3-colorable planar graphs whose triangles have pairwise distance at least 3; thus they have at most $n/4$ triangles yet we only know $\alpha(G) \geq 1/\chi(G) = 1/4$ (see Fig. 2).

Acknowledgements. I am indebted to Zdeněk Dvořák for helpful remarks, and an anonymous reviewer who suggested considering treewidth over pathwidth.

References

1. Albertson, M., Bollobas, B., Tucker, S.: The independence ratio and maximum degree of a graph. In: Proceedings of the Seventh Southeastern Conference on Combinatorics, Graph Theory and Computing, pp. 43–50. Congressus Numerantium, No. XVII. Utilitas Math., Winnipeg, Man. (1976)
2. Appel, K., Haken, W.: Every planar map is four colorable. Bull. Amer. Math. Soc. **82**(5), 711–712 (1976)
3. Berge, C.: Graphs and Hypergraphs, revised edn. North-Holland Publishing Co., Amsterdam (1976)
4. Bodlaender, H.L.: Open problems in parameterized and exact computation. Technical report UU-CS-2008-017, Utrecht University (2008)
5. Brooks, R.L.: On colouring the nodes of a network. Proc. Camb. Philos. Soc. **37**, 194–197 (1941)
6. Cai, L., Juedes, D.: On the existence of subexponential parameterized algorithms. J. Comput. Syst. Sci. **67**(4), 789–807 (2003)
7. Cranston, D.W., Rabern, L.: Planar graphs are 9/2-colorable and have independence ratio at least 3/13 (2015). http://arxiv.org/abs/1410.7233
8. Crowston, R., Fellows, M., Gutin, G., Jones, M., Rosamond, F., Thomassé, S., Yeo, A.: Simultaneously satisfying linear equations over $\mathbb{F}_2$: MaxLin2 and Max-r-Lin2 parameterized above average. In: Proceedings of FSTTCS 2011, pp. 229–240 (2011)
9. Crowston, R., Jones, M., Mnich, M.: Max-cut parameterized above the Edwards-Erdős bound. In: Czumaj, A., Mehlhorn, K., Pitts, A., Wattenhofer, R. (eds.) ICALP 2012, Part I. LNCS, vol. 7391, pp. 242–253. Springer, Heidelberg (2012)
10. Cygan, M., Fomin, F., Jansen, B., Kowalik, Ł., Lokshtanov, D., Marx, D., Pilipczuk, M., Pilipczuk, M., Saurabh, S.: Open problems from the Bedlewo school on parameterized algorithms and complexity (2014). http://fptschool.mimuw.edu.pl/opl.pdf
11. Cygan, M., Fomin, F.V., Kowalik, Ł., Lokshtanov, D., Marx, D., Pilipczuk, M., Pilipczuk, M., Saurabh, S.: Parameterized Algorithms. Springer, New York (2015)

12. Cygan, M., Pilipczuk, M., Pilipczuk, M., Wojtaszczyk, J.O.: On multiway cut parameterized above lower bounds. ACM Trans. Comput. Theory 5(1), 3:1–3:11 (2013)

13. Dvořák, Z., Mnich, M.: Large independent sets in triangle-free planar graphs. In: Schulz, A.S., Wagner, D. (eds.) ESA 2014. LNCS, vol. 8737, pp. 346–357. Springer, Heidelberg (2014)

14. Dvořák, Z., Sereni, J.-S.S., Volec, J.: Subcubic triangle-free graphs have fractional chromatic number at most 14/5. J. London Math. Soc. 89(3), 641–662 (2014)

15. Eppstein, D.: Subgraph isomorphism in planar graphs and related problems. J. Graph Algorithms Appl. 3(3), 27 (electronic) (1999)

16. Faria, L., Klein, S., Stehlík, M.: Odd cycle transversals and independent sets in fullerene graphs. SIAM J. Discrete Math. 26(3), 1458–1469 (2012)

17. Fellows, M.R., Guo, J., Marx, D., Saurabh, S.: Data reduction and problem kernels (Dagstuhl Seminar 12241). Dagstuhl Reports 2(6), 26–50 (2012)

18. Fleischner, H., Sabidussi, G., Sarvanov, V.I.: Maximum independent sets in 3- and 4-regular Hamiltonian graphs. Discrete Math. 310(20), 2742–2749 (2010)

19. Garey, M.R., Johnson, D.S.: Computers and Intractability. Freeman, New York (1979)

20. Giannopoulou, A.C., Kolay, S., Saurabh, S.: New lower bound on MAX CUT of hypergraphs with an application to r-SET SPLITTING. In: Fernández-Baca, D. (ed.) LATIN 2012. LNCS, vol. 7256, pp. 408–419. Springer, Heidelberg (2012)

21. Grohe, M.: Local tree-width, excluded minors, and approximation algorithms. Combinatorica 23(4), 613–632 (2003)

22. Grötzsch, H.: Zur Theorie der diskreten Gebilde. VII. Ein Dreifarbensatz für dreikreisfreie Netze auf der Kugel. Wiss. Z. Martin-Luther-Univ. Halle-Wittenberg. Math.-Nat. Reihe, 8, 109–120 (1958/1959)

23. Gutin, G., Jones, M., Yeo, A.: Kernels for below-upper-bound parameterizations of the hitting set and directed dominating set problems. Theoret. Comput. Sci. 412(41), 5744–5751 (2011)

24. Gutin, G., Kim, E.J., Mnich, M., Yeo, A.: Betweenness parameterized above tight lower bound. J. Comput. System Sci. 76(8), 872–878 (2010)

25. Gutin, G., van Iersel, L., Mnich, M., Yeo, A.: Every ternary permutation constraint satisfaction problem parameterized above average has a kernel with a quadratic number of variables. J. Comput. System Sci. 78(1), 151–163 (2012)

26. Heckman, C.C., Thomas, R.: Independent sets in triangle-free cubic planar graphs. J. Combin. Theory Ser. B 96(2), 253–275 (2006)

27. Kammer, F., Tholey, T.: Approximate tree decompositions of planar graphs in linear time. In: Proceedings of SODA 2012, pp. 683–698 (2012)

28. King, A.D., Lu, L., Peng, X.: A fractional analogue of Brooks' theorem. SIAM J. Discrete Math. 26(2), 452–471 (2012)

29. Lu, L., Peng, X.: The fractional chromatic number of triangle-free graphs with $\Delta \leq 3$. Discrete Math. 312(24), 3502–3516 (2012)

30. Mahajan, M., Raman, V.: Parameterizing above guaranteed values: MaxSat and MaxCut. J. Algorithms 31(2), 335–354 (1999)

31. Mahajan, M., Raman, V., Sikdar, S.: Parameterizing above or below guaranteed values. J. Comput. System Sci. 75(2), 137–153 (2009)

32. Mnich, M.: Algorithms in moderately exponential time. Ph.D. thesis, TU Eindhoven (2010)

33. Mnich, M., Zenklusen, R.: Bisections above tight lower bounds. In: Golumbic, M.C., Stern, M., Levy, A., Morgenstern, G. (eds.) WG 2012. LNCS, vol. 7551, pp. 184–193. Springer, Heidelberg (2012)

34. Molloy, M., Reed, B.: Graph Colouring and the Probabilistic Method. Algorithms and Combinatorics, vol. 23. Springer, Berlin (2002)
35. Niedermeier, R.: Invitation to Fixed-parameter Algorithms. Oxford Lecture Series in Mathematics and its Applications, vol. 31. OUP, Oxford (2006)
36. Robertson, N., Sanders, D., Seymour, P., Thomas, R.: The four-colour theorem. J. Combin. Theory Ser. B **70**(1), 2–44 (1997)
37. Scheinerman, E.R., Ullman, D.H.: Fractional Graph Theory. Dover Publications Inc., Mineola (2011)
38. Sikdar, S.: Parameterizing from the extremes. Ph.D. thesis, The Institute of Mathematical Sciences, Chennai (2010)

As Close as It Gets

Mike Behrisch[1], Miki Hermann[2], Stefan Mengel[2], and Gernot Salzer[1(✉)]

[1] Technische Universität Wien, Vienna, Austria
{behrisch,salzer}@logic.at
[2] LIX (UMR CNRS 7161), École Polytechnique, Palaiseau, France
hermann@lix.polytechnique.fr, mengel@cril.fr

Abstract. We study the minimum Hamming distance between distinct satisfying assignments of a conjunctive input formula over a given set of Boolean relations (MinSolutionDistance, MSD). We present a complete classification of the complexity of this optimization problem with respect to the relations admitted in the formula. We give polynomial time algorithms for several classes of constraint languages. For all other cases we prove hardness or completeness with respect to poly-APX, or NPO, or equivalence to a well-known hard optimization problem.

1 Introduction

We study the following optimization problem related to Boolean constraint satisfaction problems (CSPs): Given a formula built from atomic constraint relations by means of conjunction and variable identification, the task is to produce two satisfying assignments having minimal Hamming distance among all distinct pairs in the solution space of the CSP instance represented by the formula (MinSolutionDistance, MSD). Note that the dual problem MaxHammingDistance has been studied in [9].

As usual our problem is parametrized by the set of atomic constraints allowed to occur in the conjunctive formulas. With respect to this parametrization we completely classify the complexity of the minimization problem MSD: It turns out that it is either polynomial-time solvable, or that it is complete for a well-known optimization class, or else equivalent to some classical hard optimization problem.

Restricting the allowed relations to affine Boolean relations, our problem MSD becomes the well-known problem MinDistance of computing the minimum distance of a linear code. As this quantity determines the number of errors such a code can detect and correct, it is of central importance in coding theory. Our work can thus be seen as a generalization of these questions from affine to arbitrary relations.

M. Behrisch and G. Salzer— Supported by Austrian Science Fund (FWF) grant I836-N23.

M. Hermann— Supported by ANR-11-ISO2-003-01 Blanc International grant ALCOCLAN.

S. Mengel— Supported by QUALCOMM grant. Now at CRIL (UMR CNRS 8188), Lens, France.

M. Kaykobad and R. Petreschi (Eds.): WALCOM 2016, LNCS 9627, pp. 222–235, 2016.
DOI: 10.1007/978-3-319-30139-6_18

In the course of investigations it appears that MSD lacks compatibility with existential quantification, preventing classical clone theory from being applicable. Consequently, we are lead to weak co-clones that need only be closed under conjunction and equality. To deal with such structures we make use of the theory established in [15], as well as the minimal weak bases of Boolean co-clones described in [13].

This paper is part of a more general program to understand the Hamming distance between solutions of constraint satisfaction problems. The results of this program up to now, including those from this paper and some on other problems, can be found in [4,5].

2   Preliminaries

An n-ary *Boolean relation* R is a subset of $\{0,1\}^n$; its elements $(b_1, \ldots, b_n)$ are also written as $b_1 \cdots b_n$. Let $\mathcal{V}$ be a set of variables. An *atomic constraint*, or an *atom*, is an expression $R(\boldsymbol{x})$, where R is an n-ary relation and $\boldsymbol{x}$ is an n-tuple of variables from $\mathcal{V}$. Let $\mathfrak{L}$ be the collection of all non-empty finite sets of Boolean relations, also called *constraint languages*. A (conjunctive) Γ-*formula* is a finite conjunction of atoms $R_1(\boldsymbol{x_1}) \wedge \cdots \wedge R_k(\boldsymbol{x_k})$, where the R_i are relations from $\Gamma \in \mathfrak{L}$ and the $\boldsymbol{x_i}$ are variable tuples of suitable arity.

An *assignment* is a mapping $m \colon \mathcal{V} \to \{0,1\}$ assigning a Boolean value $m(x)$ to each variable $x \in \mathcal{V}$. If we arrange the variables in some arbitrary but fixed order, say as a vector $(x_1, \ldots, x_n)$, then the assignments can be identified with vectors from $\{0,1\}^n$. The i-th component of a vector m is denoted by $m[i]$ and corresponds to the value of the i-th variable, i.e., $m[i] = m(x_i)$. The *Hamming weight* $\mathrm{hw}(m) = |\{i \mid m[i] = 1\}|$ of m is the number of 1s in the vector m. The *Hamming distance* $\mathrm{hd}(m, m') = |\{i \mid m[i] \neq m'[i]\}|$ of m and m' is the number of coordinates on which the vectors disagree. The complement $\overline{m}$ of a vector m is its pointwise complement, $\overline{m}[i] = 1 - m[i]$.

Table 1. Some relevant Boolean co-clones with bases

iD_2	$\{x \oplus y, x \to y\}$	iN	$\{\mathrm{dup}^3\}$
iL	$\{\mathrm{even}^4\}$	iN_2	$\{\mathrm{nae}^3\}$
iL_2	$\{\mathrm{even}^4, \neg x, x\}$	iI	$\{\mathrm{even}^4, x \to y\}$
iV_2	$\{x \vee y \vee \neg z, \neg x, x\}$	iI_0	$\{\mathrm{even}^4, x \to y, \neg x\}$
iE_2	$\{\neg x \vee \neg y \vee z, \neg x, x\}$	iI_1	$\{\mathrm{even}^4, x \to y, x\}$

An assignment m satisfies the constraint $R(x_1, \ldots, x_n)$ if $(m(x_1), \ldots, m(x_n)) \in R$ holds. It satisfies the formula φ if it satisfies all of its atoms; m is said to be a *model* or *solution* of φ in this case. We use $[\varphi]$ to denote the Boolean relation containing all models of φ. In sets of relations represented this way we usually omit the brackets. A *literal* is a variable v, or its negation $\neg v$. Assignments m are extended to literals by defining $m(\neg v) = 1 - m(v)$.

The following Boolean functions and relations are of particular relevance to us: we write $x \oplus y$ for addition modulo 2 and $x \equiv y$ for $x \oplus y \oplus 1$. Further, we define the relations $\text{nae}^3 := \{0,1\}^3 \smallsetminus \{000, 111\}$, $\text{dup}^3 := \{0,1\}^3 \smallsetminus \{010, 101\}$ and $\text{even}^4 := \{(a_1, a_2, a_3, a_4) \in \{0,1\}^4 \mid \oplus_{i=1}^4 a_i = 0\}$.

Throughout the text we refer to different types of Boolean constraint relations following Schaefer's terminology [14] (see also [6,8]). A Boolean relation R is (1) *1-valid* if $1 \cdots 1 \in R$ and it is *0-valid* if $0 \cdots 0 \in R$, (2) *Horn* (*dual Horn*) if R can be represented by a formula in conjunctive normal form (CNF) having at most one unnegated (negated) variable in each clause, (3) *monotone* if it is both Horn and dual Horn, (4) *bijunctive* if it can be represented by a CNF having at most two variables in each clause, (5) *affine* if it can be represented by an affine system of equations $Ax = b$ over $\mathbb{Z}_2$, (6) *complementive* if for each $m \in R$ also $\overline{m} \in R$. A set Γ of Boolean relations is called 0-valid (1-valid, Horn, dual Horn, monotone, affine, bijunctive, complementive) if *every* relation in Γ has the respective property.

We denote by $\langle \Gamma \rangle$ the set of all relations that can be expressed using relations from $\Gamma \cup \{=\}$, conjunction, variable identification (and permutation), cylindrification and existential quantification. The set $\langle \Gamma \rangle$ is called the *co-clone* generated by Γ. A *base* of a co-clone $\mathcal{B}$ is a set of relations Γ such that $\langle \Gamma \rangle = \mathcal{B}$. The set of all co-clones constitutes a lattice with regard to set inclusion. Their bases were studied in [7]; those relevant in this paper are listed in Table 1. In particular the sets of relations being 0-valid, 1-valid, complementive, Horn, dual Horn, affine, and bijunctive each form a co-clone denoted by iI_0, iI_1, iN_2, iE_2, iV_2, iL_2, and iD_2, respectively.

We will also use a weaker closure than $\langle \Gamma \rangle$, called *conjunctive closure* and denoted by $\langle \Gamma \rangle_\wedge$, where the constraint language Γ is closed under conjunctive definitions, but not under existential quantification or addition of explicit equality constraints.

Minimal weak bases of co-clones are bases with certain additional properties. Since we rely on only some of them, we shall not define this notion but refer the reader to [13,15].

Theorem 1 (Schnoor & Schnoor [15]). *If Γ is a minimal weak base of a co-clone, then $\Gamma \subseteq \langle \Gamma' \rangle_\wedge$ for any base Γ'.*

Lagerkvist computed minimal weak bases for all Boolean co-clones in [13]. From there we infer that $\{[\text{even}_4(x_1, x_2, x_3, x_4) \wedge (x_1 \wedge x_4 \equiv x_2 \wedge x_3)]\}$ constitutes a minimal weak base of the co-clone iN; likewise $\{[(x_1 \equiv x_2 \wedge x_3) \wedge (\neg x_4 \equiv \neg x_2 \wedge \neg x_3)]\}$ is one of iI.

We assume that the reader has a basic knowledge of approximation algorithms and complexity theory, see e.g. [2,8]. For reductions among decision problems we use polynomial-time many-one reduction denoted by $\leq_{\text{m}}$. Many-one equivalence between decision problems is written as $\equiv_{\text{m}}$. For reductions among optimization problems we employ approximation preserving reductions (AP-reductions), represented by $\leq_{\text{AP}}$, while AP-equivalence of optimization problems is stated as $\equiv_{AP}$. Besides, the following approximation complexity classes in the hierarchy PO $\subseteq$ APX $\subseteq$ poly-APX $\subseteq$ NPO occur.

An optimization problem $\mathcal{P}_1$ AP-*reduces* to another optimization problem $\mathcal{P}_2$ if there are two polynomial-time computable functions f, g, and a constant $\alpha \geq 1$ such that for all $r > 1$ on any input x for $\mathcal{P}_1$ the following holds:

- $f(x)$ is an instance of $\mathcal{P}_2$;
- for any solution y of $f(x)$, the result $g(x, y)$ is a solution of x;
- if y is an r-approximate solution for the instance $f(x)$, then the solution $g(x, y)$ is $(1 + (r-1)\alpha + o(1))$-approximate for x.

If $\mathcal{P}_1$ AP-reduces to $\mathcal{P}_2$ with constant $\alpha \geq 1$ and $\mathcal{P}_2$ has an $f(n)$-approximation algorithm, then there is an $\alpha f(n)$-approximation algorithm for $\mathcal{P}_1$.

To relate our problem to well-known optimization problems we make the following convention: For optimization problems $\mathcal{P}$ and $\mathcal{Q}$ we say that $\mathcal{P}$ is $\mathcal{Q}$-*complete* if $\mathcal{P} \equiv_{AP} \mathcal{Q}$. We use this notion in particular with respect to the following well-studied problem.

Problem MinDistance. Given a matrix $A \in \mathbb{Z}_2^{k \times l}$ any non-zero vector $x \in \mathbb{Z}_2^l$ with $Ax = 0$ is considered a solution. The objective is to minimize the Hamming weight $\mathrm{hw}(x)$.

MinDistance is known to be NP-hard to approximate within a factor $2^{\Omega(\log^{1-\varepsilon}(n))}$ for every $\varepsilon > 0$, see [10]. Thus if a problem $\mathcal{P}$ is equivalent to it, then $\mathcal{P} \notin$ APX unless P = NP.

We also use the classic satisfiability problem $\mathsf{SAT}(\Gamma)$, asking for a conjunctive formula φ over a constraint language Γ, if φ is satisfiable. Schaefer presented in [14] a complete classification of complexity for SAT. His dichotomy theorem proves that $\mathsf{SAT}(\Gamma)$ is polynomial-time decidable if Γ is 0-valid ($\Gamma \subseteq \mathrm{iI}_0$), 1-valid ($\Gamma \subseteq \mathrm{iI}_1$), Horn ($\Gamma \subseteq \mathrm{iE}_2$), dual Horn ($\Gamma \subseteq \mathrm{iV}_2$), bijunctive ($\Gamma \subseteq \mathrm{iD}_2$), or affine ($\Gamma \subseteq \mathrm{iL}_2$); otherwise it is NP-complete. Moreover, we need the decision problem $\mathsf{AnotherSAT}(\Gamma)$, whose complexity was completely classified in [12]. Given a conjunctive formula φ and a satisfying assignment m, it asks if there exists another satisfying assignment $m' \neq m$ for φ.

3   Results

The input to our problem is a conjunctive formula over a constraint language. The satisfying assignments of the formula, being its solutions, constitute the codewords of the associated code. The minimization target is the distance of any two distinct solutions.

Problem MinSolutionDistance(Γ), MSD(Γ)
Input: A conjunctive formula φ over relations from Γ.
Solution: Two satisfying truth assignments $m \neq m'$ to the variables occurring in φ.
Objective: Minimum Hamming distance $\mathrm{hd}(m, m')$.

Theorem 2. *For any constraint language Γ the optimization problem* $\mathsf{MSD}(\Gamma)$ *is*

(i) *in* PO *if Γ is*
  (a) *bijunctive* $(\Gamma \subseteq \mathrm{iD}_2)$ *or*
  (b) *Horn* $(\Gamma \subseteq \mathrm{iE}_2)$ *or*
  (c) *dual Horn* $(\Gamma \subseteq \mathrm{iV}_2)$;
(ii) $\mathsf{MinDistance}$-*complete if Γ is exactly affine* $(\mathrm{iL} \subseteq \langle \Gamma \rangle \subseteq \mathrm{iL}_2)$;
(iii) *in poly-*APX *if Γ is both 0-valid and 1-valid, but does not contain an affine relation* $(\mathrm{iN} \subseteq \langle \Gamma \rangle \subseteq \mathrm{iI})$, *where* $\mathsf{MSD}(\Gamma)$ *is n-approximable but not $(n^{1-\varepsilon})$-approximable unless* P = NP; *and*
(iv) NPO-*complete otherwise* $(\mathrm{iN}_2 \subseteq \langle \Gamma \rangle$ *or* $\mathrm{iI}_0 \subseteq \langle \Gamma \rangle$ *or* $\mathrm{iI}_1 \subseteq \langle \Gamma \rangle)$.

Proof. The proof is split into several propositions presented in the remainder of the paper.

(i) See Propositions 7 and 8.
(ii) See Proposition 14.
(iii) For $\Gamma \subseteq \mathrm{iI}$, every formula φ over Γ has at least two solutions since it is both 0-valid and 1-valid. Thus $\mathsf{2SolutionSAT}(\Gamma)$ is in P, and Proposition 13 yields that $\mathsf{MSD}(\Gamma)$ is n-approximable. By Proposition 18 this approximation is indeed tight.
(iv) According to [12], $\mathsf{AnotherSAT}(\Gamma)$ is NP-hard for $\mathrm{iI}_0 \subseteq \langle \Gamma \rangle$, or $\mathrm{iI}_1 \subseteq \langle \Gamma \rangle$. By Lemma 10 it follows that $\mathsf{2SolutionSAT}(\Gamma)$ is NP-hard, too. For $\mathrm{iN}_2 \subseteq \langle \Gamma \rangle$ we can reduce the NP-hard problem $\mathsf{SAT}(\Gamma)$ to $\mathsf{2SolutionSAT}(\Gamma)$. Hence $\mathsf{MSD}(\Gamma)$ is NPO-complete in all three cases. □

The optimization problem can be transformed into a decision problem $\mathsf{MSD}^{\mathrm{d}}$ by adding a bound $k \in \mathbb{N}$ to the input and asking if $\mathrm{hd}(m, m') \leq k$. We obtain the following dichotomy:

Corollary 3. $\mathsf{MSD}^{\mathrm{d}}(\Gamma)$ *is in* P *if $\Gamma \in \mathfrak{L}$ is bijunctive, Horn, or dual-Horn, and it is* NP-*complete otherwise.*

Proof. This follows immediately from Theorem 2: All cases in PO become polynomial-time decidable, whereas the other cases, which are APX-hard, become NP-complete. According to Post's lattice this classification covers all finite sets Γ of relations. □

4   Duality and Inapplicability of Clone Closure

As the optimization problem MSD is not compatible with existential quantification, we cannot prove an AP-equivalence result between any two MSD parametrized by constraint languages generating the same co-clone. Yet, similar results hold for weak co-clones.

Proposition 4. *We have* $\mathsf{MSD}^d(\Gamma') \leq_m \mathsf{MSD}^d(\Gamma)$ *and* $\mathsf{MSD}(\Gamma') \leq_{AP} \mathsf{MSD}(\Gamma)$ *for* $\Gamma, \Gamma' \in \mathfrak{L}$ *satisfying* $\Gamma' \subseteq \langle \Gamma \rangle_\wedge$.

Proof. For similarity it suffices to prove that $\Gamma' \subseteq \langle \Gamma \rangle_\wedge$ implies $\mathsf{MSD}(\Gamma') \leq_{AP} \mathsf{MSD}(\Gamma)$.

Let a Γ'-formula φ be an instance of $\mathsf{MSD}(\Gamma')$. Since $\Gamma' \subseteq \langle \Gamma \rangle_\wedge$, every constraint $R(x_1, \ldots, x_k)$ of φ can be written as a conjunction of constraints upon relations from Γ. Substitute the latter into φ, obtaining φ'. Now φ' is an instance of $\mathsf{MSD}(\Gamma)$, where φ' is only polynomially larger than φ. As φ and φ' have the same variables and hence the same models, also the closest distinct models of φ and φ' are the same. □

For a relation $R \subseteq \{0,1\}^n$, its *dual* relation is $\mathrm{dual}(R) = \{\overline{m} \mid m \in R\}$, i.e., the relation containing the complements of tuples from R. We naturally extend this to sets of relations Γ by putting $\mathrm{dual}(\Gamma) = \{\mathrm{dual}(R) \mid R \in \Gamma\}$. Since taking complements is involutive, duality is a symmetric relation. By inspecting the bases of co-clones in Table 1, we deduce that many co-clones are duals of each other, e.g. iE$_2$ and iV$_2$.

We now show that it suffices to consider only one half of Post's lattice of co-clones.

Lemma 5. *For any set* Γ *of Boolean relations we have* $\mathsf{MSD}^d(\Gamma) \equiv_m \mathsf{MSD}^d(\mathrm{dual}(\Gamma))$ *and* $\mathsf{MSD}(\Gamma) \equiv_{AP} \mathsf{MSD}(\mathrm{dual}(\Gamma))$.

Proof. For a Γ-formula φ and an assignment m to φ we construct a $\mathrm{dual}(\Gamma)$-formula φ' by substitution of every atom $R(\boldsymbol{x})$ by $\mathrm{dual}(R)(\boldsymbol{x})$. Then m satisfies φ if and only if $\overline{m}$ satisfies φ', $\overline{m}$ being the complement of m. Moreover, $\mathrm{hd}(m, m') = \mathrm{hd}(\overline{m}, \overline{m'})$. □

5 Finding the Minimal Distance Between Solutions

5.1 Polynomial-Time Cases

We use the following result based on a previous theorem of Baker and Pixley [3], showing that it suffices to consider binary relations when studying bijunctive constraint languages.

Proposition 6 (Jeavons et al. [11]). *Any bijunctive constraint* $R(x_1, \ldots, x_n)$ *is equivalent to* $\bigwedge_{1 \leq i \leq j \leq n} R_{ij}(x_i, x_j)$, *where* R_{ij} *is the projection of* R *to the coordinates* i *and* j.

Proposition 7. *If* Γ *is bijunctive* ($\Gamma \subseteq \mathrm{iD}_2$) *then* $\mathsf{MSD}(\Gamma)$ *is polynomial-time solvable.*

By Proposition 6, an algorithm for bijunctive Γ can be restricted to at most binary clauses. We extend the algorithm of Aspvall, Plass, and Tarjan [1].

Algorithm

Input: An iD_2-formula φ viewed as a collection of one- or two-element sets of literals.

Output: "≤ 1 model" or the minimal Hamming distance of any two distinct models of φ.

Method: Let $\mathcal{V}$ be the set of variables occurring in φ, let $\mathcal{L} = \{v, \neg v \mid v \in \mathcal{V}\}$ be the set of corresponding literals, and let $\bar{u}$ denote the complementary literal to $u \in \mathcal{L}$.

- Construct the relation $R := \{(\bar{u}, v), (u, \bar{v}) \mid \{u, v\} \in \varphi\} \cup \{(\bar{u}, u) \mid \{u\} \in \varphi\}$.
  Let $\leq$ be the reflexive and transitive closure of R, i.e. the least preorder on $\mathcal{L}$ extending R.
  Let $\sim := \{(u, v) \in \mathcal{L}^2 \mid u \leq v \wedge v \leq u\}$ be the associated equivalence relation. If $v \sim \neg v$ holds for some variable v, then return "≤ 1 model" (φ is unsatisfiable).
- Otherwise, let $\mathcal{V}_0 := \{v \in \mathcal{V} \mid v \leq \neg v\}$ and $\mathcal{V}_1 := \{v \in \mathcal{V} \mid \neg v \leq v\}$ be sets of variables being false and true, respectively, in every model of φ.
  If $\mathcal{V}_0 \cup \mathcal{V}_1 = \mathcal{V}$ holds, then return "≤ 1 model" (φ has only one model).
- Otherwise, construct the sets
  $F_0 := \{L \in \mathcal{L}/\sim \mid \exists v \in \mathcal{V}_0 \colon L \leq [v]_\sim\} \cup \{L \in \mathcal{L}/\sim \mid \exists v \in \mathcal{V}_1 \colon L \leq [\neg v]_\sim\}$ and
  $F_1 := \{L \in \mathcal{L}/\sim \mid \exists v \in \mathcal{V}_0 \colon [\neg v]_\sim \leq L\} \cup \{L \in \mathcal{L}/\sim \mid \exists v \in \mathcal{V}_1 \colon [v]_\sim \leq L\}$.
  F_0 (F_1) is the set of equivalence classes of literals whose value is forced to false (to true) by backward and forward propagation from variables in $\mathcal{V}_0$ and $\mathcal{V}_1$.
  Let $P := (\mathcal{L}/\sim) \setminus (F_0 \cup F_1)$ be the set of remaining equivalence classes.
  Return $\min\{|L| \mid L \in P\}$ as minimal Hamming distance.

Complexity. The size of $\mathcal{L}$ is linear in the number of variables, the reflexive closure can be computed in linear time in $|\mathcal{L}|$, the transitive closure in cubic time in $|\mathcal{L}|$, see [16]. The equivalence relation $\sim$ is the intersection of $\leq$ and its inverse (quadratic in $|\mathcal{L}|$); from it we can obtain the partition $\mathcal{L}/\sim$ in linear time in $|\mathcal{L}|$, and combining this with the preorder $\leq$ we can compute the order on $\mathcal{L}/\sim$ in polynomial time, as well. Similarly, the remaining sets from the proof can be computed with polynomial time complexity.

Correctness. The pairs in R arise from understanding the atomic constraints in φ as implications. Therefore, by transitivity of implication, in every model of φ, literals $u, v \in \mathcal{L}$ satisfying $u \leq v$ have to be evaluated so that $(m(u), m(v))$ does not violate the Boolean order relation, i.e. $[x \to y]$. Hence, literals $u \sim v$ from one equivalence class have to share the same value in any model m. Moreover, since literals and their negations have to take on opposite values, we must have $m(v) = 0$ for all $v \in \mathcal{V}_0$ and $m(v) = 1$ for all $v \in \mathcal{V}_1$. This proves that the algorithm gives a correct answer in case there do not exist feasible solutions. Furthermore, by transitivity, we see that literals in equivalence classes in F_i must be evaluated to $i \in \{0, 1\}$. So any two models can only differ on the literals belonging to members of P. Therefore, clearly, the return value of the algorithm is a lower bound for the minimal solution distance. To prove the converse, we shall exhibit two models $m_0 \neq m_1$ of φ having the least cardinality of equivalence classes in P as their Hamming distance.

Let $L \in P$ be a class of minimum cardinality. Define $m_0(u) := 0$ for all literals $u \in L$ and likewise, $m_1(u) := 1$. We extend this by $m_1(w) := m_0(w) := 0$ for all $w \in \mathcal{L}$ such that $w \leq u$ for some $u \in L$, and by $m_0(w) := m_1(w) := 1$ for all $w \in \mathcal{L}$ such that $u \leq w$ for some $u \in L$. For variables $v \in \mathcal{V}$ satisfying $v \leq \neg v$ or $\neg v \leq v$ we have $[v]_\sim \notin P$; in other words, for $[v]_\sim \in P$ the classes $[v]_\sim$ and $[\neg v]_\sim$ are incomparable. Thus, so far, we have not defined m_0 and m_1 on a variable $v \in \mathcal{V}$ and on its negation $\neg v$ at the same time. Of course, fixing a value for a negative literal $\neg v$ to some value implicitly means that we bind the assignment for $v \in \mathcal{V}$ to the opposite value. We complete the definition of m_0 and m_1 by setting them to 0 on every $v \in \mathcal{V}$ where $[v]_\sim \in P$ and they are not yet defined. Moreover, for $v \in \mathcal{V}$ where $[v]_\sim \in F_i$ we put $m_j(v) := i$ for $i, j \in \{0, 1\}$. Obviously, m_0 differs from m_1 only in the variables corresponding to the literals in L, so their Hamming distance is $|L|$ as desired. Besides, both assignments respect the order constraints in $(\mathcal{L}, \leq)$. As these faithfully reflect all original atomic constraints, m_0 and m_1 are indeed models of φ.

Proposition 8. MSD(Γ) *is in* PO *for* $\Gamma \in \mathfrak{L}$ *satisfying* $\Gamma \subseteq \mathrm{iE}_2$ *or* $\Gamma \subseteq \mathrm{iV}_2$.

We only discuss the Horn case ($\Gamma \subseteq \mathrm{iE}_2$), dual-Horn ($\Gamma \subseteq \mathrm{iV}_2$) being symmetric.

Algorithm
Input: A Horn formula φ viewed as a set of Horn clauses.
Output: "≤ 1 model" or the minimal Hamming distance of any two distinct models of φ.
Method:

Step 1: For each variable x in φ, add the clause $(\neg x \vee x)$. Apply the following rules to φ until no more clauses and literals can be removed and no new clauses can be added.

– Unit resolution and unit subsumption: Let $\bar{u}$ denote the complement of a literal u. If the clause set contains a unit clause u, remove all clauses containing the literal u and remove all literals $\bar{u}$ from the remaining clauses.
– Hyper-resolution with binary implications: Resolve all negative literals of a clause simultaneously with binary implications possessing identical premises.

$$\frac{(\neg x \vee y_1) \cdots (\neg x \vee y_k) \ (\neg y_1 \vee \cdots \vee \neg y_k \vee z)}{(\neg x \vee z)} \qquad \frac{(\neg x \vee y_1) \cdots (\neg x \vee y_k) \ (\neg y_1 \vee \cdots \vee \neg y_k)}{(\neg x)}$$

Let $\mathcal{D}$ be the final result of this step. If $\mathcal{D}$ is empty or contains the empty clause, return "≤ 1 model".

Step 2: Let $\mathcal{V}$ be the set of variables occurring in $\mathcal{D}$, and let $\sim \subseteq \mathcal{V}^2$ be the relation defined by $x \sim y$ if $\{\neg x \vee y, \neg y \vee x\} \subseteq \mathcal{D}$. Note that $\sim$ is an equivalence, since the tautological clauses ensure reflexivity and resolution of implications computes their transitive closure. We say that a variable z depends on variables

$y_1, \ldots, y_k$, if $\mathcal{D}$ contains the clauses $\neg y_1 \vee \cdots \vee \neg y_k \vee z$, $\neg z \vee y_1$, $\ldots$, $\neg z \vee y_k$ and $z \not\sim y_i$ holds for all $i = 1, \ldots, k$.

Return $\min\{|X| \mid X \in \mathcal{V}/\!\sim, X$ does not contain dependent variables$\}$ as minimal Hamming distance.

Complexity. The run-time of the algorithm is polynomial in the number of clauses and the number of variables in φ: Unit resolution/subsumption can be applied at most once for each variable, and hyper-resolution has to be applied at most once for each variable x and each clause $\neg y_1 \vee \cdots \vee \neg y_k \vee z$ and $\neg y_1 \vee \cdots \vee \neg y_k$.

Correctness. Let $\mathcal{U}$ be the set of unit clauses removed by subsumption. Adding resolvents and removing subsumed clauses maintains logical equivalence, therefore $\mathcal{D} \cup \mathcal{U}$ is logically equivalent to φ, i.e., both clause sets have the same models. If $\mathcal{D}$ is empty, the unit clauses in $\mathcal{U}$ define a unique model of φ. If $\mathcal{D}$ contains the empty clause, the sets $\mathcal{D}$ and φ are unsatisfiable. Otherwise $\mathcal{D}$ has at least two models, as we will show below. As each model m of $\mathcal{D}$ uniquely extends to a model of φ by defining $m(x) = 1$ for $(x) \in \mathcal{U}$ and $m(x) = 0$ for $(\neg x) \in \mathcal{U}$, the minimal Hamming distances of φ and $\mathcal{D}$ are the same.

We are thus looking for models m_1, m_2 of $\mathcal{D}$ such that the size of the difference set $\Delta(m_1, m_2) = \{x \mid m_1(x) \neq m_2(x)\}$ is minimal. In fact, since the models of Horn formulas are closed under minimum, we may assume $m_1 < m_2$, i.e., we have $m_1(x) = 0$ and $m_2(x) = 1$ for all variables $x \in \Delta(m_1, m_2)$. Indeed, given two models m_2 and m_2' of $\mathcal{D}$, $m_1 = m_2 \wedge m_2'$ is also a model. Since $\mathrm{hd}(m_1, m_2) \leq \mathrm{hd}(m_2, m_2')$ holds, the minimal Hamming distance will occur between models m_1 and m_2 satisfying $m_1 < m_2$.

Note the following facts regarding the equivalence relation $\sim$ and dependent variables.

– If $x \sim y$ then the two variables must have the same value in every model of $\mathcal{D}$ in order to satisfy the implications $\neg x \vee y$ and $\neg y \vee x$. This means that for all models m of $\mathcal{D}$ and all $X \in \mathcal{V}/\!\sim$, we have either $m(x) = 0$ for all $x \in X$ or $m(x) = 1$ for all $x \in X$.

– The dependence of variables is acyclic: If z_i depends on z_{i+1} for $i = 1, \ldots, l$ and z_l depends on z_1, then we have a cycle of binary implications between the variables and thus $z_i \sim z_j$ for all i, j, contradicting the definition of dependence.

– If a variable z depending on y_1, $\ldots$, y_k belongs to a difference set $\Delta(m_1, m_2)$, then at least one of the y_is also has to belong to $\Delta(m_1, m_2)$: $m_2(z) = 1$ implies $m_2(y_j) = 1$ for all $j = 1, \ldots, k$ (because of the clauses $\neg z \vee y_i$), and $m_1(z) = 0$ implies $m_1(y_i) = 0$ for at least one i (because of the clause $\neg y_1 \vee \cdots \vee \neg y_k \vee z$). Therefore $\Delta(m_1, m_2)$ is the union of at least two sets in $\mathcal{V}/\!\sim$, namely the equivalence class of z and the one of y_i.

Hence the difference between any two models cannot be smaller than the cardinality of the smallest set in $\mathcal{V}/\!\sim$ without dependent variables. It remains to show that we can indeed find two such models.

Let X be a set in $\mathcal{V}/\!\sim$ which has minimal cardinality among the sets without dependent variables, and let m_1, m_2 be interpretations defined as follows: (1) $m_1(y) = 0$ and $m_2(y) = 1$ if $y \in X$; (2) $m_1(y) = 1$ and $m_2(y) = 1$ if $y \notin X$ and

$(\neg x \vee y) \in \mathcal{D}$ for some $x \in X$; (3) $m_1(y) = 0$ and $m_2(y) = 0$ otherwise. We have to show that m_1 and m_2 satisfy all clauses in $\mathcal{D}$. Let m be any of these models. $\mathcal{D}$ contains two types of clauses.

Type 1: Horn clauses with a positive literal $\neg y_1 \vee \cdots \vee \neg y_k \vee z$. If $m(y_i) = 0$ for any i, we are done. So suppose $m(y_i) = 1$ for all $i = 1, \ldots, k$; we have to show $m(z) = 1$. The condition $m(y_i) = 1$ means that either $y_i \in X$ (for $m = m_2$) or that there is a clause $(\neg x_i \vee y_i) \in \mathcal{D}$ for some $x_i \in X$. We distinguish the two cases $z \in X$ and $z \notin X$.
Let $z \in X$. If $z \sim y_i$ for any i, we are done for we have $m(z) = m(y_i) = 1$. So suppose $z \not\sim y_i$ for all i. As the elements in X, in particular z and the x_is, are equivalent and the binary clauses are closed under resolution, $\mathcal{D}$ contains the clause $\neg z \vee y_i$ for all i. But this would mean that z is a variable depending on the y_is, contradicting the assumption $z \in X$. Let $z \notin X$, and let $x \in X$. As the elements in X are equivalent and the binary clauses are closed under resolution, $\mathcal{D}$ contains $\neg x \vee y_i$ for all i. Closure under hyper-resolution with the clause $\neg y_1 \vee \cdots \vee \neg y_k \vee z$ means that $\mathcal{D}$ also contains $\neg x \vee z$, whence $m(z) = 1$.

Type 2: Horn clauses with only negative literals $\neg y_1 \vee \cdots \vee \neg y_k$. If $m(y_i) = 0$ for any i, we are done. It remains to show that the assumption $m(y_i) = 1$ for all $i = 1, \ldots, k$ leads to a contradiction. The condition $m(y_i) = 1$ means that either $y_i \in X$ (for $m = m_2$) or that there is a clause $(\neg x_i \vee y_i) \in \mathcal{D}$ for some $x_i \in X$. Let x be some particular element of X. Since the elements in X are equivalent and the binary clauses are closed under resolution, $\mathcal{D}$ contains the clause $\neg x \vee y_i$ for all i. But then a hyper-resolution step with the clause $\neg y_1 \vee \cdots \vee \neg y_k$ would yield the unit clause $\neg x$, which by construction does not occur in $\mathcal{D}$. Therefore at least one y_i is neither in X nor part of a clause $\neg x \vee y_i$ with $x \in X$, i.e., $m(y_i) = 0$.

5.2   Hard Cases

Two Solution Satisfiability. In this section we study the feasibility problem of $\mathsf{MSD}(\Gamma)$ which is, given a Γ-formula φ, to decide if φ has two distinct solutions.

Problem: 2SolutionSAT(Γ)
Input: Conjunctive formula φ over the relations from Γ.
Question: Are there two satisfying assignments $m \neq m'$ of φ?

A priori it is not clear that the tractability of 2SolutionSAT is fully characterized by co-clones. The problem is that the implementation of relations of some language Γ by another language Γ' might not be parsimonious, that is, in the implementation one solution to a constraint might be blown up into several ones in the implementation. Fortunately we can still determine the tractability frontier for 2SolutionSAT by combining the corresponding results for SAT and AnotherSAT.

Lemma 9. *For $\Gamma \in \mathfrak{L}$ where* SAT(Γ) *is* NP-*hard,* 2SolutionSAT(Γ) *is* NP-*hard.*

Proof. Since SAT(Γ) is NP-hard, there must be a relation R in Γ having more than one tuple, because every relation containing only one tuple is at the same time Horn, dual Horn, bijunctive, and affine. Given an instance φ for SAT(Γ), construct φ' as $\varphi \wedge R(y_1, \ldots, y_\ell)$ where ℓ is the arity of R and $y_1, \ldots, y_\ell$ are new variables not appearing in φ. Obviously, φ has a solution if and only if φ' has at least two solutions. Hence, we have proved $\mathsf{SAT}(\Gamma) \leq_m 2\mathsf{SolutionSAT}(\Gamma)$.   □

Lemma 10. *If $\Gamma \in \mathfrak{L}$ and* $\mathsf{AnotherSAT}(\Gamma)$ *is* NP-*hard,* $2\mathsf{SolutionSAT}(\Gamma)$ *is* NP-*hard.*

Proof. Let a formula φ and a satisfying assignment m be an instance of $\mathsf{AnotherSAT}(\Gamma)$. Then φ has a solution other than m if and only if it has two distinct solutions.   □

Lemma 11. *If* $\mathsf{SAT}(\Gamma)$ *and* $\mathsf{AnotherSAT}(\Gamma)$ *are in* P *for* $\Gamma \in \mathfrak{L}$, *then the same holds for* $2\mathsf{SolutionSAT}(\Gamma)$.

Proof. Let φ be an instance of $2\mathsf{SolutionSAT}(\Gamma)$. All polynomial-time decidable cases of $\mathsf{SAT}(\Gamma)$ are constructive, i.e., whenever that problem is polynomial-time decidable, there exists a polynomial-time algorithm computing a satisfying assignment. Thus we can compute in polynomial time a satisfying assignment m of φ. Now use the algorithm for $\mathsf{AnotherSAT}(\Gamma)$ on the instance (φ, m) to decide if there is a second solution to φ.   □

Corollary 12. *For $\Gamma \in \mathfrak{L}$, the problem* $2\mathsf{SolutionSAT}(\Gamma)$ *is polynomial-time decidable if both* $\mathsf{SAT}(\Gamma)$ *and* $\mathsf{AnotherSAT}(\Gamma)$ *are polynomial-time decidable. Otherwise,* $2\mathsf{SolutionSAT}(\Gamma)$ *is* NP-*hard.*

Proposition 13. *For $\Gamma \in \mathfrak{L}$ such that* $2\mathsf{SolutionSAT}(\Gamma)$ *is in* P, *there is a polynomial-time n-approximation algorithm for* $\mathsf{MSD}(\Gamma)$, *where n is the number of variables.*

Proof. Since $2\mathsf{SolutionSAT}(\Gamma)$ is in P, both $\mathsf{SAT}(\Gamma)$ and $\mathsf{AnotherSAT}(\Gamma)$ must be in P by Corollary 12. Since $\mathsf{SAT}(\Gamma)$ is in P, we can compute a model m of the input φ in polynomial time if it exists. Now we check the $\mathsf{AnotherSAT}(\Gamma)$-instance (φ, m). If it has a solution $m' \neq m$, it is also polynomial time computable, and we return (m, m'). If we fail somewhere in this process, then $\mathsf{MSD}(\Gamma)$ does not have feasible solutions; otherwise, $\mathrm{hd}(m, m') \leq n \leq n \cdot \mathrm{OPT}(\varphi)$.   □

MinDistance-Equivalent Cases. In this section we show that, as for the Nearest Other Solution problem (see [4,5]), the affine cases of MSD are MinDistance-complete.

Proposition 14. $\mathsf{MSD}(\Gamma)$ *is* MinDistance-*complete if $\Gamma \in \mathfrak{L}$ satisfies* $\mathrm{iL} \subseteq \langle \Gamma \rangle \subseteq \mathrm{iL}_2$.

Proof. We prove $\mathsf{MSD}(\Gamma) \equiv_{\mathrm{AP}} \mathsf{NearestOtherSolution}(\Gamma)$, which is MinDistance-complete by [4]. As $\Gamma \subseteq \mathrm{iL}_2 = \langle \{\mathrm{even}^4, [x], [\neg x]\} \rangle$, any Γ-formula ψ is expressible as $\exists y (A_1 x + A_2 y = c)$. The projection of the affine solution space is again an affine space, so it can be understood as solutions of a system $Ax = b$. If (ψ, m_0) is an instance of $\mathsf{NSol}(\Gamma)$, then ψ is an $\mathsf{MSD}(\Gamma)$-instance, and a feasible solution $m_1 \neq m_2$ satisfying ψ gives a feasible solution $m_3 := m_0 + (m_2 - m_1)$ for (ψ, m_0), where $\mathrm{hd}(m_0, m_3) = \mathrm{hd}(m_2, m_1)$. Conversely, a solution $m_3 \neq m_0$ to (ψ, m_0) yields a feasible answer to the MSD-instance ψ. Thus, $\mathrm{OPT}(\psi) = \mathrm{OPT}(\psi, m_0)$ and so $\mathsf{NSol}(\Gamma) \leq_{\mathrm{AP}} \mathsf{MSD}(\Gamma)$. The other way round, if ψ is an MSD-instance, then solve the system $Ax = b$ defined by it; let m_0 be a model of ψ. As above we conclude $\mathrm{OPT}(\psi) = \mathrm{OPT}(\psi, m_0)$, and therefore, $\mathsf{MSD}(\Gamma) \leq_{\mathrm{AP}} \mathsf{NSol}(\Gamma)$. □

Tightness Results. It will be convenient to consider the following decision problem, already studied in [5].

Problem: $\mathsf{AnotherSAT}_{<n}(\Gamma)$

Input: A conjunctive formula φ over relations from Γ and an assignment m satisfying φ.

Question: Is there another satisfying assignment m' of φ, different from m, such that $\mathrm{hd}(m, m') < n$, where n is the number of variables of φ?

Note that $\mathsf{AnotherSAT}_{<n}(\Gamma)$ is not compatible with existential quantification. Let $\varphi(y, x_1, \ldots, x_n)$ with the satisfying assignment m be an instance of $\mathsf{AnotherSAT}_{<n}(\Gamma)$ and m' its solution satisfying $\mathrm{hd}(m, m') < n + 1$. Let m_1 and m_1' be the corresponding vectors to m and m', respectively, with the first coordinate truncated. When we existentially quantify the variable y in φ, producing $\varphi_1(x_1, \ldots, x_n) = \exists y \, \varphi(y, x_1, \ldots, x_n)$, then both m_1 and m_1' are solutions of φ', but we cannot guarantee that $\mathrm{hd}(m_1, m_1') < n$. Hence we need the equivalent of Proposition 4 for this problem, whose proof is analogous.

Proposition 15 (Behrisch et al. [4,5]). *For* $\Gamma, \Gamma' \in \mathfrak{L}$ *with* $\Gamma' \subseteq \langle \Gamma \rangle_\wedge$ *we have the reduction* $\mathsf{AnotherSAT}_{<n}(\Gamma') \leq_{\mathrm{m}} \mathsf{AnotherSAT}_{<n}(\Gamma)$.

The following proposition presents only a partial result for $\mathsf{AnotherSAT}_{<n}(\Gamma)$ already proved in [5]. An exhaustive complexity classification of the problem $\mathsf{AnotherSAT}_{<n}(\Gamma)$ has been performed in [4].

Proposition 16 (Behrisch et al. [4,5]). $\mathsf{AnotherSAT}_{<n}(\Gamma)$ *is* NP-*complete for* $\Gamma \in \mathfrak{L}$ *such that* $\mathrm{iN} \subseteq \langle \Gamma \rangle \subseteq \mathrm{iI}$.

Remark 17. It is easy to see that $\mathsf{AnotherSAT}_{<n}(\Gamma)$ is NP-complete for $\mathrm{iI}_0 \subseteq \langle \Gamma \rangle$ and $\mathrm{iI}_1 \subseteq \langle \Gamma \rangle$, since already $\mathsf{AnotherSAT}(\Gamma)$ is NP-complete for these cases, as it was proved in [12]. It is also clear that $\mathsf{AnotherSAT}_{<n}(\Gamma)$ is polynomial-time decidable if Γ is Horn ($\Gamma \subseteq \mathrm{iE}_2$), dual Horn ($\Gamma \subseteq \mathrm{iV}_2$), bijunctive ($\Gamma \subseteq \mathrm{iD}_2$), or affine ($\Gamma \subseteq \mathrm{iL}_2$), just for the same reason as for $\mathsf{AnotherSAT}(\Gamma)$. In all these four Schaefer cases, for each variable x_i we flip the value of $m[i]$, substitute $\overline{m}(x_i)$ for x_i, and construct another satisfying assignment if it exists. Consider now the solutions which we get for every variable x_i. Either there is no solution for any variable, then $\mathsf{AnotherSAT}_{<n}(\Gamma)$ has no solution; or there are only the solutions which are the complement of m, then $\mathsf{AnotherSAT}_{<n}(\Gamma)$ has no solution as well; or else we get a solution m' with $\mathrm{hd}(m, m') < n$, then $\mathsf{AnotherSAT}_{<n}(\Gamma)$ also has a solution. Hence, there is an easy to prove dichotomy result also for $\mathsf{AnotherSAT}_{<n}(\Gamma)$.

We prove that Proposition 13 is essentially tight.

Proposition 18. *For $\Gamma \in \mathfrak{L}$ such that* $\text{iN} \subseteq \langle \Gamma \rangle \subseteq \text{iI}$ *and any* $\varepsilon > 0$ *there is no polynomial-time $n^{1-\varepsilon}$-approximation algorithm for* $\text{MSD}(\Gamma)$, *unless* $\text{P} = \text{NP}$.

Proof. We show that any polynomial time $n^{1-\varepsilon}$-approximation algorithm for $\text{MSD}(\Gamma)$ would also allow to decide $\text{AnotherSAT}_{<n}(\Gamma)$, being NP-complete by Proposition 16, in polynomial time.

The algorithm works as follows. Given an instance (φ, m) for $\text{AnotherSAT}_{<n}(\Gamma)$, the algorithm accepts if m is not a constant assignment. Since Γ is 0-valid (and 1-valid), this output is correct. If φ has only one variable, reject because φ has only two models; otherwise, proceed as follows.

For each variable x of φ, we construct a new formula φ'_x as follows. Let k be the smallest integer greater than $1/\varepsilon$. Introduce $n^k - n$ new variables x^i for $i = 1, \ldots, n^k - n$. For every $i \in \{1, \ldots, n^k - n\}$ and every constraint $R(y_1, \ldots, y_\ell)$ in φ, such that $x \in \{y_1, \ldots, y_\ell\}$, construct a new constraint $R(z_1^i, \ldots, z_\ell^i)$ by $z_j^i = x^i$ if $y_j = x$ and $z_j^i = y_j$ otherwise; add all the newly constructed constraints to φ in order to get φ'_x. Note, that we can extend models s of φ to models s' of φ'_x by setting $s'(x^i) = s(x)$. Now run the $n^{1-\varepsilon}$-approximation algorithm for $\text{MSD}(\Gamma)$ on φ'_x. If for every x the answer is a pair (m_1, m_2) with $m_2 = \overline{m_1}$, then reject, otherwise accept.

This procedure is a correct polynomial-time algorithm for $\text{AnotherSAT}_{<n}(\Gamma)$. For polynomial runtime is clear, it remains to show correctness. If φ has only constant models, then the same is true for every φ'_x. Thus each approximation must result in a pair of complementary constant assignments, and the output is correct. Assume now that there is a model s of φ different from $\mathbf{0}$ and $\mathbf{1}$. Hence, there exists a variable x such that $s(x) = m(x)$. It follows that φ'_x has a model s' for which $\text{hd}(s', m') < n$ holds, where n is the number of variables of φ. But then the approximation algorithm must find two distinct models $m_1 \neq m_2$ of φ'_x satisfying $\text{hd}(m_1, m_2) < n \cdot (n^k)^{1-\varepsilon} = n^{k(1-\varepsilon)+1}$. Since the inequality $k > 1/\varepsilon$ holds, it follows that $\text{hd}(m_1, m_2) < n^k$. Consequently, we have $m_2 \neq \overline{m_1}$ and the output of our algorithm is again correct. $\square$

6 Concluding Remarks

Our problem is in PO for constraints, which are bijunctive, or Horn, or dual Horn. The next complexity stage of the solution structure is characterized by affine constraints. In fact, these constraints represent the error correcting codes used in real-word applications. If we search for arbitrary two satisfying assignments with minimum distance, we can apply standard linear algebra techniques and perform an affine transformation, where we can enforce one of the assignments to be the zero-vector. This is not surprising, since in linear algebra many problems in an affine space can be transformed to the same problems in the corresponding vector space. The penultimate stage of solution structure complexity is represented by constraints, for which the existence of a solution is guaranteed by their definition, but we do not have any other exploitable information. For MSD we need a guarantee of at least two solutions. Our problem belongs to the class poly-APX for these constraints. We can exactly pinpoint the polynomial (n, i.e. arity of the formula) for which we can get a polynomial-time approximation. Our complexity results indicate moreover that we cannot get a suitable approximation for these types of the considered optimization problem. All other cases are not polynomial-time approximable at all. It is interesting to see that our results differ considerably from those of [9] for MaxHammingDistance, asking to produce two satisfying assignments having maximal Hamming distance, even if the two problems are dual.

References

1. Aspvall, B., Plass, M.R., Tarjan, R.E.: A linear-time algorithm for testing the truth of certain quantified Boolean formulas. Inf. Process. Lett. **8**(3), 121–123 (1979)
2. Ausiello, G., Crescenzi, P., Gambosi, G., Kann, V., Marchetti-Spaccamela, A., Protasi, M.: Complexity and Approximation: Combinatorial Optimization Problems and Their Approximability Properties. Springer, New York (1999)
3. Baker, K.A., Pixley, A.F.: Polynomial interpolation and the Chinese Remainder Theorem for algebraic systems. Mathematische Zeitschrift **143**(2), 165–174 (1975)
4. Behrisch, M., Hermann, M., Mengel, S., Salzer, G.: Minimal distance of propositional models (2015). abs/1502.06761
5. Behrisch, M., Hermann, M., Mengel, S., Salzer, G.: Give me another one!. In: Elbassioni, K., Makino, K. (eds.) ISAAC 2015. LNCS, vol. 9472, pp. 664–676. Springer, Heidelberg (2015). doi:10.1007/978-3-662-48971-0_56
6. Böhler, E., Creignou, N., Reith, S., Vollmer, H.: Playing with Boolean blocks, part II: constraint satisfaction problems. SIGACT News **35**(1), 22–35 (2004)
7. Böhler, E., Reith, S., Schnoor, H., Vollmer, H.: Bases for Boolean co-clones. Inf. Process. Lett. **96**(2), 59–66 (2005)
8. Creignou, N., Khanna, S., Sudan, M.: Complexity Classifications of Boolean Constraint Satisfaction Problems. SIAM Monographs on Discrete Mathematics and Applications, vol. 7. SIAM, Philadelphia (2001)
9. Crescenzi, P., Rossi, G.: On the Hamming distance of constraint satisfaction problems. Theor. Comput. Sci. **288**(1), 85–100 (2002)
10. Dumer, I., Micciancio, D., Sudan, M.: Hardness of approximating the minimum distance of a linear code. IEEE Trans. Inf. Theory **49**(1), 22–37 (2003)
11. Jeavons, P., Cohen, D., Gyssens, M.: Closure properties of constraints. J. Assoc. Comput. Mach. **44**(4), 527–548 (1997)
12. Juban, L.: Dichotomy theorem for the generalized unique satisfiability problem. In: Ciobanu, G., Păun, G. (eds.) FCT 1999. LNCS, vol. 1684, pp. 327–337. Springer, Heidelberg (1999)
13. Lagerkvist, V.: Weak bases of Boolean co-clones. Inf. Process. Lett. **114**(9), 462–468 (2014)
14. Schaefer, T.J.: The complexity of satisfiability problems. In: Proceedings of the Tenth Annual ACM Symposium on Theory of Computing, STOC 1978, San Diego, California, pp. 216–226. ACM, New York (1978). http://dx.doi.org/10.1145/800133.804350
15. Schnoor, H., Schnoor, I.: Partial polymorphisms and constraint satisfaction problems. In: Creignou, N., Kolaitis, P.G., Vollmer, H. (eds.) Complexity of Constraints. LNCS, vol. 5250, pp. 229–254. Springer, Heidelberg (2008)
16. Warshall, S.: A theorem on Boolean matrices. J. Assoc. Comput. Mach. **9**(1), 11–12 (1962)

Shortest Reconfiguration of Sliding Tokens on a Caterpillar

Takeshi Yamada[1] and Ryuhei Uehara[1(✉)]

School of Information Science, JAIST, Nomi, Ishikawa, Japan
{tyama,uehara}@jaist.ac.jp

Abstract. For given two independent sets I_b and I_r of a graph, the SLIDING TOKEN problem is to determine if there exists a sequence of independent sets which transforms I_b into I_r so that each independent set in the sequence results from the previous one by sliding exactly one token along an edge in the graph. The SLIDING TOKEN problem is one of the reconfiguration problems that attract the attention from the viewpoint of theoretical computer science. These problems tend to be PSPACE-complete in general, and some polynomial time algorithms are shown in restricted cases. Recently, the problems for finding a *shortest* reconfiguration sequence are investigated. For the 3SAT reconfiguration problem, a trichotomy for the complexity of finding the shortest sequence has been shown; it is in P, NP-complete, or PSPACE-complete in certain conditions. Even if it is polynomial time solvable to decide whether two instances are reconfigured with each other, it can be NP-complete to find a shortest sequence between them. We show nontrivial polynomial time algorithms for finding a shortest sequence between two independent sets for some graph classes. As far as the authors know, one of them is the first polynomial time algorithm for the SHORTEST SLIDING TOKEN PROBLEM that requires detours of tokens.

1 Introduction

Recently, the *reconfiguration problems* attract the attention from the viewpoint of theoretical computer science. The problem arises when we wish to find a step-by-step transformation between two feasible solutions of a problem such that all intermediate results are also feasible and each step abides by a fixed reconfiguration rule. The reconfiguration problems have been studied extensively for several well-known problems, including SATISFIABILITY [7,13], INDEPENDENT SET [8–11,15], SET COVER, CLIQUE, MATCHING [10], and so on.

The reconfiguration problem can be seen as a natural "puzzle" from the viewpoint of recreational mathematics. The *15 puzzle* is one of the most famous classic puzzles, that had the greatest impact on American and European society of any mechanical puzzle the word has ever known in 1880 (Fig. 1; see [18] for its history). It is well known that it has a parity; for any two placements, we can decide if they are reconfigurable or not by the parity. Thus, we can solve the reconfiguration problem in linear time just by checking their parities. Moreover,

© Springer International Publishing Switzerland 2016
M. Kaykobad and R. Petreschi (Eds.): WALCOM 2016, LNCS 9627, pp. 236–248, 2016.
DOI: 10.1007/978-3-319-30139-6_19

Fig. 1. The 15 puzzle, Dad's puzzle, and its Chinese variant.

the distance between any two reconfigurable placements is $O(n^3)$, i.e., we can reconfigure from one to the other in $O(n^3)$ sliding pieces, where the board is of size $n \times n$. However, surprisingly, for these two reconfigurable placements, finding a shortest path is NP-complete in general [17]. That is, although we know that it is $O(n^3)$, finding a shortest one is NP-complete. Another interesting property of the 15 puzzle is in another way of generalization. We have the other famous classic puzzles that can be seen as a generalization from this viewpoint. Namely, while every piece is a unit square in the 15 puzzle, when rectangles are allowed, we have the other classic puzzles, called "Dad puzzle" and its variants (Fig. 1). In 1964, Gardner said that "These puzzles are very much in want of a theory" [6], and Hearn and Demaine gave it after 40 years [8]; they proved that these puzzles are PSPACE-complete using their nondeterministic constraint logic model [9]. That is, the sliding block puzzle is PSPACE-complete in general decision problem, and it is linear time solvable for unit square pieces. However, finding a shortest reconfiguration for unit square pieces is NP-complete. In other words, we can characterize these three complexity classes using the model of the sliding block puzzle.

From the viewpoint of theoretical computer science, one of the most important problems is the 3SAT problem. Recently, for the 3SAT problem, a similar trichotomy for the complexity of finding a shortest sequence has been shown; that is, for the reconfiguration problem, finding a shortest sequence between two satisfiable assignments is in P, NP-complete, or PSPACE-complete in certain conditions [14]. In general, the reconfiguration problems tend to be PSPACE-complete, and some polynomial time algorithms are shown in restricted cases. However, finding a shortest sequence can be a new trend in theoretical computer science because it has a potential to characterize the class NP and gives us a new insight into this class.

Beside the 3SAT problem, one of the most important problems in theoretical computer science is the independent set problem. For this notion, the natural reconfiguration problem is called the SLIDING TOKEN problem introduced by Hearn and Demaine [8]: Suppose that we are given two independent sets $\mathbf{I}_b$ and $\mathbf{I}_r$ of a graph $G = (V, E)$ and imagine that a token is placed on each vertex in $\mathbf{I}_b$. Then, the SLIDING TOKEN problem asks if there exists a sequence $\langle \mathbf{I}_1, \mathbf{I}_2, \ldots, \mathbf{I}_\ell \rangle$ of independent sets of G such that (a) $\mathbf{I}_1 = \mathbf{I}_b$, $\mathbf{I}_\ell = \mathbf{I}_r$, and $|\mathbf{I}_b| = |\mathbf{I}_i|$ for all i with $1 \leq i \leq \ell$; and (b) for each i, $2 \leq i \leq \ell$, there is an edge $\{u, v\}$ in E

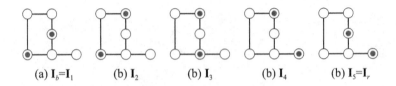

(a) $\mathbf{I}_b = \mathbf{I}_1$ (b) $\mathbf{I}_2$ (b) $\mathbf{I}_3$ (b) $\mathbf{I}_4$ (b) $\mathbf{I}_5 = \mathbf{I}_r$

Fig. 2. A sequence $\langle \mathbf{I}_1, \mathbf{I}_2, \ldots, \mathbf{I}_5 \rangle$ of independent sets of the same graph, where the vertices in independent sets are depicted by small black circles (tokens).

such that $\mathbf{I}_{i-1} \setminus \mathbf{I}_i = \{u\}$ and $\mathbf{I}_i \setminus \mathbf{I}_{i-1} = \{v\}$. Figure 2 illustrates a sequence $\langle \mathbf{I}_1, \mathbf{I}_2, \ldots, \mathbf{I}_5 \rangle$ of independent sets which transforms $\mathbf{I}_b = \mathbf{I}_1$ into $\mathbf{I}_r = \mathbf{I}_5$. Hearn and Demaine proved that the SLIDING TOKEN problem is PSPACE-complete for planar graphs.

For the SLIDING TOKEN problem, some polynomial time algorithms are investigated as follows: Linear time algorithms have been shown for cographs [11] and trees [3]. Polynomial time algorithms are shown for bipartite permutation graphs [5] and claw-free graphs [1]. On the other hand, PSPACE-completeness is also shown for graphs of bounded tree-width [16] and planar graphs [9].

In this context, we investigate for finding a shortest sequence of the SLIDING TOKEN problem, which is called the SHORTEST SLIDING TOKEN problem defined as follows:

Input: A graph $G = (V, E)$ and two independent sets $\mathbf{I}_b, \mathbf{I}_r$ with $|\mathbf{I}_b| = |\mathbf{I}_r|$.

Output: A *shortest* reconfiguration sequence $\mathbf{I}_b = \mathbf{I}_1, \mathbf{I}_2, \ldots, \mathbf{I}_\ell = \mathbf{I}_r$ such that $\mathbf{I}_i$ can be obtained from $\mathbf{I}_{i-1}$ by sliding exactly one token on a vertex $u \in \mathbf{I}_{i-1}$ to its adjacent vertex v along $\{u, v\} \in E$ for each i, $2 \le i \le \ell$.

We note that ℓ is not necessarily in polynomial of $|V|$; this is an issue how we formalize the problem, and if we do not know that ℓ is in polynomial or not. If the length k is given as a part of input, we may be able to decide if $\ell \le k$ in polynomial time even if ℓ itself is not in polynomial. However, if we have to output the sequence itself, it cannot be solved in polynomial time if ℓ is not in polynomial.

In this paper, we will show that the SHORTEST SLIDING TOKEN problem is solvable in polynomial time for the following graph classes [2]:

Proper interval graphs: We first prove that every proper interval graph with two independent sets $\mathbf{I}_b$ and $\mathbf{I}_r$ is a yes-instance if $|\mathbf{I}_b| = |\mathbf{I}_r|$. Furthermore, we can find the ordering of tokens to be slid in a shortest sequence in $O(n)$ time (implicitly), even though there exists an infinite family of independent sets on paths for which any sequence requires $\Omega(n^2)$ length.

Trivially perfect graphs: We then give an $O(n)$-time algorithm for trivially perfect graphs which actually finds a shortest sequence if such a sequence exists. In contrast to proper interval graphs, any shortest sequence is of length $O(n)$ for trivially perfect graphs. Note that trivially perfect graphs form a subclass of cographs, and hence its decision problem can be solved in polynomial time [11].

Caterpillars: We finally give an $O(n^2)$-time algorithm for caterpillars for the SHORTEST SLIDING TOKEN problem. To make self-contained, we first show a linear time algorithm for decision problem that asks if two independent sets can be transformed into each other. (We note that more general result for a tree has been shown [3].) For a yes-instance, we next show an algorithm that finds a shortest sequence between two independent sets.

We here remark that, since the problem is PSPACE-complete in general, an instance of the SLIDING TOKEN problem may require the exponential number of independent sets to transform. In such a case, tokens should make detours to avoid violating to be independent (as shown in Fig. 2). As we will see, caterpillars certainly require to make detours to transform. Therefore, it is remarkable that any yes-instance on a caterpillar requires a sequence of token-slides of polynomial length. This is still open even for a tree; in a tree, we can determine if two independent sets are reconfigurable in linear time due to [3], however, we do not know if the length of the sequence is in polynomial or not.

As far as the authors know, this is the first polynomial time algorithm for the shortest sliding token problem for a graph class that requires detours.

Due to the lack of space, some proofs are omitted, and available in a draft on arXiv [19].

2   Preliminaries

In this section, we introduce some basic terms and notations. In the SLIDING TOKEN problem, we may assume without loss of generality that graphs are simple and connected.

SLIDING TOKEN: For two independent sets I_i and I_j in a graph $G = (V, E)$, if there exists exactly one edge $\{u, v\}$ in G such that $I_i \setminus I_j = \{u\}$ and $I_j \setminus I_i = \{v\}$, then we say that I_j can be obtained from I_i by *sliding* a token on the vertex $u \in I_i$ to its adjacent vertex v along the edge $\{u, v\}$, and denote it by $I_i \vdash I_j$.

A *reconfiguration sequence* between two independent sets I_1 and I_ℓ of G is a sequence $\langle I_1, I_2, \ldots, I_\ell \rangle$ of independent sets of G such that $I_{i-1} \vdash I_i$ for $i = 2, 3, \ldots, \ell$. We denote by $I_1 \vdash^* I_\ell$ if there exists a reconfiguration sequence between I_1 and I_ℓ. We note that a reconfiguration sequence is *reversible*, that is, we have $I_1 \vdash^* I_\ell$ iff $I_\ell \vdash^* I_1$. Thus we say that two independent sets I_1 and I_ℓ are *reconfigurable* into each other if $I_1 \vdash^* I_\ell$. The *length* of a reconfiguration sequence $\mathcal{S}$ is defined as the number of independent sets contained in $\mathcal{S}$. For example, the length of the reconfiguration sequence in Fig. 2 is 5.

The SLIDING TOKEN problem is to determine if two given independent sets I_b and I_r are reconfigurable into each other. We may assume without loss of generality that $|I_b| = |I_r|$; otherwise the answer is clearly "no." In this paper, we will consider the SHORTEST SLIDING TOKEN problem that computes the length of a shortest reconfiguration sequence between two independent sets. Note that the length of a reconfiguration sequence may not be in polynomial of the size of the graph when the sequence may contain detours of tokens.

We always denote by $\mathbf{I}_b$ and $\mathbf{I}_r$ the initial and target independent sets of G, respectively, as an instance of the (SHORTEST) SLIDING TOKEN problem; we wish to slide tokens on the vertices in $\mathbf{I}_b$ to the vertices in $\mathbf{I}_r$. We sometimes call the vertices in $\mathbf{I}_b$ *blue*, and the vertices in $\mathbf{I}_r$ *red*; each vertex in $\mathbf{I}_b \cap \mathbf{I}_r$ is blue *and* red.

Target-assignment: We here give another notation of the SLIDING TOKEN problem to explain our algorithm. Let $\mathbf{I}_b = \{b_1, b_2, \ldots, b_k\}$ be an initial independent set of a graph G. For the sake of convenience, we label the tokens on the vertices in $\mathbf{I}_b$; let t_i be the token placed on b_i for each i, $1 \le i \le k$. Let $\mathcal{S}$ be a reconfiguration sequence between $\mathbf{I}_b$ and an independent set $\mathbf{I}$ of G, and hence $\mathbf{I}_b \vdash^* \mathbf{I}$. Then, for each token t_i, $1 \le i \le k$, we denote by $f_{\mathcal{S}}(t_i)$ the vertex in $\mathbf{I}$ on which the token t_i is placed via the reconfiguration sequence $\mathcal{S}$. Notice that $\{f_{\mathcal{S}}(t_i) \mid 1 \le i \le k\} = \mathbf{I}$.

Let $\mathbf{I}_r$ be a target independent set of G, which is not necessarily reconfigurable from $\mathbf{I}_b$. Then, we call a mapping $g : \mathbf{I}_b \to \mathbf{I}_r$ a *target-assignment* between $\mathbf{I}_b$ and $\mathbf{I}_r$. The target-assignment g is said to be *proper* if there exists a reconfiguration sequence $\mathcal{S}$ such that $f_{\mathcal{S}}(t_i) = g(b_i)$ for all i, $1 \le i \le k$. Note that there is no proper target-assignment between $\mathbf{I}_b$ and $\mathbf{I}_r$ if $\mathbf{I}_b \nvdash^* \mathbf{I}_r$. Therefore, the SLIDING TOKEN problem can be seen as the problem of determining whether there exists at least one proper target-assignment between $\mathbf{I}_b$ and $\mathbf{I}_r$.

Interval graphs and subclasses: The *neighborhood* of a vertex v in a graph $G = (V, E)$ is the set of all vertices adjacent to v, and denoted by $N(v) = \{u \in V \mid \{u, v\} \in E\}$. Let $N[v] = N(v) \cup \{v\}$. Two vertices u and v are called *strong twins* if $N[u] = N[v]$, and *weak twins* if $N(u) = N(v)$. We only consider the graphs without strong twins since only one of them can be used by a token for strong twins.

A graph $G = (V, E)$ with $V = \{v_1, v_2, \ldots, v_n\}$ is an *interval graph* if there exists a set $\mathcal{I}$ of intervals $I_1, I_2, \ldots, I_n$ such that $\{v_i, v_j\} \in E$ iff $I_i \cap I_j \ne \emptyset$ for each i and j with $1 \le i, j \le n$.[1] We call the set $\mathcal{I}$ of intervals an *interval representation* of the graph, and sometimes identify a vertex $v_i \in V$ with its corresponding interval $I_i \in \mathcal{I}$. We denote by $L(I)$ and $R(I)$ the left and right endpoints of an interval $I \in \mathcal{I}$, respectively. That is, we always have $L(I) \le R(I)$ for any interval $I = [L(I), R(I)]$.

We suppose that an interval graph $G = (V, E)$ is given as an input by its interval representation using $O(n)$ space, where $n = |V|$. (An interval representation of G can be found in $O(n + m)$ time [12], where $m = |E|$.) More precisely, G is given by a string of length $2n$ over alphabets $\{L(I_1), L(I_2), \ldots, L(I_n), R(I_1), R(I_2), \ldots, R(I_n)\}$.

An interval graph is *proper* if it has an interval representation such that no interval properly contains another. An interval graph is *trivially perfect* if it has an interval representation such that the relationship between any two intervals is either disjoint or inclusion.

---

[1] In this paper, a bold $\mathbf{I}$ denotes an "independent set," an italic I denotes an "interval," and calligraphy $\mathcal{I}$ denotes "a set of intervals.".

A *caterpillar* $G = (V, E)$ is a tree that consists of two subsets S and L of V as follows. The vertex set S induces a path $(s_1, \ldots, s_k)$ in G, each vertex v in L has degree 1, and its unique neighbor is in S. We call the path $(s_1, \ldots, s_k)$ *spine*, and each vertex in L *leaf*. In this paper, without loss of generality, we assume that $k \geq 2$, $\deg(s_1) \geq 2$, and $\deg(s_k) \geq 2$. It is easy to see that the class of caterpillars is a proper subset of the class of interval graphs.

3 Proper Interval Graphs

We show the main theorem in this section for proper interval graphs. Firstly, the answer of SLIDING TOKEN is always "yes" for connected proper interval graphs. We give a constructive proof of the claim, and it certainly finds a shortest sequence in linear time.

Theorem 1. *For a connected proper interval graph* $G = (V, E)$, *any two independent sets* $\mathbf{I}_b$ *and* $\mathbf{I}_r$ *with* $|\mathbf{I}_b| = |\mathbf{I}_r|$ *are reconfigurable into each other, i.e.,* $\mathbf{I}_b \vdash^* \mathbf{I}_r$. *Moreover, the shortest reconfiguration sequence can be found in polynomial time.*

We give an algorithm which actually finds a shortest reconfiguration sequence between any two independent sets $\mathbf{I}_b$ and $\mathbf{I}_r$ of a connected proper interval graph G. A connected proper interval graph $G = (V, E)$ has a unique interval representation (up to reversal), and we can assume that each interval is of unit length in the representation [4]. Therefore, by renumbering the vertices, we can fix an interval representation $\mathcal{I} = \{I_1, I_2, \ldots, I_n\}$ of G so that $L(I_i) < L(I_{i+1})$ (and $R(I_i) < R(I_{i+1})$) for each i, $1 \leq i \leq n - 1$, and each interval $I_i \in \mathcal{I}$ corresponds to the vertex $v_i \in V$.

Let $\mathbf{I}_b = \{b_1, b_2, \ldots, b_k\}$ and $\mathbf{I}_r = \{r_1, r_2, \ldots, r_k\}$ be any given initial and target independent sets of G, respectively. W.l.o.g., we assume that the blue vertices $b_1, b_2, \ldots, b_k$ are labeled from left to right (according to the unique interval representation $\mathcal{I}$ of G), that is, $L(b_i) < L(b_j)$ if $i < j$; similarly, we assume that the red vertices $r_1, r_2, \ldots, r_k$ are labeled from left to right. Then, we define a target-assignment $g : \mathbf{I}_b \to \mathbf{I}_r$, as follows: for each blue vertex $b_i \in \mathbf{I}_b$

$$g(b_i) = r_i. \tag{1}$$

To prove Theorem 1, it suffices to show that g is proper, and each token takes no detours.

String representation: By traversing the interval representation $\mathcal{I}$ of a connected proper interval graph G from left to right, we can obtain a string $S = s_1 s_2 \cdots s_{2k}$ which is a superstring of both $b_1 b_2 \cdots b_k$ and $r_1 r_2 \cdots r_k$, that is, each letter s_i in S is one of the vertices in $\mathbf{I}_b \cup \mathbf{I}_r$ and s_i appears in S before s_j if $L(s_i) < L(s_j)$. We may assume without loss of generality that $s_1 = b_1$ since the reconfiguration rule is symmetric. If a vertex is in $\mathbf{I}_b \cap \mathbf{I}_r$ as b_i and r_j, then we define that it appears as $b_i r_j$ in S. Then, for each i, $1 \leq i \leq 2k$, we define the *height* $h(i)$ *at* i by the number of blue vertices appeared in the substring $s_1 s_2 \cdots s_i$ minus the number of red vertices appeared in $s_1 s_2 \cdots s_i$. For the sake of notational convenience, we define $h(0) = 0$. Then $h(i)$ can be recursively computed as follows:

$$h(i) = \begin{cases} 0 & \text{if } i = 0; \\ h(i-1) + 1 & \text{if } s_i \text{ is blue}; \\ h(i-1) - 1 & \text{if } s_i \text{ is red.} \end{cases} \tag{2}$$

Note that $h(2k) = 0$ for any string S since $|\mathbf{I}_b| = |\mathbf{I}_r|$.

Using the notion of height, we split the string S into substrings $S_1, S_2, \ldots, S_h$ at every point of height 0, that is, in each substring $S_j = s_{2p+1} s_{2p+2} \cdots s_{2q}$, we have $h(2q) = 0$ and $h(i) \neq 0$ for all i, $2p + 1 \leq i \leq 2q - 1$. Then, the substrings $S_1, S_2, \ldots, S_h$ form a partition of S, and each substring S_j contains the same number of blue and red tokens. We call such a partition the *partition of S at height 0*.

Lemma 1. *Let $S_j = s_{2p+1} s_{2p+2} \cdots s_{2q}$ be a substring in the partition of the string S at height 0. Then, (a) the blue vertices $b_{p+1}, b_{p+2}, \ldots, b_q$ appear in S_j, and their corresponding red vertices $r_{p+1}, r_{p+2}, \ldots, r_q$ appear in S_j; (b) if S_j starts with the blue vertex b_{p+1}, then each blue vertex b_i, $p + 1 \leq i \leq q$, appears in S_j before its corresponding red vertex r_i; and (c) if S_j starts with the red vertex r_{p+1}, then each blue vertex b_i, $p + 1 \leq i \leq q$, appears in S_j after its corresponding red vertex r_i.*

Algorithm: Recall that we have fixed the unique interval representation $\mathcal{I} = \{I_1, I_2, \ldots, I_n\}$ of a connected proper interval graph G so that $L(I_i) < L(I_{i+1})$ for each i, $1 \leq i \leq n - 1$, and each interval $I_i \in \mathcal{I}$ corresponds to the vertex $v_i \in V$. Since all intervals in $\mathcal{I}$ have unit length, the following proposition clearly holds.

Proposition 1. *For two vertices v_i and v_j in G such that $i < j$, there is a path P in G which passes through only intervals (vertices) contained in $[L(I_i), R(I_j)]$. Furthermore, if $I_{i'} \cap I_i = \emptyset$ for some index i' with $i' < i$, no vertex in $v_1, v_2, \ldots, v_{i'}$ is adjacent to any vertex in P. If $I_j \cap I_{j'} = \emptyset$ for some index j' with $j < j'$, no vertex in $v_{j'}, v_{j'+1}, \ldots, v_n$ is adjacent to any vertex in P.*

Let S be the string of length $2k$ obtained from two given independent sets $\mathbf{I}_b$ and $\mathbf{I}_r$ of a connected proper interval graph G, where $k = |\mathbf{I}_b| = |\mathbf{I}_r|$. Let $S_1, S_2, \ldots, S_h$ be the partition of S at height 0. The following lemma shows that the tokens in each substring S_j can always reach their corresponding red vertices. (Note that we sometimes denote simply by S_j the set of all vertices appeared in the substring S_j, $1 \leq j \leq h$.)

Lemma 2. *Let $S_j = s_{2p+1} s_{2p+2} \cdots s_{2q}$ be a substring in the partition of S at height 0. Then, there exists a reconfiguration sequence between $\mathbf{I}_b \cap S_j$ and $\mathbf{I}_r \cap S_j$ such that tokens are slid along edges only in the subgraph of G induced by the vertices contained in $[L(s_{2p+1}), R(s_{2q})]$.*

Proof of Theorem 1. We now give an algorithm for sliding all tokens on the vertices in $\mathbf{I}_b$ to the vertices in $\mathbf{I}_r$. Recall that $S_1, S_2, \ldots, S_h$ are the substrings in the partition of S at height 0. Intuitively, the algorithm repeatedly picks up one substring S_j, and slides all tokens in $\mathbf{I}_b \cap S_j$ to $\mathbf{I}_r \cap S_j$. By Lemma 2 it works locally in each substring S_j, but it should be noted that a token in S_j may be adjacent to another token in S_{j-1} or S_{j+1} at the boundary of the substrings. To avoid this, we define a partial order over the substrings $S_1, S_2, \ldots, S_h$, as follows.

Consider any two consecutive substrings S_j and S_{j+1}, and let $S_j = s_{2p+1}s_{2p+2}\cdots s_{2q}$. Then, the first letter of S_{j+1} is s_{2q+1}. We first consider the case where both s_{2q} and s_{2q+1} are the same color. Then, since s_{2q} and s_{2q+1} are both in the same independent set of G, they are not adjacent. Therefore, by Proposition 1 and Lemma 2, we can deal with S_j and S_{j+1} independently. In this case, we do not define the ordering between S_j and S_{j+1}. We next consider the case where s_{2q} and s_{2q+1} have different colors; in this case, we have to define their ordering. Suppose that s_{2q} is blue and s_{2q+1} is red; then we have $s_{2q} = b_q$ and $s_{2q+1} = r_{q+1}$. By Lemma 2 the token t_q on s_{2q} is slid to left, and the token t_{q+1} will reach r_{q+1} from right. Therefore, the algorithm has to deal with S_j before S_{j+1}. Note that, after sliding all tokens $t_{p+1}, t_{p+2}, \ldots, t_q$ in S_j, they are on the red vertices $r_{p+1}, r_{p+2}, \ldots, r_q$, respectively, and hence the tokens in S_{j+1} are not adjacent to any of them. By the symmetric argument, if s_{2q} is red and s_{2q+1} is blue, S_{j+1} should be dealt with before S_j.

Such an ordering is defined only for two consecutive substrings S_j and S_{j+1}, $1 \le j \le h - 1$. Therefore, the partial order over the substrings is acyclic, and hence there exists a total order consistent with the partial order. The algorithm certainly slides all tokens from $\mathbf{I}_b$ to $\mathbf{I}_r$ according to this total order. Therefore, the target-assignment g defined in Eq. (1) is proper, and hence $\mathbf{I}_b \vdash^* \mathbf{I}_r$.

We now discuss the length of reconfiguration sequences between $\mathbf{I}_b$ and $\mathbf{I}_r$, together with the running time of our algorithm.

Proposition 2. *For two given independent sets $\mathbf{I}_b$ and $\mathbf{I}_r$ of a connected proper interval graph G with n vertices, (1) the ordering of tokens to be slid in a shortest reconfiguration sequence between them can be computed in $O(n)$ time and $O(n)$ space, and (2) a shortest reconfiguration sequence between them can be output in $O(n^2)$ time and $O(n)$ space.*

This proposition also completes the proof of Theorem 1. $\square$

It is remarkable that there exists an infinite family of instances for which any reconfiguration sequence requires $\Omega(n^2)$ length. Simple example is: G is a path $(v_1, v_2, \ldots, v_{8k})$ of length $n = 8k$ for any positive integer k, $\mathbf{I}_b = \{v_1, v_3, v_5, \ldots, v_{2k-1}\}$, and $\mathbf{I}_r = \{v_{6k+2}, v_{6k+4}, \ldots, v_{8k}\}$. In this instance, each token t_i must be slid $\Theta(n)$ times, and hence it requires $\Theta(n^2)$ time to output them all.

4   Caterpillars

The main result of this section is the following theorem.

Theorem 2. *The* SLIDING TOKEN *problem for a caterpillar* $G = (V, E)$ *and two independent sets* $\mathbf{I}_b$ *and* $\mathbf{I}_r$ *of* G *can be solved in* $O(n)$ *time and* $O(n)$ *space, where* $n = |V|$. *Moreover, for a yes-instance, a shortest reconfiguration sequence between them can be output in* $O(n^2)$ *time and* $O(n)$ *space.*

Let $G = (S \cup L, E)$ be a caterpillar with spine S which induces the path $(s_1, \ldots, s_m)$, and leaf set L. We assume that $m \geq 2$, $\deg(s_1) \geq 2$, and $\deg(s_m) \geq 2$. First we show that we can assume that each spine vertex has at most one leaf without loss of generality.

Lemma 3. *For any given caterpillar* $G = (S \cup L, E)$ *and two independent sets* $\mathbf{I}_b$ *and* $\mathbf{I}_r$ *on* G, *there is a linear time reduction from them to another caterpillar* $G' = (S' \cup L', E')$ *and two independent sets* $\mathbf{I}'_b$ *and* $\mathbf{I}'_r$ *such that (1)* G *with* $\mathbf{I}_b$ *and* $\mathbf{I}_r$ *is a yes-instance of the* SLIDING TOKEN *problem if and only if* G' *with* $\mathbf{I}'_b$ *and* $\mathbf{I}_r$ *is a yes-instance of the* SLIDING TOKEN *problem, (2) the maximum degree of* G' *is at most 3, and (3)* $\deg(s_1) = \deg(s_m) = 2$, *where* $m = |S'|$. *In other words, the* SLIDING TOKEN *problem on a caterpillar is sufficient to consider only caterpillars of maximum degree 3.*

Hereafter, we only consider the caterpillars stated in Lemma 3, and we denote the unique leaf of s_i by ℓ_i if it exists. We here introduce a key notion of the problem on these caterpillars that is named *locked* path. Let G and $\mathbf{I}$ be a caterpillar and an independent set of G, respectively. A path $P = (p_1, p_2, \ldots, p_k)$ on G is *locked* by $\mathbf{I}$ iff (a) k is odd and greater than 2, (b) $\mathbf{I} \cap P = \{p_1, p_3, p_5, \ldots, p_k\}$, (c) $\deg(p_1) = \deg(p_k) = 1$ (in other words, they are leaves), and $\deg(p_3) = \deg(p_5) = \cdots = \deg(p_{k-2}) = 2$. This notion is simplified version of a *locked* tree used in [3]. Using the discussion in [3], we obtain the condition for the immovable independent set on a caterpillar:

Theorem 3 ([3]). *Let* G *and* $\mathbf{I}$ *be a caterpillar and an independent set of* G, *respectively. Then we cannot slide any token in* $\mathbf{I}$ *on* G *at all if and only if there exist a set of locked paths* $P_1, \ldots, P_h$ *for some* h *such that* $\mathbf{I}$ *is a union of them.*

The proof can be found in [3], and omitted here. Intuitively, for any caterpillar G and its independent set $\mathbf{I}$, if $\mathbf{I}$ contains a locked path P, we cannot slide any token through the vertices in P. Therefore, P splits G into two subgraphs, and we obtain two completely separated subproblems. Therefore, we obtain the following lemma:

Lemma 4. *For any given caterpillar* $G = (S \cup L, E)$ *and two independent sets* $\mathbf{I}_b$ *and* $\mathbf{I}_r$ *on* G, *there is a linear time reduction from them to another caterpillar* $G' = (S' \cup L', E')$ *and two independent sets* $\mathbf{I}'_b$ *and* $\mathbf{I}'_r$ *such that (1)* G, $\mathbf{I}_b$, *and* $\mathbf{I}_r$ *are a yes-instance of the* SLIDING TOKEN *problem if and only if* G', $\mathbf{I}'_b$, *and* $\mathbf{I}_r$ *are a yes-instance of the* SLIDING TOKEN *problem, and (2) both of* $\mathbf{I}'_b$ *and* $\mathbf{I}'_r$ *contain no locked path.*

Hereafter, without loss of generality, we assume that the caterpillar G with two independent sets $\mathbf{I}_b$ and $\mathbf{I}_r$ satisfies the conditions in Lemmas 3 and 4.

That is, each spine vertex s_i has at most one leaf ℓ_i, s_1 and s_m have one leaf ℓ_1 and ℓ_m, respectively, both of $\mathbf{I}_b$ and $\mathbf{I}_r$ contain no locked path, and $|\mathbf{I}_b| = |\mathbf{I}_r|$. Then, by the result in [3], this is a yes-instance. Thus, it is sufficient to show an $O(n^2)$ time algorithm that finds a shortest reconfiguration sequence between $\mathbf{I}_b$ and $\mathbf{I}_r$.

It is clear that each pair (s_i, ℓ_i) can have at most one token. Therefore, without loss of generality, we can assume that the blue vertices $b_1, b_2, \ldots, b_k$ in $\mathbf{I}_b$ (and the red vertices $r_1, r_2, \ldots, r_k$) are labeled from left to right according to the order (s_1, ℓ_1), (s_2, ℓ_2), ..., (s_m, ℓ_m) of G. Then, by a similar argument for the proper interval graphs, we have $g(b_i) = r_i$ for each i. To prove Theorem 2, it suffices to show that we can move tokens with fewest detours by case analysis.

Now we introduce *direction* of a token t denoted by $dir(t)$ as follows: when t moves from $v_i \in \{s_i, \ell_i\}$ in $\mathbf{I}_b$ to $v_j \in \{s_j, \ell_j\}$ in $\mathbf{I}_r$ with $i < j$, the direction of t is said to be R and denoted by $dir(t) = R$. If $i > j$, it is said to be L and denoted by $dir(t) = L$. If $i = j$, the direction of t is said to be C and denoted by $dir(t) = C$.

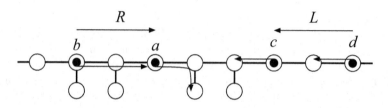

Fig. 3. The most right R token a has to precede the most left L token c.

We first consider a simple case: all directions are either R or L. In this case, we can use the same idea appearing in the algorithm for a proper interval graph in Sect. 3. We can introduce a partial order over the tokens, and move them straightforwardly using the same idea in Sect. 3. Intuitively, a sequence of R tokens are moved from left to right, and a sequence of L tokens are moved from right to left, and we can define a partial order over the sequences of different directions. The only additional considerable case is shown in Fig. 3. That is, when the token a moves to ℓ_i from left and the other token c moves to s_{i+1} from right, a should precede c. It is not difficult to see that this (and its symmetric case) is the only exception than the algorithm in Sect. 3 when all tokens move to right or left.

We next suppose that $\mathbf{I}_b$ (and hence $\mathbf{I}_r$) contains some token t with $dir(t) = C$. In other words, t is put on s_i or ℓ_i for some i in both of $\mathbf{I}_b$ and $\mathbf{I}_r$. We have five cases. Among them, here we consider the most complicated case that t is put on s_i in $\mathbf{I}_b$ and $\mathbf{I}_r$, and ℓ_i does not exist (the other cases are simpler, and omitted here). By assumption, $1 < s < m$ (since ℓ_1 and ℓ_m exist). Without loss of generality, we suppose t is the leftmost spine with the condition. We first observe that $|\mathbf{I}_b \cap \{s_{i-1}, \ell_{i-1}, s_{i+1}, \ell_{i+1}\}|$ is at most 1. Clearly, we have no token on s_{i-1} and s_{i+1}. When we have two tokens on ℓ_{i-1} and ℓ_{i+1}, the path $(\ell_{i-1}, s_{i-1}, s_i, s_{i+1}, \ell_{i+1})$ is a locked path, which contradicts the assumption. We also have $|\mathbf{I}_r \cap \{s_{i-1}, \ell_{i-1}, s_{i+1}, \ell_{i+1}\}| \leq 1$ by the same argument.

Now we consider the most serious case since the other cases are simpler and easier than this case. The most serious case is that $\mathbf{I}_b$ contains ℓ_{i-1} and $\mathbf{I}_r$ contains ℓ_{i+1}. Since any token cannot bypass the other, $\mathbf{I}_b$ contains an L token on ℓ_{i-1}, and $\mathbf{I}_r$ contains an L token on ℓ_{i+1}. In this case, by the L token on ℓ_{i-1}, first, t should make a detour to right, and by the L token in $\mathbf{I}_r$, t next should make a detour to left twice after the first detour. It is clear that this three slides should not be avoided, and this ordering of three slides cannot be violated. Therefore, t itself should slide at least four times to return to the original position, and t can done it in four slides. During this slides, since t is the leftmost spine with this condition, the tokens on $s_1, \ell_1, s_2, \ell_2, \ldots, s_{i-1}, \ell_{i-1}$ do not make any detours. Thus we focus on the tokens on $s_{i+1}, \ell_{i+1}, \ldots$ Let t' be the token that should be on ℓ_{i+1} in $\mathbf{I}_r$. Since t is on s_i, t' is not on $\{s_{i+1}, \ell_{i+1}\}$. If t' is on one of $\ell_{i+2}, s_{i+3}, \ell_{i+3}, s_{i+4}, \ldots$ in $\mathbf{I}_b$, we have nothing to do; just make a detour for only t. The problem occurs when t' is on s_{i+2} in $\mathbf{I}_b$. If there exists ℓ_{i+2}, we first slide t' to it, and it is not difficult to see that this detour for t' is unavoidable. If ℓ_{i+2} does not exist, we have to slide t' to s_{i+3} before slide of t. This can be done immediately except the similar situation that the only considerable case is that we have another L or S token t'' on s_{i+3}. We can repeat this process and confirm that each detour is unavoidable. Since G with $\mathbf{I}_b$ and $\mathbf{I}_r$ contains no locked path, this process will halts. (More precisely, this process will be stuck if and only if these sequence of tokens forms a locked path on G, which contradicts the assumption.) Therefore, traversing this process, we can construct the shortest reconfiguration sequence.

Proof of Theorem 2. Using the algorithm in [3], we can decide if the input is a yes-instance. For a yes-instance, a shortest sequence can be constructed from the case analysis above. For each token, the number of detours made by the token is bounded above by $O(n)$, the number of slides of the token is also bounded above by $O(n)$, and the computation for the token can be done in $O(n)$ time. Therefore, the algorithm runs in $O(n^2)$ time, and the length of the shortest sequence is $O(n^2)$. (We note that, as shown before, there is a simple instance of the problem that requires a shortest sequence of length $\Theta(n^2)$.) □

5   Concluding Remarks

In this paper, we showed that the SHORTEST SLIDING TOKEN problem can be solved in polynomial time for three subclasses of interval graphs. The computational complexity of the problem for chordal graphs, interval graphs, and trees are still open. Especially, tree seems to be the next target from the viewpoint of finding a shortest sequence. We can decide if two independent sets are reconfigurable in linear time [3], then can we find a shortest sequence for a yes-instance? As in the 15-puzzle, finding a shortest one can be NP-hard even if the decision problem is polynomial time solvable. (In fact, the exact analysis for the 15-puzzle has been done up to 4×4 so far.) From the viewpoint of finding any sequence, the next target can be interval graphs. In general, it is an interesting open question whether there is any instance on some graph classes whose reconfiguration sequence requires super-polynomial length.

References

1. Bonsma, P., Kamiński, M., Wrochna, M.: Reconfiguring independent sets in claw-free graphs. In: Ravi, R., Gørtz, I.L. (eds.) SWAT 2014. LNCS, vol. 8503, pp. 86–97. Springer, Heidelberg (2014). arXiv:1403.0359
2. Brandstädg, A., Le, V.B., Spinrad, J.P.: Graph Classes: A Survey. SIAM, Philadelphia (1999)
3. Demaine, E.D., Demaine, M.L., Fox-Epstein, E., Hoang, D.A., Ito, T., Ono, H., Otachi, Y., Uehara, R., Yamada, T.: Linear-time algorithm for sliding tokens on trees. Theor. Comput. Sci. **600**, 132–142 (2015)
4. Deng, X., Hell, P., Huang, J.: Linear-time representation algorithms for proper circular-arc graphs and proper interval graphs. SIAM J. Comput. **25**, 390–403 (1996)
5. Fox-Epstein, E., Hoang, D.A., Otachi, Y., Uehara, R.: Sliding token on bipartite permutation graphs. In: Elbassioni, K., Makino, K. (eds.) ISAAC 2015. LNCS, vol. 9472, pp. 237–247. Springer, Heidelberg (2015). doi:10.1007/978-3-662-48971-0_21
6. Gardner, M.: The hypnotic fascination of sliding-block puzzles. Sci. Am. **210**, 122–130 (1964)
7. Gopalan, P., Kolaitis, P.G., Maneva, E.N., Papadimitriou, C.H.: The connectivity of Boolean satisfiability: computational and structural dichotomies. SIAM J. Computing **38**, 2330–2355 (2009)
8. Hearn, R.A., Demaine, E.D.: PSPACE-completeness of sliding-block puzzles and other problems through the nondeterministic constraint logic model of computation. Theor. Comput. Sci. **343**, 72–96 (2005)
9. Hearn, R.A., Demaine, E.D.: Games, Puzzles, and Computation. A K Peters, Natick (2009)
10. Ito, T., Demaine, E.D., Harvey, N.J.A., Papadimitriou, C.H., Sideri, M., Uehara, R., Uno, Y.: On the complexity of reconfiguration problems. Theor. Comput. Sci. **412**, 1054–1065 (2011)
11. Kamiński, M., Medvedev, P., Milanič, M.: Complexity of independent set reconfigurability problems. Theor. Comput. Sci. **439**, 9–15 (2012)
12. Korte, N., Möhring, R.: An incremental linear-time algorithm for recognizing interval graphs. SIAM J. Computing **18**, 68–81 (1989)
13. Makino, K., Tamaki, S., Yamamoto, M.: An exact algorithm for the Boolean connectivity problem for k-CNF. Theor. Comput. Sci. **412**, 4613–4618 (2011)
14. Mouawad, A.E., Nishimura, N., Pathak, V., Raman, V.: Shortest reconfiguration paths in the solution space of boolean formulas. In: Halldórsson, M.M., Iwama, K., Kobayashi, N., Speckmann, B. (eds.) ICALP 2015. LNCS, vol. 9134, pp. 985–996. Springer, Heidelberg (2015)
15. Mouawad, A.E., Nishimura, N., Raman, V., Simjour, N., Suzuki, A.: On the parameterized complexity of reconfiguration problems. In: Gutin, G., Szeider, S. (eds.) IPEC 2013. LNCS, vol. 8246, pp. 281–294. Springer, Heidelberg (2013)
16. Mouawad, A.E., Nishimura, N., Raman, V., Wrochna, M.: Reconfiguration over tree decompositions. In: Cygan, M., Heggernes, P. (eds.) IPEC 2014. LNCS, vol. 8894, pp. 246–257. Springer, Heidelberg (2014)

17. Ratner, R., Warmuth, M.: Finding a shortest solution for the $N \times N$-extension of the 15-puzzle is intractable. J. Symb. Comp. **10**, 111–137 (1990)
18. Slocum, J.: The 15 Puzzle Book: How it Drove the World Crazy. Slocum Puzzle Foundation, Beverly Hills (2006)
19. Yamada, T., Uehara, R.: Shortest Reconfiguration of Sliding Tokens on a Caterpillar, 1 November 2015. arxiv:1511.00243

Approximation Algorithms

Fast and Simple Local Algorithms for 2-Edge Dominating Sets and 3-Total Vertex Covers

Toshihiro Fujito[1(✉)] and Daichi Suzuki[1]

Department of Computer Science and Engineering,
Toyohashi University of Technology, Toyohashi 441-8580, Japan
fujito@cs.tut.ac.jp

Abstract. A local algorithm is a deterministic (i.e., non-randomized) distributed algorithm in an anonymous port-numbered network running in a constant number of synchronous rounds, and this work studies the approximation performance of such algorithms. The problems treated are b-edge dominating set (b-EDS) that is a multiple domination version of the edge dominating set (EDS) problem, and t-total vertex cover (t-TVC) that is a variant of the vertex cover problem with a clustering property. After observing that EDS and 2-TVC are approximable within 4 and 3, respectively, using a single run of the local algorithm for finding a maximal matching in a bicolored graph, it will be seen that running the maximal matching local algorithm for bicolored graph twice, 2-EDS and 3-TVC can be approximated within factors 2 and 3, respectively.

1 Introduction

In the era of big data, it is almost mandatory to compute solutions an order of magnitude faster than ever before, and sublinear or constant time algorithms are urgently wanted in various areas of computation. It is fortunate meanwhile that the high computation power has become relatively easily accessible nowadays, and it is typically provided by computer networks of large scale. Distributed algorithms of high efficiency can be regarded as lying at the crossing of these demands and supplies, and this paper focuses on such algorithms running in constant time.

A *local algorithm* is a distributed algorithm, under the message-passing model of computation, that runs in a constant number of synchronous communication rounds (An excellent survey on local algorithms can be found in [26]). Here, the same computer network, called *communication graph* $G = (V, E)$, is both the input and the system for solving the problem. Each node of the communication graph is a computational entity having an unlimited computing power. The computation proceeds in rounds, in each of which each node can send and receive messages of unbounded length to and from all of its neighboring nodes (although

T. Fujito—Supported in part by the Kayamori Foundation of Informational Science Advancement and a Grant in Aid for Scientific Research of the Ministry of Education, Science, Sports and Culture of Japan.

M. Kaykobad and R. Petreschi (Eds.): WALCOM 2016, LNCS 9627, pp. 251–262, 2016.
DOI: 10.1007/978-3-319-30139-6_20

the algorithms to be presented use only messages of $O(1)$ length). There are some variants in the communication graph models, and we assume a very weak one among them throughout the paper. It is assumed that a *port numbering* is assigned in G, which means that the edges incident to a node $u \in V$ are uniquely labeled and u can use those labels to choose which neighbors of u it sends messages to and receives from, for all the nodes u in G. While no other information such as *unique identifiers* are available to any nodes, it is also assumed in this paper that G is a graph of bounded degree, and there is a constant Δ such that any node in G has at most Δ neighbors. In this case, every node in G is initially given Δ as the only local input, and must produce the local output of its own, by running some algorithm common to all the other nodes in G. The computing power of distributed algorithms of this sort can be said to be severely limited, and independent sets or matchings, for instance, that can be computed in cycles are empty (vertex or edge) sets only [19]. Nevertheless, some nontrivial results, both positive and negative, are getting accumulated in recent years and the lists of those results can be found in [26].

The main problem treated in the paper is a graph covering problem called *edge dominating set*. In an undirected graph an edge is said to *dominate* itself and all the edges adjacent to it, and a set of edges is an *edge dominating set* (henceforth an *eds*) if the edges in it collectively dominate all the edges in a graph. The *edge dominating set* problem (henceforth *EDS*) asks to find an eds of minimum cardinality. The EDS problem is one of classic NP-complete graph problems, and it was proven to be so even if graphs are planar or bipartite of maximum degree 3 by Yannakakis and Gavril [28]. While the problem was later shown to remain NP-hard under various classes of restricted graphs [17], some polynomially solvable special cases have been also discovered [17,21,24]. Computing the minimum size edge dominating set is equivalent to that of the *minimum maximal matching*, and simply computing any maximal matching is a 2-approximation algorithm for them. Whereas EDS is known to admit a PTAS (polynomial time approximation scheme) for some special cases [4,18], no better approximation algorithm has been found in the general case, and some nontrivial approximation lower bounds have been obtained (under some likely complexity hypothesis) [8,9,23]. The parameterized complexity of EDS has also been exten- sively studied [7,9,10,13,27].

The *b-edge dominating set* problem (henceforth *b-EDS*) is a multi-domination version of EDS, and it is a natural extension of EDS such as the (multi)set multicover and multi-dominating set problems. Here, each edge e of an input graph is associated with an integer $b(e)$, and a solution is required to dominate each e $b(e)$ times (and hence, the ordinary EDS corresponds to the case when $b(e) \equiv 1, \forall e \in E$). Typically two versions of b-EDS can be considered, depending on the types of feasible solutions, where a solution can be an edge multiset (called an *mb-eds* henceforth) in one case, and it has to be an ordinary edge set (called an *sb-eds* henceforth) in the other. The former is named *multiple b-edge dominating set* (henceforth *mb-EDS*) and the latter *simple b-edge dominating set* (henceforth *sb-EDS*). Whereas 8/3-approximation is known possible for the most general type of b-EDS [5], mb-EDS was shown approximable within 2 in linear time [6], and sb-EDS within 2 when $b(e) \leq 3, \forall e \in E$ [14]. The b-EDS problem treated in this paper is 2-EDS, that is the case when $b(e) \equiv 2, \forall e \in E$.

The EDS problem itself has some interesting applications, especially in view of its close relation to minimum maximal matchings, such as telephone switching networking as described in [28], and b-EDS plays an important role in any application of EDS when the fault tolerance and/or robustness need to be taken into account. Another aspect of an eds is that it induces a *vertex cover* where a vertex cover $C \subseteq V$ is a set of nodes such that every edge in G is incident to some node in C; namely, an edge set $D \subseteq E$ is an eds for a graph $G = (V, E)$ if and only if the set of endnodes of the edges in D, denoted $V(D)$, is a vertex cover for G. It is perhaps worth pointing out here that the vertex set $V(D)$ thus induced from an eds D is not a mere vertex cover but with a *clustering* property. A vertex set $C \subseteq V$ is said to be a t-*total vertex cover* $(t \geq 1)$, henceforth a t-*tvc*, for a connected graph G if it is a vertex cover for G such that each connected component of the subgraph of G induced by C has at least t nodes. Hence, if C is a t-tvc, C is a vertex cover and each member of C belongs to a "cluster" containing at least t members of C. The problem of computing a minimum t-tvc is named t-TVC (thus, 1-TVC is the ordinary vertex cover problem). It was introduced in [12,20], and was further studied in [11]. Having such clustering properties could be desirable or required in some applications, and variants with such properties enforced are considered in other combinatorial optimization problems as well, such as r-*gatherings* [1]. It is known that the t-TVC problem is NP-hard, not approximable within $10\sqrt{5} - 21 - \epsilon$ (unless P=NP), and approximable within 2, for each $t \geq 1$ [12].

1.1 Previous Work and Ours

Not so many works are known for EDS in the area of distributed algorithms, and it could be partially due to the fact that, at least under the model of local algorithms considered (i.e., deterministic distributed algorithms in anonymous port-numbered networks running in a constant number of rounds), the case is in a sense settled. It was shown by Suomela that EDS can be approximated within $4-2/\Delta$ in $O(\Delta^2)$ rounds, and the matching lower bound for approximation ratios was obtained at the same time [25]. Moreover, the same lower bound was shown to hold even if each node is provided with a unique identifier [15]. The vertex cover problem is known to be approximable within 2 by a local algorithm [2,3], but nothing is known about the t-TVC problem for $t \geq 2$.

This work is mainly concerned with local algorithms for approximating the 2-EDS problem. It will be shown that, after observing in passing that EDS is approximable within 4 in only 2Δ rounds, m2-EDS is within 2 in the same running time. We then present a local algorithm for s2-EDS, designed by extending that for m2-EDS. Interestingly, approximation becomes easier in either version of 2-EDS than in EDS, and s2-EDS will be shown approximable within 2 in $4\Delta + 2$ rounds. Local algorithms for 2-TVC and 3-TVC are considered as well. It follows from the way vertex covers are constructed by the 3-approximation algorithm of Polishchuk and Suomela [22] that 2-TVC can be approximated within 3 in $2\Delta + 1$ rounds. It will be seen that 3-TVC can be approximated equally well, within the same factor of 3. A 3-tvc is obtained from an s2-eds computed by the previous algorithm, and it will be shown to become no larger than thrice the minimum vertex cover size despite the fact that the s2-eds used is, as constructed by extending an eds, in general larger than the eds used to 3-approximate 2-TVC.

2   Preliminaries

For an edge set $F \subseteq E$ in a graph $G = (V, E)$, $V(F)$ denotes the set of nodes induced by the edges in F (i.e., the set of all the endnodes of the edges in F). For a node set $S \subseteq V$ let $\delta(S)$ denote the set of edges incident to a node in S. When S is an edge set, we let $\delta(S) = \delta(\cup_{e \in S} e)$ where edge e is a set of two nodes; then, $\delta(S)$ also denotes the set of edges dominated by S. When S is a singleton set $\{s\}$, $\delta(\{s\})$ is abbreviated to $\delta(s)$. For a node set $U \subseteq V$, $N(U)$ denotes the set of neighboring nodes of those in U (i.e., $N(U) = \{v \in V \mid \{u, v\} \in E$ for some $u \in U\}$), and $N(u)$ means $N(\{u\})$.

An edge set in G is a *simple 2-matching* if at most two edges in it are incident to any node in G.

3   A Local Algorithm for EDS, M2-EDS, and 2-TVC

For a graph $G = (V, E)$ let $G_D = (V_L \cup V_R, E_D)$ denote the *bipartite double cover* of G, where $V_L = \{u_L \mid u \in V\}, V_R = \{u_R \mid u \in V\}$, and $E_D = \{\{u_L, v_R\}, \{u_R, v_L\} \mid \{u, v\} \in E\}$. Thus, there exist exactly two edges, $\{u_L, v_R\}$ and $\{u_R, v_L\}$, in G_D corresponding to any edge $\{u, v\}$ in G. Let $p : E_D \to E$ denote the function mapping each of $\{u_L, v_R\}$ and $\{u_R, v_L\}$ to $\{u, v\}$.

For any maximal matching $M_D \subseteq E_D$ computed in G_D, let $\tilde{M}$ denote the mapping of M_D into E; that is, $\tilde{M} = \{p(e) \mid e \in M_D\}$. The *multiplicity* of an edge $e \in \tilde{M}$ is the number of edges in M_D corresponding to e, and it is defined by the function $m : \tilde{M} \to \mathbb{N}$ such that $m(e) \overset{\text{def}}{=} \mid p^{-1}(e) \cap M_D \mid$. Clearly, $m(e) \in \{1, 2\}$ for all $e \in \tilde{M}$.

It is rather straightforward to verify that (1) $\tilde{M} \subseteq E$ is a simple 2-matching in G, and (2) $V(\tilde{M}) \subseteq V$ is a vertex cover for G [26]. While $\tilde{M}$ is not necessarily a maximal simple 2-matching in G, this means that $\tilde{M}$ is an edge dominating set for G as well, and we can say more:

Lemma 1. *For any maximal matching M_D in the bipartite double cover G_D of G,*

1. *$\tilde{M} \subseteq E$ is an eds and $(\tilde{M}, m)$ is an m2-eds for G, and*
2. $\displaystyle\sum_{e' \in \delta(e) \cap \tilde{M}} m(e') \leq 4$ *for any $e \in E$,*

where $\tilde{M} = p(M_D)$ *and* $m(e) = | \, p^{-1}(e) \cap M_D \, |$.

Proof. 1. Clearly, each of $\{u_L, v_R\}$ and $\{u_R, v_L\}$ is dominated by M_D in G_D for any edge $\{u, v\}$ of G as M_D is a maximal matching in G_D. Moreover, any edge dominating $\{u_L, v_R\}$ cannot simultaneously dominate $\{u_R, v_L\}$ in G_D, and vice versa, from the way G_D is constructed, for any $\{u, v\} \in E$. Therefore, there exist two different edges in M_D dominating $\{u_L, v_R\}$ and $\{u_R, v_L\}$ in G_D, and both of them appear in $(\tilde{M}, m)$, either as two edges or as a single edge with multiplicity of 2, and hence, $(\tilde{M}, m)$ is an m2-eds for G.

2. Observe that $\displaystyle\sum_{e' \in \delta(e) \cap \tilde{M}} m(e')$ denote the number of edges in M_D dominating either $\{u_L, v_R\}$ or $\{u_R, v_L\}$ for $e = \{u, v\} \in E$. Any edge in G_D is dominated by at most two edges of the matching M_D, and hence, the number of edges in M_D dominating either $\{u_L, v_R\}$ or $\{u_R, v_L\}$ is at most 4 for any $\{u, v\} \in E$. $\square$

An eds $\tilde{M}$ can be computed by the following technique which has been often used in designing local algorithms for various graph problems.

1. A key component of this technique is a simple local algorithm of Hańćkowiak et al. for computing a maximal matching in a bounded-degree bipartite graph G, with color classes L and R, where each node of G is informed of which color class of G it belongs to by the local input [16]. Port numberings are assumed but unique node identifies are not. The algorithm repeatedly performs the following steps for $i = 1, \cdots, \Delta$:
   (a) Any unmatched left node (in L) sends a proposal to its ith neighbor.
   (b) If any unmatched right node (in R) receives a proposal, it accepts the proposal, becomes matched, and informs the proposal sender of its acceptance. In case more than one proposal arrives simultaneously, it accepts the one received from a neighbor with the smallest port number.
   (c) If an unmatched left node (in L) receives a reply of acceptance from its ith neighbor, it becomes matched and halts (Otherwise, it goes on by returning to Step (1a)).
   As Steps (1a) and (1c) can be executed in a single round, a maximal matching in a bipartite graph with the local inputs of color classes can be computed in 2Δ rounds.

2. Observe now that, by simulating the algorithm above on G for the problem of finding a maximal matching in a bipartite graph, one can compute a maximal matching M_D in the bipartite double cover G_D; each node u of G simulates the behavior of both of its copies, the left node u_L and the right node u_R, both inheriting the port numbering of the original node u.
   Once M_D is computed, $\tilde{M}$ is available almost immediately as $\{u, v\} \in \tilde{M}$ iff $\{u_L, v_R\}$ or $\{u_R, v_L\} \in M_D$. The multiplicity of each $e \in \tilde{M}$ is easy to compute as well. For each $u \in V$ matched by $\tilde{M}$, check if both of u_L and u_R are matched by M_D, and if so, check if their mates are the same ($m(e) = 2$ in this case) or not ($m(e) = 1$ in this case).

Mapping M_D to $\tilde{M}$ and setting the multiplicity of each edge in $\tilde{M}$ require no additional communication round, and hence, both $\tilde{M}$ and the multiset $(\tilde{M}, m)$ can be computed in 2Δ rounds.

To analyze the quality of an m2-eds $(\tilde{M}, m)$ computed by the algorithm above, let us consider an integer program formulation of the mb-EDS problem:

$$\min\{x(E) \mid x(\delta(e)) \geq b(e) \text{ and } x_e \in \mathbb{Z}_+, \forall e \in E\},$$

where $x(F) = \sum_{e \in F} x_e$ for $F \subseteq E$, and $\delta(e) = \{e\} \cup \{e' \in E \mid e' \text{is adjacent to} e\}$ for $e \in E$. Replacing the integrality constraints by linear constraints $0 \leq x_e$, we obtain an LP and its dual LP in the following forms:

LP: (P_{eds}) $\min z_P(x) = x(E)$ LP: (D_{eds}) $\max z_D(y) = \sum_{e \in E} b(e) y_e$

subject to: $x(\delta(e)) \geq b(e), \quad \forall e \in E$ subject to: $y(\delta(e)) \leq 1, \quad \forall e \in E$

$\qquad\qquad\quad x_e \geq 0, \quad \forall e \in E$ $\qquad\qquad\quad y_e \geq 0, \quad \forall e \in E$

Let $\tilde{y} \in \mathbb{R}^E$ denote a vector of dual variables for the multiset $(\tilde{M}, m)$ such that

$$\tilde{y}_e = \begin{cases} m(e)/4 & \text{if } e \in \tilde{M} \\ 0 & \text{otherwise} \end{cases}$$

We are ready to show the performance of the algorithm above for approximating EDS and m2-EDS problems:

Theorem 1. *The local algorithm given above computes a 4-approximation to EDS and a 2-approximation to m2-EDS, in 2Δ rounds.*

Proof. By Lemma 1.2 $\tilde{y}$ is dual feasible in LP:(D_{eds}). In case of EDS, $b(e) \equiv 1, \forall e \in E$, and hence, its objective value is

$$z_D(\tilde{y}) = \sum_{e \in \tilde{M}} y_e = \sum_{e \in \tilde{M}} (m(e)/4) \geq \left| \tilde{M}/4 \right|,$$

whereas it is

$$z_D(\tilde{y}) = \sum_{e \in \tilde{M}} 2 y_e = \sum_{e \in \tilde{M}} (m(e)/2) = \left(\sum_{e \in \tilde{M}} m(e) \right) / 2$$

in case of m2-EDS where $b(e) \equiv 2, \forall e \in E$. Therefore, the optimum of EDS is lower bounded by $|\tilde{M}/4|$ and that of m2-EDS by $\left(\sum_{e \in \tilde{M}} m(e) \right) / 2$. $\square$

Clearly, the vertex set $V(\tilde{M})$ is a 2-tvc for G, and it can be computed by each node checking if it is matched by $\tilde{M}$ after $\tilde{M}$ is computed. It is then exactly the algorithm of Polishchuk and Suomela [22], who showed that $V(\tilde{M})$ is no larger than thrice the minimum vertex cover size, and hence,

Corollary 1. *The 2-TVC problem can be approximated within 3 in 2Δ rounds.*

Remark: Better algorithms are known for the vertex cover problem [2,3] as stated in Sect. 1, but their outputs are not necessarily 2-tvc's.

4   A Local Algorithm for S2-EDS and 3-TVC

As was seen already, $\tilde{M} \subseteq E$ computed by the algorithm of Sect. 3 is a simple 2-matching as well as an eds for $G = (V, E)$. It is not necessarily a maximal simple 2-matching, and even if it is so, it doesn't have to be a simple 2-eds.

As observed in the proof of Lemma 1.1, there exist two different edges, say e_1 and e_2, in M_D dominating $\{u_L, v_R\}$ and $\{u_R, v_L\}$ in G_D, for any $\{u, v\} \in E$. When M_D is mapped to $\tilde{M}$, however, these two might become one resulting in a single domination of $\{u, v\}$ in G. More precisely, when $\{u_L, v_R\}$ (or $\{u_R, v_L\}$) is dominated by these two edges e_1 and e_2 in G_D, $\tilde{M}$ dominates $\{u, v\}$ twice as $p(e_1) \neq p(e_2)$. Therefore, $\{u, v\}$ is dominated only once by $\tilde{M}$ if and only if e_1 (e_2, respectively) is the only edge of M_D dominating $\{u_L, v_R\}$ ($\{u_R, v_L\}$, respectively) in G_D and $p(e_1) = p(e_2)$. Formally, let $\tilde{M}$ be divided into $\tilde{M}_1$ and $\tilde{M}_2$ such that $\tilde{M}_2 = \{e \in \tilde{M} \mid p^{-1}(e) \subseteq M_D\}$ and $\tilde{M}_1 = \tilde{M} \setminus \tilde{M}_2 = \{e \in \tilde{M} \mid |p^{-1}(e) \cap M_D| = 1\}$. We can then restate the above argument as follows:

Lemma 2. *An edge e is dominated only once by $\tilde{M}$ in G iff e is dominated only by a single edge of $\tilde{M}_2$ in G.*

It thus suffices to dominate those edges specified in Lemma 2, on top of $\tilde{M}$, to construct an s2-eds. For this purpose, let $V_2 = V(\tilde{M}_2)$ and then, the set of edges subject to additional dominations is exactly $\tilde{M}_2 \cup E_2$, where $E_2 = \{\{u, v\} \in E \mid u \in V_2, v \notin V(\tilde{M})\}$.

To describe the algorithm for dominating those edges in $\tilde{M}_2 \cup E_2$, consider the bipartite graph $G_B = (V_2 \cup F, E_2)$ that we can find once $\tilde{M}$ is computed by the algorithm of Sect. 3, where $F = N(V_2) \setminus V(\tilde{M})$, the set of nodes in the neighborhood of V_2 and unmatched by $\tilde{M}$.

1. Compute a simple 2-matching $\tilde{M} \subseteq E$ by running the algorithm of Sect. 3 on $G = (V, E)$.
2. Compute a maximal matching M_B in G_B with color classes V_2 and F. To do so, we once again use the local algorithm of Hańćkowiak et al. [16]. Each node of G knows if it belongs to $V_2 = V(\tilde{M}_2)$ immediately after $\tilde{M}$ is computed in Step 1, and any node unmatched by $\tilde{M}$ can know if it belongs to F by checking if any of its neighbors belongs to $V(\tilde{M}_2)$ using one additional round.

Clearly, any edge in E_2 is dominated by M_B. On the other hand, there are three cases for $e = \{u, v\} \in \tilde{M}_2$ to consider: (1) $\{u, v\} \subseteq V(M_B)$ (i.e., both u and v matched by M_B), (2) $|\{u, v\} \cap V(M_B)| = 1$ (i.e., only one of them matched), and (3) $\{u, v\} \cap V(M_B) = \emptyset$ (i.e., neither matched). For each $\{u, v\} \in \tilde{M}_2$, u and v can check which is the case, by exchanging messages between them in one round. In cases (1) or (2) $\{u, v\}$ is successfully dominated twice by $\tilde{M} \cup M_B$, whereas it is still dominated only once (by $\{u, v\}$ itself) otherwise. So, we need to pick one additional edge to dominate $\{u, v\}$ in case (3), but picking exactly one edge among those incident to either u or v requires the symmetry breaking in general, and it is hard to do in an anonymous network. Therefore, instead of trying to do so, we let each of u and v to add one edge incident to it, other than $\{u, v\}$, to M_B while dropping $\{u, v\}$ from $\tilde{M}_2$.

3. For any $\{u, v\} \in \tilde{M}_2$, if $\{u, v\} \cap V(M_B) = \emptyset$, each of u and v picks an edge incident to it other than $\{u, v\}$, and adds it to $\tilde{M}_2$ while dropping $\{u, v\}$ from $\tilde{M}_2$. In case when u or v cannot pick any edge other than $\{u, v\}$, then keep it in $\tilde{M}_2$.

Let $\tilde{M}'_2 \subseteq E$ denote the edge set resulting from modifying $\tilde{M}_2$ in Step 3 above, and $\tilde{M}_2$ the original subset of $\tilde{M}$. It is then clear at this point that every edge in $\tilde{M}_2 \cup E_2$ is dominated twice by $\tilde{M}'_2 \cup M_B$, and hence, the output $\tilde{M}_1 \cup \tilde{M}'_2 \cup M_B$ of the algorithm is a valid s2-eds for G, which is computed in $4\Delta + 2$ rounds in total.

It remains to analyze the performance of this algorithm, and it will be based again on the the dual LP:(D_{eds}) of the LP relaxation for m2-EDS. Recall the vector $\tilde{y} \in \mathbb{R}^E$ of dual variables defined for the multiset $(\tilde{M}, m)$ in Sect. 3, and we also use it here as defined in terms of $\tilde{M}_1$ and $\tilde{M}_2$ such that

$$\tilde{y}_e = \begin{cases} 1/4 & \text{if } e \in \tilde{M}_1 \\ 1/2 & \text{if } e \in \tilde{M}_2 \\ 0 & \text{otherwise} \end{cases}$$

By the same reasoning as the one used for an m2-eds $(\tilde{M}, m)$, it can be seen that $\tilde{y}$ is dual feasible in LP:(D_{eds}), and moreover, the solution size $|\tilde{M}_1 \cup \tilde{M}'_2 \cup M_B|$ would be bounded above by twice the objective value of $\tilde{y}$, which is $z_D(\tilde{y}) = \sum_{e \in \tilde{M}} 2y_e$, if it is the case that $|\tilde{M}'_2 \cup M_B| \leq 2|\tilde{M}_2|$. Among the three cases considered earlier for $e = \{u, v\} \in \tilde{M}_2$, two edges of $\tilde{M}'_2 \cup M_B$ can be distinctively associated with e in cases (2) and (3). In case of (1), however, where both u and v are matched by M_B, three edges (two from M_B and $e \in \tilde{M}'_2$) must be balanced with e. To deal with such a case, $\tilde{y}$ is modified as follows. Suppose that both u and v are matched by M_B for $\{u, v\} \in \tilde{M}_2$, and let e_1 and e_2 denote those two edges in M_B matching u and v, respectively. Replace $e = \{u, v\}$ in $\tilde{M}_2$ by these two edges e_1 and e_2, and do this operation for every $e \in \tilde{M}_2$ corresponding to case (1). Each of these operations can be seen to be an augmentation of the matching $\tilde{M}_2$ along an alternating path of length 3, and hence, the resulting edge set $\tilde{M}''_2$ remains as a matching in G_B. Moreover, no edge in M_B touches a node in $V(\tilde{M}_1)$, and therefore, when $\tilde{y}$ is altered to $\tilde{y}'$ such that

$$\tilde{y}'_e = \begin{cases} 1/4 & \text{if } e \in \tilde{M}_1 \\ 1/2 & \text{if } e \in \tilde{M}''_2 \\ 0 & \text{otherwise} \end{cases}$$

it remains dual feasible in LP:(D_{eds}). Since each edge $e \in \tilde{M}_2$ corresponding to case (1) is replaced by e_1 and e_2 in $\tilde{M}''_2$ distinctively, each with the dual value of $1/2$, those three edges associated with e, namely e_1 and e_2 in M_B and e itself, can be accounted for by the values of y_{e_1} and y_{e_2}, in bounding the solution size within a factor 2 of the optimum; or in other words,

$$|\tilde{M}_1 \cup \tilde{M}_2' \cup M_B| \leq 2z_D(\tilde{y}').$$

We may thus conclude:

Theorem 2. *The local algorithm given above computes a 2-approximation to s2-EDS in $4\Delta + 2$ rounds.*

Let us turn our attention to the 3-TVC problem. Each component of the subgraph of G induced by any s2-eds S for G contains at least two edges, and hence, $V(S)$ is always a 3-tvc for G. Therefore, attaching the following step, which requires no additional round of communication, to the above algorithm at the end enables it to compute a 3-tvc for G:

4. For each $u \in V$ check if any edge incident to it belongs to the previous output of $\tilde{M}_1 \cup \tilde{M}_2' \cup M_B$. Set the local output of u as "yes, I'm in a solution" if it does, and "no, I'm not in a solution" otherwise.

So the output of this algorithm is $V(\tilde{M}_1 \cup \tilde{M}_2' \cup M_B)$, and it remains to estimate its size. To do so, consider the following LP relaxation of the vertex cover problem and its dual LP:

LP: (P_{vc}) $\min \sum_{v \in V} x_v$ 　　　　　 LP: (D_{vc}) $\max \sum_{e \in E} y_e$

subject to: $x_u + x_v \geq 1,$ $\forall \{u, v\} \in E$ 　　 subject to: $y(\delta(v)) \leq 1,$ $\forall v \in V$

　　　　　$x_v \geq 0,$ $\forall v \in V$ 　　　　　　　　　　　　$y_e \geq 0,$ $\forall e \in E$

where $y(F) = \sum_{e \in F} y_e$ for $F \subseteq E$.

Recall now the feasible solution $\tilde{y}' \in \mathbb{R}^E$ of LP:(D_{eds}) used in lower bounding the size of a minimum s2-eds, and observe that $\tilde{y}'(\delta(v)) \leq 1/2$ for all $v \in V$. It then means that $2\tilde{y}'$ is feasible in LP:(D_{vc}).

Let us now consider $V(\tilde{M}_1)$ and $V(\tilde{M}_2' \cup M_B)$ separately:

- $\tilde{M}_1$ is a simple 2-matching consisting of paths of length at least 2 and cycles. For every component C of the subgraph induced by $V(\tilde{M}_1)$, let $V(C)$ and $\tilde{M}_1(C)$ denote the sets of nodes and edges in $\tilde{M}_1$ contained in C, respectively. Then, (1) $|\tilde{M}_1(C)| \geq 2$, and (2) $|V(C)| \leq |\tilde{M}_1(C)|+1$. Therefore, when $|V(C)|$ is compared with the duals assigned on the edges of $\tilde{M}_1(C)$, we have

$$\frac{|V(C)|}{\sum_{e \in \tilde{M}_1(C)} 2\tilde{y}'_e} = \frac{|V(C)|}{|\tilde{M}_1(C)|/2} \leq \frac{2k+2}{k} \leq 3.$$

- The edge set $\tilde{M}_2'$ is obtained from the matching $\tilde{M}_2$ by adding more edges than deleted. It should be noted, however, that $V(\tilde{M}_2' \cup M_B)$ remains the same as $V(\tilde{M}_2 \cup M_B)$ because M_B is a maximal matching in G_B. Also recall that nonzero duals are assigned, within G_B, only on the edges in $\tilde{M}_2''$. We here do the case analysis as was done earlier depending on the number of edges in M_B incident to u or v for $\{u, v\} \in \tilde{M}_2$.

Case $\{u, v\} \subseteq V(M_B)$ (i.e., both u and v matched by M_B). In this case both of u and v are matched by two edges of M_B, say e_1 and e_2. It is also the case that both e_1 and e_2 are in $\tilde{M}_2''$ (but $\{u, v\}$ is not). Therefore, the dual value of $2(1/2 + 1/2) = 2$ can be associated with those 4 nodes matched by e_1 and e_2.

Case $|\{u, v\} \cap V(M_B)| = 1$ (i.e., only one of them matched). There exists just one edge, say e, in M_B incident to either u or v. For those 3 nodes, u, v, and another one matched by e, the dual of $2 \times (1/2) = 1$ on $\{u, v\}$ can be associated.

Case $\{u, v\} \cap V(M_B) = \emptyset$ (i.e., neither matched). There are only two nodes to account for in this case, namely, u and v, and the dual of 1 on the edge $\{u, v\}$ can be associated.

In either case the number of nodes is thus bounded by thrice the corresponding dual values.

It follows that the number of nodes in a computed solution is no larger than three times the objective value of dual feasible $2\tilde{y}'$; i.e.,

$$V(\tilde{M}_1 \cup \tilde{M}_2' \cup M_B) \leq 3 \sum_{e \in E} 2\tilde{y}_e'.$$

Therefore, although the 3-tvc $V(\tilde{M}_1 \cup \tilde{M}_2' \cup M_B)$ is in general larger than the 2-tvc $V(\tilde{M})$ as the former is constructed by augmenting the latter into a 3-tvc, it is till within the range of 3-approximation of the minimum vertex cover, and hence,

Theorem 3. *The local algorithm given above computes a 3-approximation to 3-TVC in* $4\Delta + 2$ *rounds.*

Acknowledgments. The authors are very grateful to the anonymous referees for their valuable comments and suggestions.

References

1. Armon, A.: On min-max r-gatherings. Theor. Comput. Sci. **412**(7), 573–582 (2011)
2. Åstrand, M., Floréen, P., Polishchuk, V., Rybicki, J., Suomela, J., Uitto, J.: A local 2-approximation algorithm for the vertex cover problem. In: Keidar, I. (ed.) DISC 2009. LNCS, vol. 5805, pp. 191–205. Springer, Heidelberg (2009)
3. Åstrand, M., Suomela, J.: Fast distributed approximation algorithms for vertex cover and set cover in anonymous networks. In: Proceedings of the Twenty-second Annual ACM Symposium on Parallelism in Algorithms and Architectures, SPAA 2010, pp. 294–302 (2010)
4. Baker, B.S.: Approximation algorithms for NP-complete problems on planar graphs. J. ACM **41**, 153–180 (1994)
5. Berger, A., Fukunaga, T., Nagamochi, H., Parekh, O.: Approximability of the capacitated b-edge dominating set problem. Theor. Comput. Sci. **385**(1–3), 202–213 (2007)

6. Berger, A., Parekh, O.: Linear time algorithms for generalized edge dominating set problems. Algorithmica **50**(2), 244–254 (2008)
7. Binkele-Raible, D., Fernau, H.: Enumerate and measure: improving parameter budget management. In: Raman, V., Saurabh, S. (eds.) IPEC 2010. LNCS, vol. 6478, pp. 38–49. Springer, Heidelberg (2010)
8. Chlebík, M., Chlebíková, J.: Approximation hardness of edge dominating set problems. J. Comb. Optim. **11**(3), 279–290 (2006)
9. Escoffier, B., Monnot, J., Paschos, V.T., Xiao, M.: New results on polynomial inapproximabilityand fixed parameter approximability of edge dominating set. Theor. Comput. Syst. **56**(2), 330–346 (2015)
10. Fernau, H.: EDGE DOMINATING SET: Efficient enumeration-based exact algorithms. In: Bodlaender, H.L., Langston, M.A. (eds.) IWPEC 2006. LNCS, vol. 4169, pp. 142–153. Springer, Heidelberg (2006)
11. Fernau, H., Fomin, F.V., Philip, G., Saurabh, S.: The curse of connectivity: t-total vertex (edge) cover. In: Thai, M.T., Sahni, S. (eds.) COCOON 2010. LNCS, vol. 6196, pp. 34–43. Springer, Heidelberg (2010)
12. Fernau, H., Manlove, D.F.: Vertex and edge covers with clustering properties: complexity and algorithms. J. Discrete Algorithms **7**(2), 149–167 (2009)
13. Fomin, F.V., Gaspers, S., Saurabh, S., Stepanov, A.A.: On two techniques of combining branching and treewidth. Algorithmica **54**(2), 181–207 (2009)
14. Fujito, T.: On matchings and b-edge dominating sets: a 2-approximation algorithm for the 3-Edge dominating set problem. In: Ravi, R., Gørtz, I.L. (eds.) SWAT 2014. LNCS, vol. 8503, pp. 206–216. Springer, Heidelberg (2014)
15. Göös, M., Hirvonen, J., Suomela, J.: Lower bounds for local approximation. J. ACM **60**(5), 1–23 (2013)
16. Hańćkowiak, M., Karoński, M., Panconesi, A.: On the distributed complexity of computing maximal matchings. In: Proceedings of the Ninth Annual ACM-SIAM Symposium on Discrete Algorithms, pp. 219–225 (1998)
17. Horton, J.D., Kilakos, K.: Minimum edge dominating sets. SIAM J. Discrete Math. **6**(3), 375–387 (1993)
18. Hunt III, H.B., Marathe, M.V., Radhakrishnan, V., Ravi, S.S., Rosenkrantz, D.J., Stearns, R.E.: A unified approach to approximation schemes for NP- and PSPACE-hard problems for geometric graphs. In: Proceedings 2nd Annual European Symposium on Algorithms, pp. 424–435 (1994)
19. Linial, N.: Locality in distributed graph algorithms. SIAM J. Comput. **21**(1), 193–201 (1992)
20. Małafiejski, M., Żyliński, P.: Weakly cooperative guards in grids. In: Gervasi, O., Gavrilova, M.L., Kumar, V., Laganá, A., Lee, H.P., Mun, Y., Taniar, D., Tan, C.J.K. (eds.) ICCSA 2005. LNCS, vol. 3480, pp. 647–656. Springer, Heidelberg (2005)
21. Mitchell, S., Hedetniemi, S.: Edge domination in trees. In: Proceedings 8th Southeastern Conference on Combinatorics, Graph Theory, and Computing, pp. 489–509 (1977)
22. Polishchuk, V., Suomela, J.: A simple local 3-approximation algorithm for vertex cover. Inform. Process. Lett. **109**(12), 642–645 (2009)
23. Schmied, R., Viehmann, C.: Approximating edge dominating set in dense graphs. Theor. Comput. Sci. **414**(1), 92–99 (2012)
24. Srinivasan, A., Madhukar, K., Nagavamsi, P., Pandu, C., Pandu Rangan, C., Chang, M.S.: Edge domination on bipartite permutation graphs and cotriangulated graphs. Inform. Process. Lett. **56**, 165–171 (1995)

25. Suomela, J.: Distributed algorithms for edge dominating sets. In: Proceedings of the 29th ACM SIGACT-SIGOPS Symposium on Principles of Distributed Computing, PODC 2010, pp. 365–374 (2010)
26. Suomela, J.: Survey of local algorithms. ACM Comput. Surv. **45**(2), 24–40 (2013)
27. Xiao, M., Kloks, T., Poon, S.-H.: New parameterized algorithms for the edge dominating set problem. Theor. Comput. Sci. **511**, 147–158 (2013)
28. Yannakakis, M., Gavril, F.: Edge dominating sets in graphs. SIAM J. Appl. Math. **38**(3), 364–372 (1980)

Approximation Algorithms for Generalized Bounded Tree Cover

Barun Gorain[1], Partha Sarathi Mandal[2], and Krishnendu Mukhopadhyaya[1]([✉])

[1] Indian Statistical Institute, Kolkata, India
{baruniittg123,krishnendu.mukhopadhyaya}@gmail.com
[2] Indian Institute of Technology Guwahati, Guwahati, India
psm@iitg.ernet.in

Abstract. A tree cover is a collection of subtrees of a graph such that each vertex is a part of at least one subtree. The bounded tree cover problem (BTC) requires to find a tree cover with minimum number of subtrees of bounded weight. This paper considers two generalized versions of BTC. The first problem deals with graphs having multiple weight functions. Two variations called strong and weak tree cover problems are defined. In $strong$ tree cover, every subtree must be bounded with respect to all weight functions, whereas in $weak$ tree cover, each subtree must be bounded with respect to at least one weight function. We consider only metric weight functions. A 4-approximation algorithm for strong tree cover and an $O(\log n)$-approximation algorithm for weak tree cover problem has been proposed. In the second problem, the objective is to find a tree cover where the bounds of the subtrees are not necessarily same. We show that this problem cannot be approximated within a constant factor unless P=NP.

Keywords: Tree cover · TSP · k-MST · Approximation algorithms

1 Introduction

Covering the vertices of a graph with subgraphs like trees, tours or paths is a widely studied topic of research [1,6]. For a given graph $G = (V, E)$, a collection of subtrees (tours/paths) is called a tree (tour/path) cover of G if every vertex of G appears in at least one of the subtrees (tours/paths). This problem has applications in vehicle routing [1,8], where the objective is to serve a set of clients by assigning some vehicles with proper scheduling. The objectives of the problems vary depending on the applications. Two classes of problems are popular in literature. The Min-Max problems [1,2,6] aim to minimize the maximum weight of a subtree in the tree cover where the number of subtrees is given. In the mincover or bounded cover problems [1,6], the maximum weight of a subtree is given. The objective is to find a tree cover with minimum number of subtrees that covers all the vertices of the graph. Both the above tree cover problems are NP-hard [1]. In this paper, we concentrate on bounded tree cover problem.

© Springer International Publishing Switzerland 2016
M. Kaykobad and R. Petreschi (Eds.): WALCOM 2016, LNCS 9627, pp. 263–273, 2016.
DOI: 10.1007/978-3-319-30139-6_21

Definition 1 (Bounded tree cover problem (BTC) [6]). Let $G = (V, E, w)$ be an undirected graph with positive weights. For a given bound $\lambda \geq 0$, the objective is to find a tree cover of G with minimum number of subtrees such that the weight of each subtree is at most λ.

In many applications, multiple weight functions may be associated with a graph [9]. Consider an application, where geographic locations are represented as vertices of a graph. Between a pair of locations several attributes like distance, travel time and associated cost etc. may be associated. These attributes can be represented as multiple weight functions of the graph. Some other problems may also be equivalently restated as BTC of a graph with multiple weight functions. For example, let us consider the following variation of BTC. Let $G = (V, E, w)$ be a weighted graph, λ a positive real number, and p a positive integer. The objective is to find a tree cover of G such that the weight of each subtree is at most λ and each subtree can have at most p vertices. This problem can be reformulated on G by assigning it another weight function. Let w' be another weight function on G, where $w'(e) = 1$, for all $e \in E$. Now, the objective of the problem is to find a tree cover of G such that for each subtree T of the tree cover, $w(T) \leq \lambda$ and $w'(T) \leq p - 1$.

Throughout this paper, we denote the vertex and edge set of a graph G by $V(G)$ and $E(G)$, respectively. For any subgraph H and a weight function w of G, $w(H)$ denotes the sum of the edge weights of H. Without loss of generality, we assume that the input graph G is a complete graph. If G is not complete graph, it can be transformed into its shortest path metric completion $\tilde{G}$. A tree cover of G can be constructed from a tree cover of $\tilde{G}$ by replacing each edge of a subtree in $\tilde{G}$ with the shortest path between the corresponding vertices in G [6]. (v_i, v_j) denotes the edge between two vertices v_i and v_j. A path P of G is denoted by $v_1 v_2 \cdots v_j$, where $v_1, v_2, \cdots, v_j$ are consecutive vertices along P from v_1 to v_j. For any set S, $|S|$ denotes the number of elements in S.

Related Work: Many interesting results have been reported in the literature on tree cover problems. Evan et al. [2] considered the rooted version of Min-Max k-tree cover problem. In this problem a set of vertices called *roots* are given as input with the graph. The objective is to find a tree cover with k subtrees such that the weight of the maximum subtree is minimum. A 4-approximation algorithm is proposed to solve the problem. Nagamochi [7] proposed a $(3 - \frac{2}{k+1})$-approximation algorithm for the Min-Max k-tree cover problem where each subtree of the tree cover has a common root. If the underlying graph is a tree, the authors proposed a $(2 + \epsilon)$-approximation algorithm for the rooted version and a $(2 - \frac{2}{k+1})$-approximation algorithm for the unrooted version of the Min-Max k-tree cover problem, respectively. The objective of the Min-Max tree partition problem is to partition the graph into k equal size vertex sets such that the maximum weight of the minimum spanning trees of each of the vertex sets is minimized. Guttmann-Beck and Hassin [5] proposed $(2k - 1)$-approximation algorithm to solve this problem for a graph with metric weight function. Frederickson et al. [3] considered the k-TSP problem, where the objective is to cover a graph with k tours rooted at a given vertex such that the total weight of the tours is minimized. An $(e + 1 - \frac{1}{k})$-approximation algorithm was proposed, where e is the approximation ratio for the TSP problem.

Arkin et al. [1] proposed a 3-approximation algorithm for BTC. The algorithm computes paths of bounded weight and the set of paths is returned as the set of subtrees of the tree cover. For each k, $1 \leq k \leq n$, minimum spanning forest with k connected components is computed. Then paths of desired weights are calculated from the tours which are computed on each of the components by doubling the edges and shortcutting. Shortcutting is a technique to compute a tour from an Eulerian tour by eliminating duplicate entries of all vertices except the first vertex. The minimum number of subtrees over all iterations is returned as the output of the algorithm. The authors also proposed a 4-approximation algorithm for Min-Max k-tree cover problem. Khani and Salavatipour [6] proposed a 2.5-approximation algorithm for BTC. The proposed algorithm joins the smaller subtrees to reduce the number of trees in the tree cover. The authors also proposed a 3-approximation algorithm for Min-Max k-tree cover problem. These are the best known constant factor algorithms for the above two problems.

Our Contribution: In this paper we consider two variations of BTC. Strong tree cover and weak tree cover problems are introduced on a graph with multiple weight functions. A 4-approximation algorithm is proposed for strong tree cover problem. For weak tree cover problem, an $O(\log n)$-approximation algorithm is proposed for a graph having two weight functions. The same algorithm is extended to work for arbitrary numbers of weight functions. We establish an inapproximability result for BTC when weights of the subtrees of the tree cover are bounded by a set of given bounds which are not necessarily same.

2   Tree Cover for Graphs with Multiple Weight Functions

We introduce strong tree cover and weak tree cover on a graph with multiple weight functions. In the strong tree cover, the subtrees in the tree cover need to be bounded with respect to each of the weight functions. In weak tree cover, the subtrees in the tree cover need to be bounded with respect to at least one of the weight functions. The formal definitions of the problems are given below.

Definition 2 (Strong tree cover problem). *Let G be a complete graph with multiple metric weight functions $w_1, w_2, \ldots, w_l$, $(l \geq 1)$. For given bounds $\lambda_1, \lambda_2, \cdots, \lambda_l \geq 0$, the objective is to find a tree cover with minimum number of subtrees such that for each subtree T in the tree cover, $w_i(T) \leq \lambda_i$, for all $i = 1, 2, \cdots, l$.*

Definition 3 (Weak tree cover problem). *Let G be a complete graph with multiple metric weight functions $w_1, w_2, \cdots, w_l$ $(l \geq 1)$. For given bounds $\lambda_1, \lambda_2, \cdots, \lambda_l \geq 0$, the objective is to find a tree cover with minimum number of subtrees such that for each subtree T in the tree cover, there exists a j, $1 \leq j \leq l$, with $w_j(T) \leq \lambda_j$.*

Both of the above problems are NP-hard, since for $l = 1$, the problems reduce to BTC which is NP-hard [1].

2.1   Strong Tree Cover

In this section, a 4-approximation algorithm for strong tree cover problem is proposed. We define a weight function w' on $G = (V, E, w_1, w_2, \cdots, w_l)$ as follows. For any edge $e \in E(G)$,

$$
w'(e) = \begin{cases} 1 & \text{if there exists a } j, 1 \leq j \leq l, \\ & \text{such that } \lambda_j = 0 \\ \min\left\{\max\left\{\frac{w_1(e)}{\lambda_1}, \frac{w_2(e)}{\lambda_2}, \cdots, \frac{w_l(e)}{\lambda_l}\right\}, 1\right\} & \text{otherwise.} \end{cases}
$$

Note that w' is a metric as the maximum of two metric is a metric and minimum of a metric and a constant is also a metric.

Algorithm 1 (STRONGTREECOVER) is proposed to solve strong tree cover problem on G.

Algorithm 1. STRONGTREECOVER(G)

1: $SOL = \{\}$.
2: Find the minimum spanning tree Γ of G with respect to w'.
3: Delete each edge e from Γ for which $w_j(e) > \lambda_j$, for some j.
    Let $C_1, C_2, \cdots C_h$ be the connected components after deletion of the edges.
4: **for** $i = 1$ to h **do**
5:    Find a tour τ_i on C_i after doubling the edges and shortcutting.
    Let P_i be the path after deleting an edge from τ_i.
6:    **while** $|V(P_i)| > 0$ **do**
7:        $P_i := v_i^1 v_i^2 \cdots, v_i^{|V(P_i)|}$.
8:        Let v_i^j be the first vertex on P_i such that $w'(v_i^1 v_i^2 \cdots v_i^{j+1}) > 1$.
9:        $SOL = SOL \bigcup (v_i^1 v_i^2 \cdots v_i^j)$ and $P_i = P_i \setminus (v_i^1 v_i^2 \cdots v_i^j) \setminus (v_i^j, v_i^{j+1})$.
10:   **end while**
11: **end for**
12: **return** SOL.

Now we analyze the approximation factor of Algorithm 1. Let opt be the number of subtrees in an optimal tree cover. The following lemma gives a lower bound on the optimal solution.

Lemma 1. *If Γ is a minimum spanning tree of G with respect to w' then* $\lceil w'(\Gamma) \rceil \leq 2opt$.

Proof. Let $T_1, T_2, \cdots, T_{opt}$ be the subtrees in the optimal tree cover. Clearly, $w'(T_i) \leq 1$ for $i = 1, 2, \cdots, opt$. Let v_i be a vertex on T_i. Construct a spanning tree H of G by adding the set of edges $\{(v_i, v_{i+1}) | i = 1, 2, \cdots, opt - 1\}$ with $\bigcup_{i=1}^{opt} T_i$. Since $w'(v_i, v_{i+1}) \leq 1$,

$$w'(H) = \sum_{i=1}^{opt} w'(T_i) + \sum_{i=1}^{opt-1} w'(v_i, v_{i+1}) \leq opt + opt - 1 \leq 2opt - 1.$$

As Γ is a minimum spanning tree of G with respect to w', $w'(\Gamma) \leq w'(H) \leq 2opt - 1$. Hence, $\lceil w'(\Gamma) \rceil \leq 2opt$. □

Theorem 1. *The approximation factor of Algorithm 1 is 4.*

Proof. Let opt be the number of subtrees in an optimal solution and $|SOL|$ the number of subtrees calculated by Algorithm 1. Let $C_1, C_2, \cdots, C_h$ be the connected components after deleting $h - 1$ edges in step 3 from the minimum spanning tree Γ. Then,

$$w'(\Gamma) = w'(C_1) + w'(C_2) + \cdots + w'(C_h) + h - 1 \tag{1}$$

Since τ_i is a tour found after doubling the edges in C_i and shortcutting, and w' is a metric, we have

$$w'(\tau_i) \leq 2w'(C_i) \tag{2}$$

According to step 9, each time a subtree is computed from P_i, the weight of modified P_i is reduced by at least one with respect to w'. Thus the number of subtrees computed from τ_i can be at most $\lceil w'(\tau_i) \rceil$. If no edge is deleted from Γ in step 3, then $|SOL| \leq 2 \lceil w'(\Gamma) \rceil \leq 4opt$. When $h \geq 2$, i.e., at least one edge is deleted from Γ in step 3,

$$|SOL| \leq \sum_{i=1}^{h} \lceil w'(\tau_i) \rceil$$

$$\leq \sum_{i=1}^{h} w'(\tau_i) + h$$

$$\leq 2 \sum_{i=1}^{h} w'(C_i) + h \quad (From\ Eq.2)$$

$$\leq 2w'(\Gamma) \quad (From\ Eq.1)$$

$$\leq 4opt \quad (From\ Lemma1)$$

Therefore, the approximation factor of Algorithm 1 is 4. □

2.2   Weak Tree Cover

First we propose an algorithm to solve weak tree cover problem for a graph with two weight functions. Then we extend the algorithm to solve the problem for a graph with any given number of weight functions. We use the 2-approximation algorithm for k-MST [4] as a subroutine in the proposed algorithm. The definition of k-MST problem is given below.

Definition 4 (k-MST problem [4]). *Let $G = (V, E, w)$ be a weighed graph, where the edge weights are positive real numbers and k a given positive integer. The objective is to find a minimum weighted tree of G that spans any k vertices of G.*

Let w_1 and w_2 be two metric weight functions on G. We define two weight functions w_1' and w_2' as follows:

$$w_1'(e) = \begin{cases} \frac{w_1(e)}{\lambda_1} & \text{if } w_1(e) \le \lambda_1, \\ 1 & \text{otherwise.} \end{cases}$$

$$w_2'(e) = \begin{cases} \frac{w_2(e)}{\lambda_2} & \text{if } w_2(e) \le \lambda_2, \\ 1 & \text{otherwise.} \end{cases}$$

Algorithm 3 (WEAKTREECOVER) is proposed to solve weak tree cover problem on G. The algorithm calls the recursive Algorithm 2 (FINDTREE). FIND-TREE returns $O(\log n)$ subtrees, each of which is associated with an integer $i \in \{1, 2\}$. $(\Gamma_1, 1)$ means the subtree Γ_1 is calculated with respect to the weight functions w_1'. Similarly, $(\Gamma_2, 2)$ means the subtree Γ_2 is calculated with respect to the weight functions w_2'. For each of the tuples (Γ', i) returned by FINDTREE, each edge e with $w_i(e) > \lambda_i$ is deleted from Γ'. This deletion may split Γ' into several connected components. Then paths of weights at most λ_i with respect to w_i are calculated from the tours which are computed on each of the components by doubling the edges and shortcutting. This set of paths is returned as the tree cover of G.

Algorithm 2. FINDTREE(G)

1: $n' := |V(G)|$.
2: **if** $n' = 1$ **then**
3:    Return $(G, 1)$.
4: **end if**
5: Find a $\left\lceil \frac{n'}{2} \right\rceil$-MST Γ_1 of G with respect to w_1' using the 2-approximation algorithm [4]. Let G_1 be the induced subgraph with the remaining vertices.
6: Find a $\left\lceil \frac{n'}{2} \right\rceil$-MST Γ_2 of G with respect to w_2' using the 2-approximation algorithm [4]. Let G_2 be the induced subgraph with the remaining vertices.
7: **if** $w_1'(\Gamma_1) \le w_2'(\Gamma_2)$ **then**
8:    Return $(\Gamma_1, 1) \bigcup$ FINDTREE(G_1).
9: **else**
10:    Return $(\Gamma_2, 2) \bigcup$ FINDTREE(G_2).
11: **end if**

Lemma 2. *If $(\Gamma', i) \in S$, then $w_i'(\Gamma') \leq 4opt'$, where opt' is the number of subtrees in the optimal tree cover of G.*

Proof. Since (Γ', i) is returned by FINDTREE, there exists a subgraph G' of G such that Γ' is a subtree spanning $\left\lceil \frac{|V(G')|}{2} \right\rceil$ vertices of G'. Let $opt_{G'}$ be the number of subtrees in the optimal tree cover of G'. Let $T_1, T_2, \cdots, T_{opt_1}, T_{opt_1+1},$ $T_{opt_1+2}, \cdots T_{opt_{G'}}$ be the subtrees in the optimal tree cover of G', where $w_1(T_i) \leq \lambda_1$ for $i = 1, 2, \cdots, opt_1$ and $w_2(T_i) \leq \lambda_2$ for $i = opt_1 + 1, opt_1 + 2, \cdots, opt_{G'}$. Let $v_i \in V(T_i)$. Construct two subgraphs H_1 and H_2 as follows:

$$H_1 = \left(\bigcup_{i=1}^{opt_1} T_i \right) \bigcup \{(v_i, v_{i+1}) | i = 1, 2, \cdots, opt_1 - 1\}$$

$$H_2 = \left(\bigcup_{i=opt_1+1}^{opt_{G'}} T_i \right) \bigcup \{(v_i, v_{i+1}) | i = opt_1 + 1, opt_1 + 2, \cdots, opt_{G'} - 1\}$$

Now,

$$w_1'(H_1) = \sum_{i=1}^{opt_1} w_1'(T_i) + \sum_{i=1}^{opt_1-1} w_1'(v_i, v_{i+1})$$
$$\leq 2opt_1 - 1$$

Therefore, $\lceil w_1'(H_1) \rceil \leq 2opt_1 \leq 2opt_{G'}$. Similarly, $\lceil w_2'(H_2) \rceil \leq 2opt_{G'}$.

Since $\{T_1, T_2, \cdots, T_{opt_1}, T_{opt_1+1}, T_{opt_1+2}, \cdots, T_{opt_{G'}}\}$ is a tree cover of G', either $|V(H_1)| \geq \left\lceil \frac{|V(G')|}{2} \right\rceil$ or $|V(H_2)| \geq \left\lceil \frac{|V(G')|}{2} \right\rceil$.

Without loss of generality let $|V(H_1)| \geq \left\lceil \frac{|V(G')|}{2} \right\rceil$. For $j = 1, 2$, let Γ_j^{opt} be the optimal $\left\lceil \frac{|V(G')|}{2} \right\rceil$-MST of G' with respect to w_j'. Since H_1 is a spanning tree that spans at least $\left\lceil \frac{|V(G')|}{2} \right\rceil$ vertices of G', therefore,

$$\lceil w_1'(\Gamma_1^{opt}) \rceil \leq \lceil w_1'(H_1) \rceil \leq 2opt_{G'}$$

Let Γ_1 and Γ_2 be the $\left\lceil \frac{|V(G')|}{2} \right\rceil$-MST of G' with respect to w_1' and w_2', respectively computed using 2-approximation algorithm [4] (Ref. step 5 and step 6 of FINDTREE). Then, $w_1'(\Gamma_1) \leq 2w_1'(\Gamma_1^{opt})$.

Therefore,

$$\lceil w_i'(\Gamma') \rceil = \min\{\lceil w_1(\Gamma_1) \rceil, \lceil w_2(\Gamma_2) \rceil\} \leq 2 \lceil w_1'(\Gamma_1^{opt}) \rceil \leq 4opt_{G'}$$

Since $opt_{G'} \leq opt'$, therefore $\lceil w_i'(\Gamma') \rceil \leq 4opt'$. $\square$

Algorithm 3. WEAKTREECOVER(G, w_1, w_2)

1: S=FINDTREE(G).
2: $SOL'=\{\}$.
3: **for** each $(\Gamma', p) \in S$ **do**
4:    Delete every edge e from Γ' for which $w_p(e) > \lambda_p$.
5:    Let $C_1, C_2, \cdots C_h$ be the connected components after deletion of the edges.
6:    **for** $i = 1$ to h **do**
7:       Find a tour τ_i from C_i after doubling the edges and shortcutting.
8:       Let P_i be the path after deleting any edge from τ_i.
9:       **while** $|V(P_i)| > 0$ **do**
10:         $P_i := v_i^1 v_i^2 \cdots v_i^{|V(P_i)|}$.
11:         Let v_i^j be the first vertex on P_i such that $w_p(v_i^1 v_i^2 \cdots v_i^{j+1}) > \lambda_p$.
12:         $SOL' = SOL' \bigcup (v_i^1 v_i^2 \cdots v_i^j)$ and $P_i = P_i \setminus (v_i^1 v_i^2 \cdots v_i^j) \setminus (v_i^j, v_i^{j+1})$.
13:       **end while**
14:    **end for**
15: **end for**
16: Return SOL'.

Theorem 2. *The approximation factor of Algorithm 3 is $O(\log n)$, where n is the number of vertices of G.*

Proof. Let $(\Gamma', i) \in S$. By Lemma 2, $\lceil w_i'(\Gamma') \rceil \leq 4opt'$. In step 5 through step 14 of Algorithm 3, a set of subtrees are computed from Γ'. Let the total number of subtree computed from Γ' be N_1. If no edge from Γ' is deleted in step 5 of Algorithm 3, i.e., $h = 1$ then $N_1 \leq 2 \left\lceil \frac{w_i(\Gamma')}{\lambda_i} \right\rceil = \lceil w_i'(\Gamma') \rceil \leq 4opt'$.

Consider the case when at least one edge is deleted from Γ', i.e., $h \geq 2$. After the deletion of $h-1$ edges, Γ' splits into h connected components $C_1, C_2, \cdots, C_h$. The corresponding tours $\tau_1, \tau_2, \cdots, \tau_h$ are computed after doubling the edges of each component and shortcutting.

Therefore, $N_1 \leq \sum\limits_{j=1}^{h} \lceil w_i'(\tau_j) \rceil \leq \sum\limits_{j=1}^{h} 2w_i'(C_j) + h$.

Also, $w_i'(\Gamma') = \sum\limits_{j=1}^{h} w_i'(C_j) + h - 1$. Therefore,

$$N_1 \leq \sum_{j=1}^{h} 2w_i'(C_j) + h$$
$$\leq 2w_i'(\Gamma')$$
$$\leq 8opt'$$

Since the total number of subtrees returned by FINDTREE is $O(\log n)$, therefore $|SOL'| \leq O(\log n) \cdot 8opt'$.

Hence, the approximation factor of Algorithm 3 is $O(\log n)$.

$\square$

Algorithm 3 can be extended to work for a graph with arbitrary number of weight functions. Let $G = (V, E, w_1, w_2, \cdots, w_l)$ be the given graph with bounds $\lambda_1, \lambda_2, \cdots, \lambda_l$. For each i, $1 \leq i \leq l$, we define w_i' as follows:

$$w_i'(e) = \begin{cases} \frac{w_i(e)}{\lambda_i} & \text{if } w_i(e) \leq \lambda_i, \\ 1 & \text{otherwise.} \end{cases}$$

Following modification on FINDTREE subroutine is made for the extended version of Algorithm 3. The spanning trees $\Gamma_1, \Gamma_2, \cdots, \Gamma_l$ are computed with respect to $w_1', w_2', \cdots, w_l'$, respectively such that each Γ_i spans any $\left\lceil \frac{|V(G)|}{l} \right\rceil$ vertices of G. FINDTREE returns $(\Gamma_j, j) \bigcup$ FINDTREE(G_j) where j $(1 \leq j \leq l)$ is the index such that $w_j'(\Gamma_j) = \min\{w_i'(\Gamma_i) | i = 1, 2, \cdots, l\}$.

Since in each call of FINDTREE, the number of vertices of the graph is reduced by a fixed fraction of $\frac{1}{l}$, FINDTREE returns $O(\log n)$ number of subtrees in S. Using arguments similar to those used in Theorem 2, one can show that the approximation factor of this modified Algorithm 3 is $O(\log n)$.

3   Tree Cover with Different Bounds

In some practical applications it may be required that the subtrees of the tree cover bounded by different limits. For example, consider a service provider having a set vehicles with different speed limits. Vehicle V_i can travel maximum distance D_i in time t. With these vehicles the service provider can provide service to a set of customers who are located at different locations. Suppose there are multiple service requests from different customers at a time instance for providing service within time t. Here the goal of the service provider is to schedule minimum number of vehicles such that all the customers will be served within that time. The problem can be formulated as BTC with different bounds. Formally, the tree cover problem with different bounds (TCDB) may be defined as:

Definition 5 (TCDB). *Let $G = (V, E, w)$ be a weighted graph and $\lambda_1, \lambda_2, \cdots, \lambda_n \geq 0$ given real numbers. The objective is to find a tree cover $\{T_{i_1}, T_{i_2}, \cdots, T_{i_p}\}$ with minimum number of subtrees such that $w(T_{i_1}) \leq \lambda_{i_1}$, $w(T_{i_2}) \leq \lambda_{i_2}$, $\cdots$, $w(T_{i_p}) \leq \lambda_{i_p}$, where $i_j \in \{1, 2, \cdots, n\}$ and $i_j \neq i_q$ for $j \neq q$ for $j, q = 1, 2, \cdots, p$.*

We prove that there does not exist any constant factor approximation algorithm for TCDB unless P=NP. Arkin et al. [1] proved the hardness of the following decision version of BTC.

Definition 6 (k-BTC [1]). *Given a graph $G = (V, E, w)$, a real number $\lambda \geq 0$ and a positive integer k, whether there is a tree cover of G with k subtrees such that the weight of each subtree is at most λ.*

Arkin et al. proved that k-BTC is NP-hard for $k = \frac{|V|}{3}$. In [5], the authors proved the hardness of the following decision version of the Min-Max tree partitioning problem.

Definition 7. *Given a graph $G = (V, E, w)$ and a real number $\lambda \geq 0$, whether G can be partitioned into two subtrees T_1 and T_2 such that $V(T_1) = V(T_2) = \frac{V(G)}{2}$ and $w(T_1) \leq \lambda$, $w(T_2) \leq \lambda$.*

It can be shown that k-BTC is NP-hard even for $k = 2$ using the same reduction technique which is used to prove the hardness of the Min-Max tree partitioning problem.

Theorem 3. *It is not possible to design any constant factor approximation algorithm for TCDB, unless P=NP.*

Proof. We propose a polynomial time reduction from an instance of 2-BTC to an instance of TCDB. We show that if there exist a constant factor approximation algorithm for TCDB then 2-BTC can be decided in polynomial time. If possible, suppose there exists a z-approximation algorithm $\mathcal{A}$ for TCDB. Without loss of generality we assume that z is a positive integer. We consider an instance of 2-BTC (G_1, λ), where $G_1 = (V_1, E_1, w_1)$ is a complete weighted graph with $n > 2z$ vertices. The weight w_1 of each edge is integer and satisfies triangle inequality. We construct a complete graph $G_2 = (V_2, E_2, w_2)$ with n^2 vertices from G_1. For each vertex $v_i \in V_1$, consider n vertices $v_i^1, v_i^2, \cdots v_i^n$ in V_2. Define weight of an edge $(v_i^l, v_j^k) \in E_2$ as $w_2(v_i^l, v_j^k) = w_1(v_i, v_j)$ for $i \neq j$ and $w_2(v_i^l, v_i^k) = \frac{1}{2n(n-1)}$.

We show that the graph G_1 has a tree cover with two subtrees having weights at most λ iff G_2 has a tree cover with two subtrees having weights at most $\lambda + \frac{1}{2}$. Let $\{T_1, T_2\}$ be a tree cover of G_1 with $w_1(T_1) \leq \lambda$ and $w_1(T_2) \leq \lambda$. Let $v_1, v_2, \cdots, v_q$ be the vertices of $V(T_1)$. We compute a subtree of T_1' of G_2 as follows. For every edge $(v_x, v_y) \in E(T_1)$, add (v_x^1, v_y^1) in T_1'. Then add the edges $\{v_f^i, v_f^{i+1} | i = 1$ to $n-1$, $v_f \in V(T_1)\}$. Similarly, T_2' can be computed from T_2. It can be easily verified that $\{T_1', T_2'\}$ is a tree cover of G_2 with $w_2(T_1') \leq \lambda + \frac{1}{2}$ and $w_2(T_2') \leq \lambda + \frac{1}{2}$.

Conversely, let $\{T_1', T_2'\}$ be a tree cover of G_2 with $w_2(T_1') \leq \lambda + \frac{1}{2}$ and $w_2(T_2') \leq \lambda + \frac{1}{2}$. A subtree T_1 of G_1 from T_1' is constructed as follows: remove all the edges of type (v_i^k, v_i^l) from T_1'. The remaining edges of T_1' are of the form (v_i^k, v_j^l) for $i \neq j$. For each and every such edge (v_i^k, v_j^l) of T_1' we consider the corresponding edge (v_i, v_j) in G_1 and construct a graph $\mathcal{G}$. Compute the minimum spanning tree T_1 of $\mathcal{G}$. Note that $w_1(T_1) \leq \lambda + \frac{1}{2}$. Since the edge weights of G_1 are integer, $w_1(T_1) \leq \lambda$. Similarly, T_2 can be constructed from T_2'.

We consider an instance of TCDB as G_2 with n^2 bounds $\{\lambda + \frac{1}{2}, \lambda + \frac{1}{2}, 0 \cdots, 0\}$. Let y be the number of subtree in the tree cover returned by algorithm $\mathcal{A}$ on G_2.

Case 1: $(y \leq 2z)]$ We show that there is a tree cover of G_1 with two subtrees having weights at most λ. In this case, among these y subtrees, $y - 2$ are singleton vertices. The other two subtrees of weight $\lambda + \frac{1}{2}$ cover at least $n^2 - 2z + 2$ vertices of G_2. Since $n^2 - 2z + 2 > n^2 - n$, at least one vertex from each of the set $\{v_i^1, v_i^2, \cdots, v_i^n\}$ of G_2 must be covered by at least one of the subtrees having weights at most $\lambda + \frac{1}{2}$. From these two subtrees, two subtrees of G_1 with weight at most λ can be computed in the similar way as explained in the previous paragraph. Hence, G_1 has a tree cover with two subtrees having weights at most λ.

Case 2: $(y > 2z)]$ As $\mathcal{A}$ is a z-approximation algorithm, we can say that there is no tree cover of G_2 with two subtrees having weight at most $\lambda + \frac{1}{2}$. This implies that there does not exist any tree cover of G_1 with two subtrees having weight at most λ.

Hence, 2-BTC can be decided in polynomial time by applying $\mathcal{A}$ on G_2, which is a contradiction unless P=NP. Hence the statement of the theorem follows. $\square$

4 Conclusion

Some variations of bounded tree cover problem have been discussed in this paper. We have introduced strong tree cover and weak tree cover for graphs with multiple weight functions. A 4-approximation algorithm is proposed for strong tree cover and an $O(\log n)$-approximation algorithm is proposed for weak tree cover. An inapproximability result is established for bounded tree cover problem where the subtrees are bounded by a set of given bounds which are not necessarily same.Designing constant factor approximation algorithm for the weak tree cover may an interesting open problem to be investigated in the future.

References

1. Arkin, E.M., Hassin, R., Levin, A.: Approximations for minimum and min-max vehicle routing problems. J. Algorithms **59**(1), 1–18 (2006)
2. Even, G., Garg, N., Könemann, J., Ravi, R., Sinha, A.: Min-max tree covers of graphs. Oper. Res. Lett. **32**(4), 309–315 (2004)
3. Frederickson, G.N., Matthew S. Hecht, Kim, C.E.: Approximation algorithms for some routing problems. In: 17th Annual Symposium on Foundations of Computer Science, pp. 216–227, October 1976
4. Garg, N.: Saving an epsilon: a 2-approximation for the k-MST problem in graphs. In: Gabow, H.N., Ronald, F. (eds). Proceedings of the 37th Annual ACM Symposium on Theory of Computing, Baltimore May 22–24, pp. 396–402. ACM (2005)
5. Guttmann-Beck, N., Hassin, R.: Approximation algorithms for min-max tree partition. J. Algorithms **24**(2), 266–286 (1997)
6. Khani, M.R., Salavatipour, M.R.: Improved approximation algorithms for the min-max tree cover and bounded tree cover problems. Algorithmica **69**(2), 443–460 (2014)
7. Nagamochi, H.: Approximating the minmax rooted-subtree cover problem. IEICE Trans. **88**(5), 1335–1338 (2005)
8. Nagarajan, V., Ravi, R.: Approximation algorithms for distance constrained vehicle routing problems. Networks **59**(2), 209–214 (2012)
9. Rocklin, M., Pinar, A.: On clustering on graphs with multiple edge types. Internet Math. **9**(1), 82–112 (2013)

Approximation Algorithms for Three Dimensional Protein Folding

Dipan Lal Shaw[1,2](✉), A.S.M. Shohidull Islam[3], Shuvasish Karmaker[1],
and M. Sohel Rahman[1]

[1] ALEDA Group, Department of CSE, BUET, Dhaka 1205, Bangladesh
dshaw003@ucr.edu, msrahman@cse.buet.ac.bd
[2] University of California, Riverside, CA 92521, USA
[3] Department of Computational Engineering and Science,
McMaster University, Hamilton, ON, Canada
sohanas@mcmaster.ca

Abstract. Predicting protein secondary structures using lattice models is one of the most studied computational problems in bioinformatics. Here the secondary structure or three dimensional structure of a protein is predicted from its amino acid sequence. The secondary structure refers to local sub-structures of a protein. Mostly founded secondary structures are alpha helix and beta sheets. Simplified energy models have been proposed in the literature on the basis of interaction of amino acid residues in proteins. Here we use well researched Hydrophobic-Polar (HP) energy model. In this paper, we propose the hexagonal prism lattice with diagonals that can overcome the problems of other lattice structures, e.g., parity problem. We give two approximation algorithms for protein folding on this lattice. Our first algorithm leads us to a similar structure of helix structure that is commonly found in a protein structure. This motivates us to propose the next algorithm with a better approximation ratio.

1 Introduction

The HP model [6] is a widely used theoretical model for determining the protein structure from its amino acid sequence. The model assumes that, there are only two types of amino acids instead of twenty different types: one type of amino acid is hydrophobic or non-polar and the other is hydrophilic or polar. In this paper, hydrophobic and hydrophilic amino acids are denoted as H and P, respectively. The model also assumes that, hydrophobicity or contacts between two hydrophobic amino acid, i.e., H-H contacts, is the dominant force in protein folding. Hence, the energy function of this model depicts the idea that for optimal embedding, our main task is to maximize the H-H contacts. Although the HP model abstracts away many details of the folding process, an optimal or near-optimal solution in the HP model can give valuable insights into the possible structure of the protein.

M.S. Rahman—Supported by a ACU Titular Fellowship.

M. Kaykobad and R. Petreschi (Eds.): WALCOM 2016, LNCS 9627, pp. 274–285, 2016.
DOI: 10.1007/978-3-319-30139-6_22

Crescenzi et al. proved that protein folding problem in HP model is NP-hard [5]. The authors used 2D square lattice and proved this by reducing the Hamiltonian cycle problem to this problem. In 1998, Leighton and Berger proved the NP-Completeness of protein folding in 3D cubic lattice [2]. Hence the number of developed approximation and heuristics algorithms using simplified lattice structures increased over decades [4,7–11,18,20]. The first approximation algorithm was proposed by Hart and Istrail [7], which gave an approximation ratio of 4 on the 2D square lattice. Mauri et al. [16] proposed a different approximation algorithm, albeit having the same approximation ratio of 4; but they argued that it worked better in practice. An improved algorithm of ratio 3 was proposed by Newman [17]. He considered the protein primary structure as a folded loop. A $\frac{8}{3}$-approximation algorithm for the problem on the cubic lattice was given by Hart and Istrail [7].

A significant drawback of the square lattice and cubic lattice is the parity problem. Parity problem refers to the phenomenon that if two residues are at even distance from one another in the sequence then they cannot be in topological contact with each other when the protein is embedded in the lattice. Agarwala et al. first suggested that triangular lattice was more suitable to remove this parity problem [1]. Using a better upper bound, they proposed two approximation algorithms, one having an approximation ration of 2 and the other $\frac{11}{6}$. Face Centered Cubic (FCC) lattice is a more generalized 3 dimensional version of the triangular lattice. Agarwala et al. gave an $\frac{5}{3}$-approximation algorithm on FCC lattice. Bockenhauer and Bongartz proposed another way to remove the parity problem. They introduced the square lattice with diagonals [3] and achieved an approximation ratio of $\frac{26}{15}$ in this lattice. Jiang and Zhu used 2D hexagonal lattice for this problem and gave an approximation algorithm of ratio 6 [14]. However, the hexagonal lattice also has the parity problem. Shaw et al. [19] removed this problem by introducing diagonals into the hexagonal lattice and gave two approximation algorithms for protein folding on hexagonal lattice with diagonals.

In this paper, we introduce the hexagonal prism lattice with diagonals for protein folding. Our prior work on the hexagonal lattice model with diagonals achieved an approximation ratio of $\frac{5}{3}$ for the primary protein structure [19]. We now focus on the hexagonal prism lattice with diagonals which can be seen as the 3 dimensional version of the hexagonal lattice with diagonals. It removes the parity problem of other 3 dimensional lattices. In hexagonal prism lattice with diagonals, contacts can occur through diagonals (see Fig. 1). Cubic lattices with diagonals and FCC lattices can also remove this problem but the main advantages of the hexagonal prism lattice with diagonals lay on its structure. One vertex of this lattice have 20 neighbours, on the other hand, the same in the cubic lattice and FCC lattice has 6 and 14 neighbours respectively. As a result, probability of having more number of contacts increases for hexagonal prism lattices with diagonals. Here, we present two novel approximation algorithms for finding protein structures. Our first algorithm has an approximation ratio of 2 for $k > 13$ where k is the number of sequences of H's in the HP string. Our next algorithm improves the approximation ratio to $\frac{9}{7}$ for $k > 132$.

2 Preliminaries

In this section, detail description of structure of the hexagonal prism lattice with diagonals is given. Also the required notion for describing the algorithm presented.

Definition 1. *The three-dimensional hexagonal prism lattice with diagonals is an infinite graph $G = (V, E)$ in the Euclidean Space with vertex set $V = R^3$ and edge set $E = \{(x, x') | x, x' \in R^3, |x - x'| \leq 2\}$, where $|.|$ denotes the Euclidean norm. The hexagonal prism lattice is composed by stacking multiple two-dimensional hexagonal lattices with diagonals on top of each other. On a hexagonal prism lattice with diagonals each two-dimensional hexagonal lattice with diagonals is called a layer. The edges connecting the two layers are called layer edges. If and only if an edge has length $|x - x'| = 1$, it is a non-diagonal edge or non-diagonal layer edge. Otherwise, it is diagonal-layer edge or diagonal edge.*

If two vertices are connected through an edge, they are called adjacent or neighbour to each other. The difference between the usual hexagonal prism and the hexagonal prism lattice with diagonals, lies in the fact that, in latter two vertices can be called adjacent, though they are diagonally apart. The number of total neighbours in hexagonal prism lattice with diagonals is 20 (see Fig. 1).

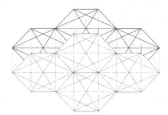

Fig. 1. A hexagonal prism lattice with diagonals. Different layers are indicated using black and red color. Connecting edges between layers are indicated using green color (Color figure online).

Fig. 2. Crossing between two binding edges. This is not possible in valid conformation.

The input to the protein folding problem is a finite string p which is often refer to as an HP string in our problem. P is a string over the alphabet $\{P, H\}$, where $p = \{P\}^* b_1 \{P\}^+ b_2 \{P\}^+ \cdots \{P\}^+ b_k \{P\}^*$. Here, $b_i \in \{H\}^+$ for $1 \leq i \leq k$. Let, $n = \sum_{i=1}^{k} |b_i|$. Here, P denotes polar amino acids and H denotes non-polar respectively. Consecutive string of H's is named as H-run and consecutive P's as P-run. So, from the notation of string p the total number of H-runs is k and total number of H is n. An H-run of odd (even) length is said to be an odd H-run (even H-run). Now, let's see the definition of a valid embedding and conformation into this lattice.

Definition 2. *Let, $G = (V, E)$ be a lattice and $p = p_1 \ldots p_t$ be an HP string of length t. An embedding into G of string p is a mapping function $f: \{1, \ldots, t\} \rightarrow V$. It assigns adjacent positions in p to adjacent vertices in G, $(f(i), f(i+1)) \in E$ for all $1 \le i \le t - 1$. The edges $(f(i), f(i+1)) \in E$ for $1 \le i \le t - 1$ are called **binding edges**. An embedding of p into G is called a conformation, if the embedding is a **self-avoiding walk** inside the grid. Self-avoiding walk means no two binding edges cross each other (see Fig. 2).*

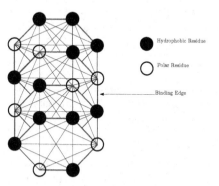

Fig. 3. Conformation of PHPHHHPH-PHPHPHPHHH on the lattice (Color figure online).

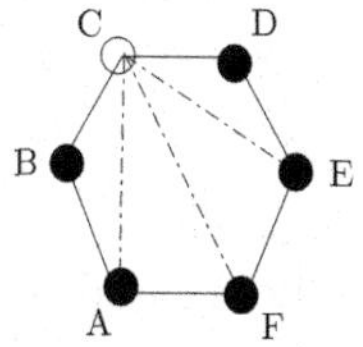

Fig. 4. (C,D) and (B,C) are alternating edges. (A,C), (C,F) and (C,E) are loss edges.

Figure 3 shows an example of a conformation. Edges coloured blue are binding edges and all other edges between residues are non-binding edges. Throughout the paper, the P-vertices are indicated by blank circles and the H-vertices are indicated by filled circle.

Definition 3. *In a conformation ϕ, if an edge (x, x') of G is not a binding edge and there exist $i, j \in \{1, \ldots, t\}$ such that $f(i) = x, f(j) = x'$, and $p_i = p_j = H$, the edge is called a **contact edge**. In a continuation, if p_i or p_j anyone contains P, the edge is named as **loss edge**. A binding edge connecting an H with a P is called an **alternating edge**. Loss edge is a non-binding edge incident to an H that is not a contact edge (see Fig. 4).*

Definition 4. *Let $e = (x, y)$ be any edge in G. Neighbourhood $N(e)$ of e can be defined as the intersection of the neighbours of its endpoints x and y.*

3 Our Approaches

3.1 Upper Bound

Based on a simple counting argument we will deduce a bound, which refers to maximum contacts possible in hexagonal prism lattice with diagonals. We will count the number of neighbours of a vertex in the lattice. Here we start our first lemma.

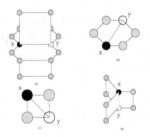

Fig. 5. (a) 12 neighbourhood of the non-diagonal edge (x, y), (b) 4 neighbourhood of the diagonal edge (x, y), (c) 2 neighbourhood of layer-diagonal edge (x, y), (d) 6 neighbourhood of layer non-diagonal edge (x, y).

Lemma 1. *Let, $G = (V, E)$ be a hexagonal lattice with diagonals and p is an HP string. If p has a conformation in G, then any H in p can have at most 18 contact edges.*

Proof: In the lattice G, every vertex has exactly 20 neighbours comprising 9 diagonal neighbours and 3 non-diagonal neighbours in one layer; 4 neighbour from upper layer and 4 neighbour from lower layer (see Fig. 1). In this conformation, there are exactly two binding edges of every H-vertex. Hence the remaining 18 edges of adjacent edges of each vertex, can be contact edges. So every vertex can have at most 18 contact edges.   □

Lemma 2. *Let, p be an input string for the problem and ϕ be a conformation of p. Let $e = (x, y)$ be a loss edge with respect to ϕ. Then there are at most four alternating edges in $N(e)$.*

Proof: From Fig. 5 if e is a non-diagonal edge, then $N(e)$ contain 12 vertices; if e is a diagonal edge, then $N(e)$ contain 4 vertices; if e is a layer-diagonal edge, then $N(e)$ contain 2 vertices; if e is a layer non-diagonal edge, then $N(e)$ contain 6 vertices. Again, each of x and y can be incident to at most two binding edges. So, there are at most four binding edges in $N(e)$. It follows immediately that there can be at most four alternating edges adjacent to e.   □

Now we are ready to present the upper bound.

Lemma 3. *Let, an HP string p with k H-runs and n H. The total number of contacts in a conformation ϕ of string p is at most $18n - \frac{1}{2}k$.*

Proof: From Lemma 1, since there are n H, total number of contacts is at most $18n$.

From Lemma 2 we know that, for every loss edge there will be at most four alternating edges in its neighbourhood. Alternatively, we can say that, there will be at least one loss edge for every four alternating edges, assuming that the alternating edges are in the neighbourhood of that loss edge. Clearly, the number of loss edges will increase, if the alternating edges are not within the neighbourhood. So, there will be at least $\frac{1}{4}$ loss edge for every alternating edge.

In total there are $2k$ alternating edges. So, the total number of loss edges will be, $\frac{1}{4} \times 2 \times k = \frac{1}{2}k$.

In a confirmation one loss edge incident to H means that it would lose one contact edge. That means $\frac{1}{2}k$ loss edges are not used to build contact. Hence, total contact is at most $18n - \frac{1}{2}k$.   □

3.2   Approximation Algorithms and Lower Bounds

In this section, for the protein folding problem, we present two novel approximation algorithms.

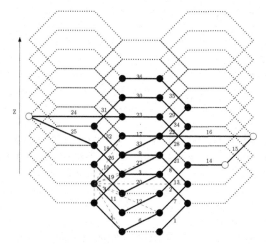

Fig. 6. Folding of HP string $H^{14}P^2H^8P^1H^{11}$ by Algorithm HelixArrangement. Dotted black line represent the lattice, solid line represent binding edge of protein, blue dashed line shows 9 contacts of a H. Binding edges are numbered sequentially. z indicates the direction of side layers of Upper layer (Color figure online).

Algorithm HelixArrangement. The idea of first algorithm is to arrange all H's of the input string in helix structure. The main difference between conventional helix structure, here we arrange P's of input string outside of the main helix structure. Figure 6 shows the way we arrange H's and P's.

 Algorithm HelixArrangement.
Input: An HP string p.

1. Arrange the H's as follows:
   (a) Starting from a layer arrange the first six H's in a hexagon. Let, called this base hexagon.
   (b) Using the layer diagonal edge climb to upper layer. In this layer arrange next six H's in a hexagon which is parallel to base hexagon.
   (c) repeat step (b) until end of string p. The hexagon where the process ended, let called that top hexagon.
2. Intermediate P-runs are arranged in the outer side of hexagon in a layer (see Fig. 6).

Approximation Ratio for Algorithm HelixArrangement. Except the H's of base hexagon and top hexagon a H can achieve at least 9 contacts. A H from its layer achieve 3 contacts, from its immediate upper layer 3 contacts and from its immediate lower layer 3 contacts. H's of base hexagon miss the contacts from lower layer and H's of top hexagon miss the contacts from upper layer. So, there is in total 12 H in base hexagon and top hexagon which miss in total $12 * 3$ *or* 36 contacts. Note that, it is possible that top hexagon is not filled 6 H's. But it does not change any computation, because there is still 6 H's in top hexagon and lower layer hexagon of top hexagon, which miss 3 contacts.

Now, if we consider the P's arrangement, we will achieve two contacts for every alternating edge. If there is k alternating edge we will achieve $2k$ contacts.

So, for n H's total number of contacts $(\mathcal{C})$ can be achieved as follows: $\mathcal{C} \geq 9n - 36 + 2k$

Hence we get the following approximation ratio A_1:

$$A_1 = \frac{18n - \frac{1}{2}k}{(9n - 36 + 2k)} \tag{1}$$

From Eq. 1 it can be seen that, A_1 tends to reach $\frac{18}{9}$ *or* 2, for large n. So we compute the value of k so that our approximation ratio is at most 2 as shown below.

$$\frac{18n - \frac{k}{2}}{(9n - 36 + 2k)} \leq \frac{18}{9}$$

$$\Rightarrow 81k \geq 18 \times 30 \times 2$$

$$\Rightarrow k \geq \frac{48}{3} \approx 16$$

So, if the total number of H-runs is greater than 16, then Algorithm HelixArrangement will achieve an approximation ratio of 2.

Theorem 1. *For any given HP string, Algorithm HelixArrangement gives a 2 approximation ratio for k > 16, where k is the total number of H-runs and n is the total number of H.*

Algorithm LayerArrangement. The idea of second algorithm is to arrange all H's occurring in the input string along the two layers. We arrange the H's in the prefix of the string up to the $\lfloor \frac{n}{2} \rfloor$-th H on the upper layer and arrange the rest of those on the lower layer. In a layer, H-runs are arranged in a spiral manner. Then we arrange the P's between the H's outside these two layers. The arrangements of the P-runs outside the two layers are shown in Fig. 7. Within a layer the arrangement is done in chains (see Fig. 7). The arrangement in the upper (lower) layer can be further divided into nine regions, namely, the left region, the right region, the up region, the down region, the inside-left region, the inside-right region, the inside-up region, the inside-down region and the middle region (see Fig. 8).

Algorithm LayerArrangement.

Input: An HP string p.

1. Set $f = \lfloor \frac{n}{2} \rfloor$.
2. Suppose $F(f)$ denotes the position in p after the f-th H. The string up to position $F(p)$ is denoted by *pref* $F(p)$ and the string after position $F(p)$ is denoted by *suff* $F(p)$. Now,
   (a) Arrange the H's in *pref* $F(p)$ in the upper layer as follows:
      i. Let, i and j are two integers that divide m_1 with reminder 0, such that the $|i-j|$ is minimal for all i and j. Let, $r = min(i, j)$, which is number of the chains in a layer. Let $s = \lfloor \frac{f}{r} \rfloor$, which is the number of residues in a chain. Suppose, $S_1, S_2, S_3, \ldots$ denote the position in p after the s-th,$2s$-th,$3s$-th$\ldots$ H respectively. Denote, $S_i(p) = p_{S_{i-1}}, \ldots p_{S_i-1}$ for $i = 1, 2, 3, \ldots$. Here S_0 is starting position.
      ii. Now arrange $S_i(p)$ in chain one by one from top to bottom for $i = 1, 2, 3, \ldots$.
      iii. Intermediate P-runs are arranged in the upper-side layers of the upper layer (see Fig. 7).
   (b) Arrange the H's in *suff* $F(p)$ along the lower layer following the same strategy spelled out in Step 2(a); intermediate P-runs are arranged in the lower-side layer of the lower layer (see Fig. 7).

Approximation Ratio for Algorithm LayerArrangement. Suppose that $m_1 = \lfloor \frac{n}{2} \rfloor$. From Algorithm LayerArrangement, the upper (lower) layer will contain m_1 (m_1 or $m_1 + 1$) H's. We consider two cases, namely, where m_1 is odd, i.e., $m_1 = 2x + 1$ and m_1 is even, i.e., $m_1 = 2x$, with an integer $x > 0$.

Now, let, i and j are two integers that divide m_1 with reminder 0, such that $|i-j|$ is minimal for all i and j. Let, $r = min(i, j)$, which is number of the chains in a layer. Now, let, $s = m_1/r$ which is number of residues in a chain. The chains are arranged spirally in a layer.

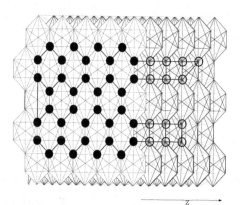

Fig. 7. Folding of HP string $H^9 P^6 H^{18} P^7 H^9$ by Algorithm LayerArrangement only in Upper layer. Z indicates the direction of side layers of Upper layer

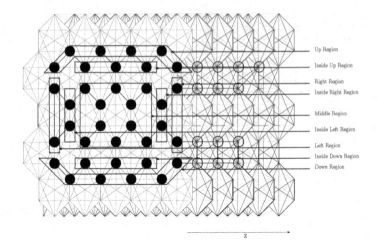

Fig. 8. Divided into 9 region. They are up region, inside up region, right region, inside right region, middle region, inside left region, left region, inside down region, down region.

Now, we will use pq-upper layer (pq-lower layer) to denote a particular region of the upper (lower) layer. So, pq could be one of the 9 options, namely, lR (left region), rR (right region), uR (up region), dR (down region), i_lR (inside-left region), i_rR (inside-right region), i_uR (inside-up region), i_dR (inside-down region) and mR (middle region). To refer the conformation given by Algorithm LayerArrangement, we use notation ϕ_{CA}.

The analysis can be easily understood with the help of Fig. 8. In ϕ_{CA}, every vertex in the lR-up layer and rR-up layer has at least 8 contacts. Every vertex in the i_lR-upper layer and the i_rR-upper layer has at least 12 contacts. For each of lR-upper layer, rR-upper layer, i_lR-upper layer and the i_rR-upper layer, there are $r - 2$ such vertices (see Fig. 8). Every vertex in the uR-upper layer and the dR-upper layer has at least 6 contacts. There are $\frac{s+3}{2}$ such vertices for each of the uR-upper layer and the dR-upper layer. Every vertex in the i_uR-upper layer and the i_dR-upper layer has at least 11 contacts. There are $(\frac{s-3}{2})$ such vertices for each of the i_uR-upper layer and the i_dR-upper layer. So there remain $(rs - 2r - 2s - 4)$ vertices in upper layer which fall to mR-upper layer, where every vertex achieved 14 contacts.

So, for all the vertices of the upper layer, the total number of contacts ($\mathcal{C}$) can be computed as follows:

$$\mathcal{C} \geq 2 \times 8 \times (r - 2) + 2 \times 12 \times (r - 2) + 2 \times 6$$
$$\times \frac{s+3}{2} + 2 \times 11 \times (\frac{s-3}{2}) + 14 \times (2x - 2r - 2s - 4)$$
$$\Rightarrow \mathcal{C} \geq 16r - 32 + 24r - 48 + 6s + 18 + 11s - 33 + 14sr - 28r - 28s - 56$$
$$\Rightarrow \mathcal{C} \geq 14sr + 12r - 11s - 151$$
$$\Rightarrow \mathcal{C} \geq 14m_1 + 12r - 11s - 151 \quad \Rightarrow \mathcal{C} \geq 7n + 12r - 11s - 151$$

because of, the upper layer is symmetric to the lower layer, both layers will have the same number of vertices if $n = 2m_1$. Hence, all the vertices of the lower layer will also have at least C contacts. So, total number of contacts will be at least $2C$ or $14n + 24r - 22s - 302$.

Now, if $n = 2m_1 + 1$, then let $n_1 = n - 1$. These n_1 vertices will have at least $14n_1 + 24r - 22s - 302$ contacts. The remaining vertex will have at least 2 contacts. So the total number of contacts will be at least $14(n - 1) + 24r - 22s - 302 + 2$ or $14n + 24r - 22s - 314$. So, combining the two cases, we get that the total number of contacts is at least $14n + 24r - 22s - 314$. Now if we consider alternating edges, for every alternating edge we get two extra contacts for the two vertices (each having one). So, we get a total of at least $14n + 24r - 22s - 314 + 2k$ contacts for n H's and k alternating edges. Approximation ratio A_2 can be calculated as follows:

$$A_2 = \frac{18n - \frac{1}{2}k}{(14n + 24r - 22s - 314 + 2k)} \tag{2}$$

From Eq. 2 it can be seen that, A_2 tends to reach $\frac{18}{14}$ for large n. So we compute the value of k so that our approximation ratio is at most $\frac{18}{14}$ as shown below.

$$\frac{18n - \frac{k}{2}}{(14n + 24r - 22s - 314 + 2k)} \leq \frac{18}{14}$$

$$\Rightarrow 14 \times 18n - \frac{k}{2} \leq \frac{18}{(14n + 24r - 22s - 314 + 2k)}$$

$$\Rightarrow 252n - 7k \leq 252n + 432r - 396s - (314 \times 18) + 36k$$

$$\Rightarrow 43k \geq 36(11s - 12r) + (314 \times 18) \quad \Rightarrow k \geq \frac{36(11s - 12r) + (314 \times 18)}{43}$$

Now, from this case if $11s = 12r$, $k \geq \frac{(314 \times 18)}{43} \approx 132$
So, the Algorithm LayerArrangement will achieve an approximation ratio of $\frac{18}{14}$ or $\frac{9}{7}$ for $11s = 12r$, if the total number of H-runs is greater than 132.

Theorem 2. *For any given HP string, Algorithm LayerArrangement gives a $\frac{9}{7}$ approximation ratio for $k > 132$, where k is the total number of H-runs and $11s = 12r$ where, $n = 2rs$ and n is the total number of H.*

Be noted that, the value of k is dependent on n and the HP string. This leads us to deduce the expected value of k for a given HP string. This can be mapped into the problem of *Integer Partitioning* which can be defined as below. Notably, for deriving an expected approximation ratio, similar problem mapping has recently been utilized in different algorithms [12, 13, 19].

From *Integer Partitioning* used in [12, 15, 19], we can say that, the expected value of H-runs, k is less than or equal to $\sqrt{n} \times \log n$ which implies that $\sqrt{n} \times \log n \geq 132$ or $n \geq 500$. Now, if $11s > 12r$, lower bound of k increases, as a result expected lower bound of n will increases. On the other side, if $11s < 12r$, expected lower bound of n will decreases. The above findings are summarized in the following theorem.

Theorem 3. *For any given HP string, Algorithm LayerArrangement is expected to achieve an approximation ratio of $\frac{9}{7}$ for $n \geq 500$ and $11s = 12r$ where, $n = 2rs$ and n is the total number of H.*

4 Conclusion

As has been already discussed above, a vertex in the SC (Simple Cubic) lattice have 6 neighbours and the same in the FCC (Face Centered Cubic) or BCC (Body Centered Cubic) lattice have 14 neighbours. On the other hand, one vertex in the hexagonal prism lattice with diagonals have 20 neighbours. This useful property has led us to find better approximation ratio. On the other hand this lattice model removes some well known problems of protein folding in SC lattices e.g., parity problem. Considering such properties of this lattice encourages us to investigate further to develop even better approximation algorithms. Also heuristics algorithm can be applied on this lattice, which can surely lead us to better results.

References

1. Agarwala, R., Batzogloa, S., Dancik, V., Decatur, S., Hannenhalli, S., Farach, M., Muthukrishnan, S., Skiena, S.: Local rules for protein folding on a triangular lattice and generalized hydrophobicity in the HP model. J. Comput. Biol. **4**(3), 276–296 (1997)
2. Berger, B., Leighton, T.: Protein folding in the hydrophobic-hydrophilic (HP) model is NP-complete. J. Comput. Biol. **5**(1), 27–40 (1998)
3. Böckenhauer, H.-J., Bongartz, D.: Protein folding in the hp model on grid lattices with diagonals. Discrete Appl. Math. **155**(2), 230–256 (2007)
4. Böckenhauer, H.-J., Dayem Ullah, A.Z.M., Kapsokalivas, L., Steinhöfel, K.: A local move set for protein folding in triangular lattice models. In: Crandall, K.A., Lagergren, J. (eds.) WABI 2008. LNCS, vol. 5251, pp. 369–381. Springer, Heidelberg (2008)
5. Crescenzi, P., Goldman, D., Papadimitriou, C.H., Piccolboni, A., Yannakakis, M.: On the complexity of protein folding. J. Comput. Biol. **5**(3), 423–465 (1998)
6. Dill, K.A.: Theory for the folding and stability of globular proteins. Biochemistry **24**, 1501–1509 (1985)
7. Hart, W., Istrail, S.: Fast protein folding in the hydrophobic-hydrophilic model within three-eighths of optimal. J. Comput. Biol. **3**(1), 53–96 (1996)
8. Hart, W., Istrail, S.: Lattice and o-lattice side chain models of protein folding: linear time structure prediction better than 86 % of optimal. Comput. Biol. **4**(3), 241–259 (1997)
9. Heun, V.: Approximate protein folding in the HP side chain model on extended cubic lattices (extended abstract). In: Nešetřil, J. (ed.) ESA 1999. LNCS, vol. 1643, pp. 212–223. Springer, Heidelberg (1999)
10. Hoque, M.T., Chetty, M., Dooley, L.S.: A hybrid genetic algorithm for 2D FCC hydrophobic-hydrophilic lattice model to predict protein folding. In: Sattar, A., Kang, B.-H. (eds.) AI 2006. LNCS, vol. 4304, pp. 867–876. Springer, Heidelberg (2006)

11. Hoque, M.T., Chetty, M., Sattar, A.: Protein folding prediction in 3D FCC HP lattice model using genetic algorithm. In: IEEE Congress on Evolutionary Computation, pp. 4138–4145 (2007)
12. Islam, A.S.M.S., Rahman, M.S.: On the protein folding problem in 2D-triangular lattices. Algorithms Mol. Biol. **8**, 30 (2013)
13. Islam, A.S.M.S., Rahman, M.S.: Protein folding in 2D-triangular lattice revisited. In: Lecroq, T., Mouchard, L. (eds.) IWOCA 2013. LNCS, vol. 8288, pp. 244–257. Springer, Heidelberg (2013)
14. Jiang, M., Zhu, B.: Protein folding on the hexagonal lattice in the HP model. J. Bioinform. Comput. Biol. **3**(1), 19–34 (2005)
15. Kessler, I., Livingston, M.: The expected number of parts in a partition of n. Monatshefte für Mathematik **81**(3), 203–212 (1976)
16. Mauri, G., Piccolboni, A., Pavesi, G.: Approximation algorithms for protein folding prediction. In: Symposium on Discrete Algorithms (SODA), pp. 945–946 (1999)
17. Newman, A.: A new algorithm for protein folding in the HP model. In: Symposium on Discrete Algorithms (SODA), pp. 876–884 (2002)
18. Newman, A., Ruhl, M.: Combinatorial problems on strings with applications to protein folding. In: Farach-Colton, M. (ed.) LATIN 2004. LNCS, vol. 2976, pp. 369–378. Springer, Heidelberg (2004)
19. Shaw, D.L., Islam, A.S.M.S., Rahman, M.S., Hasan, M.: Protein folding in HP model on hexagonal lattices with diagonals. BMC Bioinform. **15**(S–2), S7 (2014)
20. Unger, R., Moult, J.: Genetic algorithms for protein folding simulations. J. Mol. Biol. **231**, 75–81 (1993)

Parameterization of Strategy-Proof Mechanisms in the Obnoxious Facility Game

Morito Oomine, Aleksandar Shurbevski$^{(\boxtimes)}$, and Hiroshi Nagamochi

Department of Applied Mathematics and Physics, Kyoto University,
Yoshida-Honmachi, Sakyo-ku, Kyoto 606-8501, Japan
{oomine1024,shurbevski,nag}@amp.i.kyoto-u.ac.jp

Abstract. In the obnoxious facility game, a location for an undesirable facility is to be determined based on the voting of selfish agents. The design of group strategy proof mechanisms has been extensively studied, and it is known that there exists a gap between the social benefit (i.e., the sum of individual benefits) by a facility location determined by any group strategy proof mechanism and the maximum social benefit over all choices of facility locations; their ratio, called the *benefit ratio* can be 3 in the line metric space. In this paper, we investigate a trade-off between the benefit ratio and a possible relaxation of group strategy proofness, taking 2-candidate mechanisms for the obnoxious facility game in the line metric as an example. Given a real $\lambda \geq 1$ as a parameter, we introduce a new strategy proofness, called "λ-group strategy-proofness," so that each coalition of agents has no incentive to lie unless every agent in the group can increase her benefit by strictly more than λ times by doing so, where the 1-group strategy-proofness is the previously known group strategy-proofness. We next introduce "masking zone mechanisms," a new notion on structure of mechanisms, and prove that every λ-group strategy-proof (λ-GSP) mechanism is a masking zone mechanism. We then show that, for any $\lambda \geq 1$, there exists a λ-GSP mechanism whose benefit ratio is at most $1 + \frac{2}{\lambda}$, which converges to 1 as λ becomes infinitely large. Finally we prove that the bound is nearly tight: given $n \geq 1$ selfish agents, the benefit ratio of λ-GSP mechanisms cannot be better than $1 + \frac{2}{\lambda}$ when n is even, and $1 + \frac{2n-2}{\lambda n+1}$ when n is odd.

Keywords: Mechanism design · Facility game · Strategy-proof · Anonymous · Optimization

1 Introduction

1.1 Social Choice Theory

In social choice theory, we design *mechanisms* that determine a social decision based on a vote. That is, for a set Ω of voting alternatives and a set N of selfish voters with various utilities, we design a mechanism $f : \Omega^n \to \Omega$ as a collective decision making system. We call *primary benefit* the benefit that

© Springer International Publishing Switzerland 2016
M. Kaykobad and R. Petreschi (Eds.): WALCOM 2016, LNCS 9627, pp. 286–297, 2016.
DOI: 10.1007/978-3-319-30139-6_23

each voter obtains under the assumption that all votes are truthfull However, each voter may try to manipulate the decision of a mechanism by changing her voting to increase her personal utility. A voting which aims to manipulate the decision of a mechanism is called a *strategic voting*. To the effect of making a fair decision, we are interested in mechanisms in which no voter can benefit by a single-handed strategic voting. Such a mechanism is called a *strategy-proof* mechanism. Moreover, a mechanism is called a *group strategy-proof* mechanism, if there is no coalition of voters such that each member in the coalition can simultaneously benefit by their cooperative strategic-voting. Moulin [9] studied social choice theory under the condition that the set of alternatives is the one-dimensional Euclidean space and each utility function is a single peaked concave function, and gave necessary and sufficient conditions of strategy-proofness under such conditions. Following, Border and Jordan [2] extended the characterization into the multi-dimensional Euclidean space, and characterized strategy-proof mechanisms in those metrics. Schummer and Vohra [12] applied this result [2] to the case when Ω is the set of all points in a tree metric and characterized strategy-proof mechanisms in those metrics. Moreover, they characterized strategy-proof mechanisms in the case when Ω is the set of all points in a graph metric which has at least one cycle.

1.2   Facility Game

The *facility game* can be regarded as a problem in social choice theory where a location of the facility in a metric space will be decided based on locations of agents (votes by voters) and each agent tries to maximize the benefit from her utility function defined based on the distance from her location to the location of the facility.

In a facility game with a set N of agents in a space Ω, each agent reports a point in the space, and a mechanism determines a location for a facility. Each agent is selfish in the sense that she may misreport her point so that the output by the mechanism becomes more beneficial to her. The facility can be classified as either one of two types, one is *desirable* to agents (or each agent wants the facility to be located near her actual location) and the other is *obnoxious* to agents (or each agent wants the facility to be located far from her actual location).

Several extensive studies have been made on the desirable facility game, such as designing mechanisms [1,2,7,8,11,12]. Procaccia and Tennenholtz [11] proposed a group strategy-proof mechanism which returns the location of the median agent as the facility location when all agents are located on a path.

Cheng et al. [4] studied the *obnoxious facility game*. For a given mechanism, the benefit for each agent obtained under the assumption that all agents have reported their true locations is called *primary benefit*. Mechanisms which only output one of a predetermined set of k candidates for a facility location are called *k-candidate mechanisms*.

An important aspect of mechanisms of facility games is a measure of the quality of mechanisms. In general, a location of a facility that maximizes some social benefit, such as the sum of all individual benefits, is different from a location output by a strategy-proof (or group strategy-proof) mechanism. In other words, the maximum value of the social utility attained by a strategy-proof (or group strategy-proof) mechanism is smaller than that attained just by choosing the best location of the facility. This raises a problem of designing a strategy-proof (or group strategy-proof) mechanism that outputs a location of a facility that maximizes the social benefit among all strategy-proof (or group strategy-proof) mechanisms. A possible measurement of the performance for a mechanism is a *benefit-ratio*, the ratio of the social utility attained by the mechanism to that attained by a theoretically maximum possible social benefit. For example, Alon *et al.* [1] gave a complete analysis on benefit-ratios of group strategy-proof mechanisms for the desirable facility game in general graph metrics.

We review some recent results on the obnoxious facility game. Ibara and Nagamochi [5,6] presented a complete characterization of 2-candidate (group) strategy-proof mechanisms in any metric space, giving necessary and sufficient conditions for the existence of such a mechanism in a given metric, and proved that in any metric, a 2-candidate mechanism with a benefit ratio of 4 can always be designed.

For the obnoxious facility game in the line metric, Ibara and Nagamochi [5,6] showed that there exists no k-candidate group strategy-proof mechanism for any $k \geq 3$. Cheng *et al.* [4] gave a 2-candidate group strategy-proof mechanism in the line metric with a benefit ratio of 3, and showed that this is the best possible over all 2-candidate group strategy-proof mechanisms in the line metric.

1.3   Our Contribution

Since it has been shown that the best benefit ratio is 3 (see Cheng *et al.* [4]) over all 2-candidate group strategy-proof mechanisms for the obnoxious facility game in the line metric, we propound the following questions on the game:

1. Is there any way of relaxing the definition of group strategy-proofness so that the benefit ratio 3 is improved over such relaxed group strategy-proof mechanisms?
2. With an adequate parameter $\lambda \geq 1$, is there any trade-off between λ-group strategy-proofness (group strategy-proofness relaxed with λ) and the benefit ratio ρ for λ such that ρ approaches 1 as λ becomes infinitely large; and
3. For each fixed $\lambda \geq 1$, what is the benefit ratio ρ of a λ-group strategy-proof (or can tight upper and lower bounds on ρ be derived)?

This paper answers all of these questions affirmatively. First we introduce a relaxed version of (group) strategy-proofness via a parameter $\lambda \geq 1$ by assuming that an agent has no incentive to misreport her own location unless she can increase her benefit by strictly more than λ times her primary benefit. Respectively, in every group of agents, at least one agent cannot get an increase of strictly more than λ times from her primary benefit by strategically changing her report in coalition with the rest of the group. This parameterization serves as a relaxation of the notion of group strategy-proofness. Mechanisms which guarantee the above property are termed "λ-strategy-proof (λ-SP) mechanisms" and "λ-group strategy-proof (λ-GSP) mechanisms," where 1-group strategy-proofness is equivalent to the previously known group strategy-proofness.

Second, we design a λ-GSP 2-candidate mechanism whose benefit ratio ρ is at most $1 + 2/\lambda$, which approaches 1 as the parameter λ tends to ∞. This answers the first and second question.

Finally, we show that there is no λ-SP 2-candidate mechanism whose benefit ratio ρ is smaller than $1 + 2/\lambda$ for an even n and $1 + (2n - 2)/(\lambda n + 1)$ for an odd n, where n (≥ 1) is the number of agents. This is an answer to the third and second question, since our upper and lower bounds on the benefit ratio are almost tight.

The above results are obtained by introducing a new concept of mechanism design that follows naturally from the introduction of the parameter λ, called "masking zone mechanisms," which in their own right might lend interesting directions for future research.

2 Preliminaries

2.1 Notation

Let $\mathbb{R}$ and $\mathbb{R}_+$ be the sets of real and nonnegative real numbers, respectively. Let Ω be a universal set of points. For a positive integer $n \geq 1$, let N be a set of n agents. For a set $S \subseteq N$ of agents (resp., an agent $i \in N$), let $\overline{S} = N \setminus S$ (resp., $\overline{i} = N \setminus \{i\}$). Each agent $i \in N$ chooses a point $p \in \Omega$ as a reported value $\chi_i = p$. Let $\Omega_{\text{agents}} \subseteq \Omega$ denote a set of points that can be chosen by an agent. A vector $\chi \in \Omega_{\text{agents}}^n$ with reported values χ_i, $i \in N$ is called a profile of N.

A mechanism f is a function that given a profile χ of N outputs a point $t \in \Omega$. Let $\Omega_{\text{facility}} \subseteq \Omega$ denote a set of points that can be output by a mechanism, where a mechanism f is a mapping $f(\chi) : \Omega_{\text{agents}}^n \to \Omega_{\text{facility}}$. It is common in the literature, e.g., [3,4], to represent the locations reported by agents as an n-dimensional vector $\boldsymbol{x}$, where $\boldsymbol{x}_i$ is the point reported by an agent $i \in N$. Under these circumstances, the notion of anonymity plays an important role. A mechanism f is *anonymous* if $f(\boldsymbol{x}) = f(\boldsymbol{x}')$ holds for any two vectors $\boldsymbol{x}$ and $\boldsymbol{x}'$ that admit a bijection σ on N such that $\boldsymbol{x}'_i = \boldsymbol{x}_{\sigma(i)}$ for all $i \in N$.

In what follows, we treat a profile χ of N as a multiset $\{\chi_i \mid i \in N\}$ of n elements. For convenience, given a profile χ and a set $S \subseteq N$ of agents, let χ_S denote the multiset $\{\chi_i \mid i \in S\}$ of $|S|$ elements. For a subspace $\Omega' \subseteq \Omega$, the restriction $\chi|_{\Omega'}$ of a profile χ on Ω' is defined to be the multiset

$$\chi|_{\Omega'} = \{\chi_i \mid i \in N, \chi_i \in \Omega'\},$$

where $|\chi|_{\Omega'}|$ means the number of elements in $\chi|_{\Omega'}$, i.e., the number of agents in $\chi|_{\Omega'}$.

The *benefit* of an agent $i \in N$ with respect to a point $p \in \Omega_{\text{agents}}$ and a point $t \in \Omega_{\text{facility}}$ is specified by a function $\beta_i : \Omega_{\text{facility}} \times \Omega_{\text{agents}} \to \mathbb{R}$. We assume that a larger value in β_i is preferable to the agent $i \in N$. For a mechanism

$f : \Omega_{\text{agents}}^n \rightarrow \Omega_{\text{facility}}$, a point $t \in \Omega_{\text{facility}}$ is called a candidate if there is a profile $\chi \in \Omega_{\text{agents}}^n$ such that $f(\chi) = t$, and the set of all candidates of f is denoted by $C(f) \subseteq \Omega_{\text{facility}}$. A mechanism with $|C(f)| = k$ is called a k-candidate mechanism.

2.2 Strategy Proofness

In this paper, we assume that a larger value of β_i is preferable to the agent $i \in N$. We define parameterized strategy-proofness by introducing a real parameter $\lambda \geq 1$ as follows.

A mechanism f is called λ-strategy-proof (λ-SP for short) if no agent can gain strictly more than λ times her primary benefit by changing her report. Formally, for any agent $i \in N$ and any profile χ' such that $\chi'_{\bar{i}} = \chi_{\bar{i}}$, it holds that

$$\lambda \beta_i(f(\chi), \chi_i) \geq \beta_i(f(\chi'), \chi_i). \tag{1}$$

A mechanism f is called λ-group strategy-proof (λ-GSP for short) if for every group of agents, at least one agent in the group cannot gain strictly more than λ times her primary benefit by changing her report in coalition with the rest of the group. Formally, for any non-empty set $S \subseteq N$ of agents and for any profile χ' such that $\chi'_{\overline{S}} = \chi_{\overline{S}}$, there exists an agent $i \in S$ for whom

$$\lambda \beta_i(f(\chi), \chi_i) \geq \beta_i(f(\chi'), \chi_i). \tag{2}$$

By definition, any λ-GSP mechanism is λ-SP. Also, 1-strategy-proofness (resp., 1-group strategy-proofness) is equivalent to the strategy-proofness (resp., group strategy-proofness) of mechanisms as it is commonly defined in the literature [1,4–6].

2.3 Masking Zone Mechanisms

In this paper, we introduce "masking zone mechanisms," another new concept on the structure of mechanisms.

Definition 1. *Let $\mathcal{S}$ be a family of nonempty disjoint subsets of Ω, and $\overline{S} = \Omega \setminus \bigcup_{S \in \mathcal{S}}$. A mechanism f is a* masking zone mechanism *with set of masking zones $\mathcal{S}$ if it delivers the same output $f(\chi) = f(\chi')$ for any two profiles χ and χ' such that*

$$\chi|_{\overline{S}} = \chi'|_{\overline{S}} \text{ and } |\chi|_S| = |\chi'|_S| \text{ for all } S \in \mathcal{S}.$$

In other words, $f(\chi)$ of a profile χ never changes as long as a point $\chi_i \in S \in \mathcal{S}$ changes to a point in the same subset S.

2.4   Social Benefit

We introduce an objective function $\mathrm{sb}(t, \chi)$ that evaluates the quality of a point t determined based on a given profile χ. For a point $t \in \Omega_{\text{facility}}$ and a profile χ, we define the *social benefit* $\mathrm{sb}(t, \chi)$ to be the sum of individual benefits over all agents, i.e.,

$$\mathrm{sb}(t, \chi) = \sum_{i \in N} \beta_i(t, \chi_i).$$

Given a profile χ, let $\mathrm{opt}(\chi)$ denote the maximum social benefit over all choices of points $t \in \Omega_{\text{facility}}$; i.e.,

$$\mathrm{opt}(\chi) = \max_{t \in \Omega_{\text{facility}}} \{\mathrm{sb}(t, \chi)\}.$$

The *benefit ratio* $\rho_f \geq 1$ of a mechanism f is defined to be

$$\rho_f = \sup_{\chi \in \Omega_{\text{agents}}^n} \frac{\mathrm{opt}(\chi)}{\mathrm{sb}(f(\chi), \chi)}.$$

When λ becomes infinitely large, the constraints of Eqs. (1) and (2) are no longer effective. If there is no such constraint as in Eq. (1) or (2), then a λ-SP or λ-GSP mechanism can deliver a point $t \in \Omega_{\text{facility}}$ that maximizes $\mathrm{sb}(t, \chi)$ and $\rho_f = 1$ always holds in this case.

2.5   Obnoxious Facility Game in the Line Metric

In this paper, we consider the obnoxious facility game in the line metric [4]. Let (Ω, d) be a metric with a space $\Omega \subseteq \mathbb{R}$ in the line and the distance function $d : \Omega^2 \to \mathbb{R}_+$ such that

$$d(x, y) = |x - y| = \begin{cases} x - y & \text{if } x \geq y, \\ y - x & \text{otherwise.} \end{cases}$$

We assume that, for any agent $i \in N$, the benefit β_i is given by

$$\beta_i(t, p) = d(t, p) \quad t \in \Omega_{\text{facility}}, \ p \in \Omega_{\text{agents}}.$$

In interest of space and clarity, given a profile χ and a candidate location $t \in \Omega_{\text{facility}}$, henceforth we omit referring to the benefit $\beta_i(t, \chi_i)$ of agent i, and directly write $d(t, \chi_i)$. Also we let $d(t, P)$ denote $\sum_{p \in P} d(t, p)$ for a point $t \in \mathbb{R}$ and a multiset P of points in $\mathbb{R}$, where $\mathrm{sb}(c, \chi) = d(c, \chi)$ for a profile χ of N and a location $c \in \mathbb{R}$.

Recall that there is no k-candidate SP mechanism in the line metric, for any $k \geq 3$ [5,6] and that $\rho_f = 3$ for a GSP mechanism f and no GSP mechanism f attains $\rho_f < 3$ [4].

In this paper, we examine the benefit ratio of a 2-candidate λ-GSP mechanism f, and assume without loss of generality that $C(f) = \{0, 1\} = \Omega_{\text{facility}} \subseteq \Omega \subseteq \mathbb{R}$.

Given a real $\lambda \geq 1$, we define subsets I_0 and I_1 of $\mathbb{R}$ as follows:

$$I_0 = \left(\frac{-1}{\lambda - 1}, \frac{1}{\lambda + 1}\right), \; I_1 = \left(\frac{\lambda}{\lambda + 1}, \frac{\lambda}{\lambda - 1}\right) \; \text{for } \lambda > 1,$$

and $I_0 = \{p \in \mathbb{R} \mid p < \frac{1}{2}\}$ and $I_1 = \{p \in \mathbb{R} \mid p > \frac{1}{2}\}$ for $\lambda = 1$. Let $I = I_0 \cup I_1$ and $\overline{I} = \mathbb{R} \setminus I$. Then we observe the next property.

Proposition 1. *Given a real $\lambda \geq 1$, a point $p \in \mathbb{R}$ satisfies $\lambda d(0, p) < d(1, p)$ (resp., $\lambda d(1, p) < d(0, p)$) if and only if $p \in I_0$ (resp., $p \in I_1$).*

Proof. We have that $\{p \in \mathbb{R} \mid \lambda d(0, p) - d(1, p) < 0\} = \{p \in \mathbb{R} \mid \lambda^2 p^2 - (p - 1)^2 = (\lambda p + (p - 1))(\lambda p - (p - 1)) < 0\} = (\frac{-1}{\lambda - 1}, \frac{1}{\lambda + 1}) = I_0$. Analogously we see that $\{p \in \mathbb{R} \mid \lambda d(1, p) - d(0, p) < 0\} = (\frac{\lambda}{\lambda + 1}, \frac{\lambda}{\lambda - 1}) = I_1$. In the case of $\lambda = 1$, we have that $d(0, p) - d(1, p) < 0 \iff p < \frac{1}{2}$, and $d(0, p) - d(1, p) > 0 \iff p > \frac{1}{2}$, as required. $\quad\square$

In this paper, we first show that all 2-candidate λ-SP mechanisms are masking zone mechanisms.

Theorem 2. *Let $\lambda \geq 1$. Every 2-candidate λ-SP mechanism f with candidate set $C(f) = \{0, 1\}$ in the line metric is a masking zone mechanism with set of masking zones $\{I_0, I_1\}$.*

Based on this, we next design a masking zone λ-GSP mechanism whose benefit ratio is at most $1 + 2/\lambda$.

Theorem 3. *Let $\lambda \geq 1$. In the line metric, there is a 2-candidate λ-GSP mechanism f such that $\rho_f \leq 1 + \frac{2}{\lambda}$.*

Finally we examine the converse, showing that no masking zone λ-SP mechanism f attains a benefit ratio smaller than $1 + 2/\lambda$ for an even $n = |N|$ or $1 + (2n - 2)/(\lambda n + 1)$ for an odd $n = |N|$.

Theorem 4. *Let $\lambda \geq 1$ and $n = |N| \geq 1$. In the line metric, there is no 2-candidate λ-SP mechanism f such that*

$$\rho_f < \begin{cases} 1 + \frac{2}{\lambda} & \text{if } n \text{ is even,} \\ 1 + \frac{2n - 2}{\lambda n + 1} & \text{otherwise.} \end{cases}$$

3   Masking Zone Mechanisms

This section shows that any λ-SP mechanism is a masking zone mechanism. By the following lemma, we derive a necessary condition for a mechanism in the line metric to be λ-SP.

Lemma 5. *Given a real $\lambda \geq 1$, let f be a λ-SP mechanism with candidate set $C(f) = \{0,1\}$. Let χ be a profile of N such that $f(\chi) = c \in \{0,1\}$. If there is an agent i with $\chi_i \in I_c$, then the profile $\widehat{\chi}$ obtained from χ by changing χ_i to a point in I_c still satisfies $f(\widehat{\chi}) = c$, where $\widehat{\chi_{\overline{i}}} = \chi_{\overline{i}}$ and $\widehat{\chi_i} \in I_c$.*

Proof. To derive a contradiction, we assume that $f(\widehat{\chi}) = 1 - c$. Since $\chi_i \in I_c$, we know $\lambda d(c, \chi_i) < d(1 - c, \chi_i)$ by Proposition 1, i.e., $\lambda d(f(\chi), \chi_i) = \lambda d(c, \chi_i) < d(1 - c, \chi_i) = d(f(\widehat{\chi}), \chi_i)$. Since $\widehat{\chi_{\overline{i}}} = \chi_{\overline{i}}$, this contradicts that f is λ-SP.   □

We are now ready to prove Theorem 2.

Proof of Theorem 2. Let f be a λ-SP mechanism with candidate set $C(f) = \{0,1\}$. We say that two profiles χ and χ' of N are *zone-equivalent* if $\chi|_{\overline{I}} = \chi'|_{\overline{I}}$ and $|\chi|_{I_c}| = |\chi'|_{I_c}|$ for each $c \in C(f) = \{0,1\}$. It suffices to show that, for any zone-equivalent profiles χ and χ', it holds that $f(\chi) = f(\chi')$. To derive a contradiction, assume that there are zone-equivalent profiles χ and χ' with $f(\chi) \neq f(\chi')$, and let χ and χ' minimize the number $|\chi|_{\overline{I}} \setminus \chi'|_{\overline{I}}| + |\chi'|_{\overline{I}} \setminus \chi|_{\overline{I}}|$ of different locations between them among all such pairs.

Since χ and χ' are zone-equivalent, there are two distinct locations $\chi_i \in I_c$ and $\chi'_j \in I_c$ for some agents $i, j \in N$ and some $c \in \{0,1\}$. Without loss of generality $f(\chi) = c$ and $f(\chi') = 1 - c$ (if necessary we exchange the role of χ and χ'). Let $\widehat{\chi}$ be the profile obtained from χ by changing the location $\chi_i \in I_c$ of agent i to the point $\chi'_j \in I_c$. By Lemma 5 and $f(\chi) = c$, we see that $f(\widehat{\chi}) = c$.

Notice that $\widehat{\chi}$ and χ' remain zone-equivalent, and now they have a fewer number of different locations than χ and χ' have. Then by the choice of χ and χ', it must hold $f(\widehat{\chi}) = f(\chi') = 1 - c$, which contradicts that $f(\widehat{\chi}) = c$.   □

4   Upper Bounds on the Benefit Ratio

In this section, given a real $\lambda \geq 1$, we prove Theorem 3 by constructing a 2-candidate λ-GSP mechanism f whose benefit ratio ρ_f is at most $1 + 2/\lambda$.

Having in mind that for a given profile χ, the condition for λ-group strategy-proofness of Eq. (2) concerns exactly the agents $i \in N$ with $\chi_i \in I$, we define a distorted distance between a point $c \in \{0,1\}$ and a point $p \in \Omega$ of to be

$$\mu(c,p) = \begin{cases} d(c,p) & \text{if } p \in \overline{I}, \\ 0 & \text{if } p \in I_c, \\ 1 & \text{if } p \in I_{1-c}, \end{cases}$$

where clearly $-1 \leq \mu(c,p) - \mu(1 - c, p) \leq 1$ always holds. Also we let $\mu(c, P)$ denote $\sum_{p \in P} \mu(c,p)$ for a point $c \in \{0,1\}$ and a multiset P of points in $\mathbb{R}$. Then for a profile χ of N and a location $c \in \{0,1\}$, we have

$$\mu(c,\chi) = \sum_{i \in N} \mu(c,\chi_i) = d(c, \chi|_{\overline{I}}) + |\chi|_{I_{1-c}}|.$$

Based on this, we propose the following masking zone mechanism f with candidate set $C(f) = \{0,1\}$.

Mechanism $f(\chi)$: given a multiset χ, return a candidate $c \in C(f) = \{0,1\}$

$$f(\chi) = \begin{cases} 0 & \text{if } \mu(0,\chi) > \mu(1,\chi), \\ 1 & \text{if } \mu(0,\chi) \le \mu(1,\chi). \end{cases} \tag{3}$$

We claim that the mechanism f is λ-GSP.

Lemma 6. *The mechanism f of Eq. (3) is λ-GSP.*

Proof. To derive a contradiction, we assume that f is not λ-GSP; i.e., by definition, there is a non-empty subset $S \subseteq N$ which influences two profiles χ and χ' with

$$\chi'_{\overline{S}} = \chi_{\overline{S}} \text{ and } f(\chi) \ne f(\chi')$$

such that every agent $i \in S$ satisfies

$$\lambda d(f(\chi), \chi_i) < d(f(\chi'), \chi_i) \text{ or } \chi_i \in I_c \text{ for } c = f(\chi),$$

by Proposition 1, where $f(\chi) \ne f(\chi')$ must hold by $\lambda \ge 1$. Let S be minimal subject to the above condition, and j be an arbitrary agent in S.

We prove that the profile χ'' obtained from χ' just by changing χ'_j to χ_j satisfies $f(\chi') = f(\chi'')$, or equivalently

$$\mu(0,\chi') - \mu(1,\chi') \le 0 \text{ if and only if } \mu(0,\chi'') - \mu(1,\chi'') \le 0 \tag{4}$$

by the definition of f of Eq. (3). Observe that $\chi''_{\overline{j}} = \chi'_{\overline{j}}$, $\chi''_j = \chi_j$, and

$$\mu(c,\chi'') = \mu(c,\chi') - \mu(c,\chi'_j) + \mu(c,\chi_j) \text{ for each } c \in \{0,1\}$$

by the definition of μ. From this, we have

$$\mu(0,\chi'') - \mu(1,\chi'') = \mu(0,\chi') - \mu(1,\chi') + [\mu(1,\chi'_j) - \mu(0,\chi'_j)] + [\mu(0,\chi_j) - \mu(1,\chi_j)], \tag{5}$$

where we know $-1 \le \mu(1,\chi'_j) - \mu(0,\chi'_j) \le 1$ by the definition of μ. By $\chi_j \in I_c$ for $f(\chi) = c$ (i.e., $f(\chi') = 1 - c$), we also know

$$\mu(0,\chi_j) - \mu(1,\chi_j) = \begin{cases} -1 & \text{if } f(\chi') = 1 \text{ (i.e., } \mu(0,\chi') - \mu(1,\chi') \le 0 \text{ by Eq. (3))}, \\ 1 & \text{if } f(\chi') = 0 \text{ (i.e., } \mu(0,\chi') - \mu(1,\chi') > 0). \end{cases}$$

Therefore, if $\mu(0,\chi') - \mu(1,\chi')$ is nonnegative (resp., positive), then the right hand side of Eq. (5) is also nonnegative (resp., positive), implying Eq. (4).

Finally we observe that $f(\chi'') = f(\chi')$ contradicts the minimality of S.

(i) $S = \{j\}$: Since $\chi'_{\overline{S}} = \chi_{\overline{S}}$, we see that $\chi'' = \chi$ and $f(\chi'') = f(\chi) \ne f(\chi')$, a contradiction.

(ii) $S - \{j\} \ne \emptyset$: If $f(\chi'') = f(\chi')$ then the subset $T = S - \{j\}$ would satisfy $\chi''_{\overline{S}} = \chi'_{\overline{S}} = \chi_{\overline{S}}$, $\chi''_j = \chi_j$ and $\lambda d(f(\chi), \chi_j) < d(f(\chi'), \chi_j) = d(f(\chi''), \chi_j)$ for all $i \in T$, contradicting the minimality of S. $\qquad\square$

Now we derive an upper bound on the benefit ratio of the mechanism f.

Lemma 7. *The benefit ratio of the mechanism f of Eq.(3) is at most $1 + 2/\lambda$ for any real $\lambda \geq 1$.*

Proof. We use the fact that $f(\chi) = c$ for $c \in \{0,1\}$ imply $\mu(c,\chi) \geq \mu(1-c,\chi)$ in Eq.(3), i.e.,

$$d(c,\chi|_{\bar{I}}) \leq d(1-c,\chi|_{\bar{I}}) + m_c - m_{1-c}, \tag{6}$$

which is symmetric for $c \in \{0,1\}$. For notational simplicity, we consider the case of $f(\chi) = 0$, because the other case of $f(\chi) = 1$ can be treated symmetrically.

For each $c \in \{0,1\}$, define $I_c^- = \{h \in I_c \mid h < c\}$, $I_c^+ = \{h \in I_c \mid h \geq c\}$, $m_c^- = |\chi|_{I_c^-}|$, $m_c^+ = |\chi|_{I_c^+}|$ and $m_c = |\chi|_{I_c}| = m_c^- + m_c^+$. For

$$D = d(0,\chi|_{\bar{I}}) + d(0,\chi|_{I_0^-}) + d(0,\chi|_{I_1^+}) \ (\geq 0),$$

we prove

$$\mathrm{sb}(0,\chi) = d(0,\chi) \geq D + m_1^- \frac{\lambda}{\lambda+1}$$

and

$$\mathrm{opt}(\chi) \leq D + m_1^- \left(1 + \frac{1}{\lambda+1}\right),$$

which implies the desired result

$$\frac{\mathrm{opt}(\chi)}{\mathrm{sb}(0,\chi)} \leq 1 + \frac{2}{\lambda}.$$

By noting that χ is a disjoint union of five multisets $\chi|_{\bar{I}}$, $\chi|_{I_0^-}$, $\chi|_{I_0^+}$, $\chi|_{I_1^-}$ and $\chi|_{I_1^+}$, we get

$$d(0,\chi) = d(0,\chi|_{\bar{I}}) + d(0,\chi|_{I_0^-}) + d(0,\chi|_{I_0^+}) + d(0,\chi|_{I_1^-}) + d(0,\chi|_{I_1^+})$$

$$\geq D + d(0,\chi|_{I_1^-}) \geq D + m_1^- \frac{\lambda}{\lambda+1} \ (\text{by } I_1^- = (\frac{\lambda}{\lambda+1},1)).$$

On the other hand, for $\mathrm{opt}(\chi) = \mathrm{sb}(1,\chi) = d(1,\chi)$, we get

$$\mathrm{opt}(\chi) = d(1,\chi|_{\bar{I}}) + d(1,\chi|_{I_0}) + d(1,\chi|_{I_1})$$

$$< d(0,\chi|_{\bar{I}}) + m_1 - m_0 + d(1,\chi|_{I_0}) + d(1,\chi|_{I_1}) \ (\text{by Eq. (6)})$$

$$= d(0,\chi|_{\bar{I}}) + m_1 + d(0,\chi|_{I_0^-}) - d(0,\chi|_{I_0^+}) + d(0,\chi|_{I_1^+}) - m_1^+ + d(1,\chi|_{I_1^-})$$

$$\leq D + m_1 - m_1^+ + d(1,\chi|_{I_1^-})$$

$$= D + m_1^- + d(1,\chi|_{I_1^-})$$

$$\leq D + m_1^- + m_1^- \frac{1}{\lambda+1} \ (\text{by } I_1^- = (\frac{\lambda}{\lambda+1},1)),$$

as required. □

In conclusion, the results of Lemmas 6 and 7, together give a proof of Theorem 3.

In light of a previous result by Cheng *et al.* [4], who have demonstrated a strategy-proof GSP mechanism with a benefit ratio at most 3 in the line metric, we see that the result of Theorem 3 follows as a natural extension of the introduction of λ-strategy proofness, matching the result of Cheng *et al.* [4] for $\lambda = 1$.

5   Lower Bounds on the Benefit Ratio   ·

This section derives a lower bound on the benefit ratio of all 2-candidate λ-SP mechanisms in the line metric.

By Theorem 2, we only need to handle a masking zone 2-candidate λ-SP mechanism. We show that every such λ-SP mechanisms f admits a profile χ_f such that $\frac{\mathrm{opt}(\chi_f)}{\mathrm{sb}(f(\chi_f),\chi_f)}$ is not smaller than $1+2/\lambda$ if n is even; $1+(2n-2)/(\lambda n+1)$ otherwise.

The following lemma establishes a lower bound on the benefit ratio of any 2-candidate masking zone mechanisms in the line.

Lemma 8. *Given a real $\lambda \geq 1$ and a set N of n (≥ 1) agents, let f be a masking zone mechanism f with candidate set $C(f) = \{0,1\}$ and set $\{I_0, I_1\}$ of masking zones. Then for any real $\delta > 0$, there is a profile χ such that*

$$\frac{\mathrm{opt}(\chi)}{\mathrm{sb}(f(\chi),\chi)} \geq \begin{cases} 1 + \frac{2}{\lambda}(\frac{1-\delta}{1+\delta}), & \text{if } n \text{ is even} \\ 1 + \frac{2n-2}{\lambda n+1}(\frac{1-\delta}{1+\delta}), & \text{otherwise.} \end{cases}$$

The proof of Lemma 8 follows by analysis on adversarially chosen profiles, but is omitted here due to space requirements. The reader is referred to a working version of this paper [10]. By Theorem 2 and Lemma 8, we obtain Theorem 4.

6   Concluding Remarks

This paper studied a trade-off between the benefit ratio and a relaxation of group strategy proofs, taking 2-candidate mechanisms for the obnoxious facility game in the line metric as an example. As a result we introduce λ-group strategy-proofness, a parameterized strategy proofness, and demonstrated a mechanism that has a desired property of a benefit ratio of at most $1 + \frac{2}{\lambda}$, which tends to 1, as the parameter λ tends to ∞. This result was obtained via a novel view on mechanism properties, and the introduction of the concept of masking zone mechanisms, which is a necessary condition for λ-GSP mechanisms. On the other hand, we also derived lower bounds on the benefit ratio of for masking zone mechanisms: $1 + \frac{2}{\lambda}$ when $n = |N|$ is even and $1 + \frac{2n-2}{\lambda n+1}$ when $n = |N|$ is odd. The above bounds are tight when $|N|$ is even, meaning that the upper bound on the benefit ratio is the best we can hope for, but it remains an open question to the slight gap between the upper and lower bounds for the case when $|N|$ is odd.

For future work, it remains to investigate the trade-off between the benefit ratio and the λ-GSP mechanisms in other metrics such as trees, circles and Euclidean space.

References

1. Alon, N., Feldman, M., Procaccia, A.D., Tennenholtz, M.: Strategyproof approximation mechanisms for location on networks. arXiv preprint (2009). arxiv:0907.2049
2. Border, K.C., Jordan, J.S.: Straightforward elections, unanimity and phantom voters. The Rev. Econ. Stud. **50**(1), 153–170 (1983)
3. Cheng, Y., Han, Q., Yu, W., Zhang, G.: Obnoxious facility game with a bounded service range. In: Chan, T.-H.H., Lau, L.C., Trevisan, L. (eds.) TAMC 2013. LNCS, vol. 7876, pp. 272–281. Springer, Heidelberg (2013)
4. Cheng, Y., Yu, W., Zhang, G.: Mechanisms for obnoxious facility game on a path. In: Wang, W., Zhu, X., Du, D.-Z. (eds.) COCOA 2011. LNCS, vol. 6831, pp. 262–271. Springer, Heidelberg (2011)
5. Ibara, K., Nagamochi, H.: Characterizing mechanisms in obnoxious facility game. In: Lin, G. (ed.) COCOA 2012. LNCS, vol. 7402, pp. 301–311. Springer, Heidelberg (2012)
6. Ibara, K., Nagamochi, H.: Characterizing mechanisms in obnoxious facility game. Technical report 2015-006, Department of Applied Mathematics and Physics, Kyoto University (2015). http://www.amp.i.kyoto-u.ac.jp/tecrep/ps_file/2015/2015-005.pdf
7. Lu, P., Sun, X., Wang, Y., Zhu, Z.A.: Asymptotically optimal strategy-proof mechanisms for two-facility games. In: Proceedings of the 11th ACM Conference on Electronic Commerce (ACM-EC 2010), pp. 315–324. ACM (2010)
8. Lu, P., Wang, Y., Zhou, Y.: Tighter bounds for facility games. In: Leonardi, S. (ed.) WINE 2009. LNCS, vol. 5929, pp. 137–148. Springer, Heidelberg (2009)
9. Moulin, H.: On strategy-proofness and single peakedness. Public Choice **35**(4), 437–455 (1980)
10. Oomine, M., Shurbevski, A., Nagamochi, H.: Parameterization of strategy-proof mechanisms in the obnoxious facility game. Technical report 2015-006, Department of Applied Mathematics and Physics, Kyoto University (2015). http://www.amp.i.kyoto-u.ac.jp/tecrep/ps_file/2015/2015-006.pdf
11. Procaccia, A.D., Tennenholtz, M.: Approximate mechanism design without money. In: Proceedings of the 10th ACM Conference on Electronic Commerce (ACM-EC 2009), pp. 177–186. ACM (2009)
12. Schummer, J., Vohra, R.V.: Strategy-proof location on a network. J. Econ. Theory **104**(2), 405–428 (2002)

On-line Algorithms

Optimal Online Algorithms
for the Multi-objective Time Series
Search Problem

Shun Hasegawa and Toshiya Itoh$^{(\boxtimes)}$

Tokyo Institute of Technology, Yokohama 226-8502, Japan
hasegawa.s.aj@m.titech.ac.jp, titoh@ip.titech.ac.jp

Abstract. Tiedemann, et al. [Proc. of WALCOM, LNCS 8973, 2015, pp. 210–221] defined multi-objective online problems and the competitive analysis for multi-objective online problems, and presented best possible online algorithms for the multi-objective online problems with respect to several measures of competitive analysis. In this paper, we first point out that the frameworks of the competitive analysis due to Tiedemann, et al. do not necessarily capture the efficiency of online algorithms for multi-objective online problems and provide modified definitions of the competitive analysis for multi-objective online problems. Under the modified framework, we present a simple online algorithm Balanced Price Policy BPP_k for the multi-objective time series search problem, and show that the algorithm BPP_k is *best possible* with respect to any measure of the competitive analysis. For the modified framework, we derive exact values of the competitive ratio for the multi-objective time series search problem with respect to the worst component competitive analysis, the arithmetic mean component competitive analysis, and the geometric mean component competitive analysis.

Keywords: Multi-objective online algorithms · Worst component competitive ratio · Arithmetic mean component competitive ratio · Geometric mean component competitive ratio

1 Introduction

Single-objective online optimization problems are fundamental in computing, communicating, and many other practical systems. To measure the efficiency of online algorithms for single-objective online optimization problems, a notion of competitive analysis was introduced by Sleator and Tarjan [8], and since then extensive research has been made for diverse areas, e.g., online paging and caching (see [10] for a survey), metric task systems (see [6] for a survey), asset conversion problems (see [7] for a survey), buffer management of network

T. Itoh—The author gratefully acknowledges the ELC project (Grant-in-Aid for Scientific Research on Innovative Areas MEXT Japan) for encouraging the research presented in this paper.

© Springer International Publishing Switzerland 2016
M. Kaykobad and R. Petreschi (Eds.): WALCOM 2016, LNCS 9627, pp. 301–312, 2016.
DOI: 10.1007/978-3-319-30139-6_24

switches (see [4] for a survey), etc. All of these are single-objective online problems. In practice, there are many online problems of multi-objective nature, but we have no general framework of competitive analysis and no definition of competitive ratio. Tiedemann, et al. [9] first formulated a framework of multi-objective online problems as the online version of multi-objective optimization problems [2] and defined a notion of the competitive ratio for multi-objective online problems by extending the competitive ratio for single-objective online problems. To define the competitive ratio for multi-objective (k-objective) online problems, Tiedemann, et al. [9] regarded multi-objective online problems as a family of (possibly dependent) single-objective online problems and applied a monotone (continuous) function $f : \mathbf{R}^k \to \mathbf{R}$ to the family of the single-objective online problems. Given an algorithm ALG for a multi-objective (k-objective) online problem, regard ALG as a family of algorithms ALG_i for the ith objective of the input sequence. For the set of k competitive ratios $\{c_1, \ldots, c_k\}$, we say that the algorithm ALG is $f(c_1, \ldots, c_k)$-competitive with respect to a monotone continuous function $f : \mathbf{R}^k \to \mathbf{R}$. In fact, Tiedemann, et al. [9] defined the worst component competitive ratio, the arithmetic mean component competitive ratio, and the geometric mean component competitive ratio by functions

$$f_1(c_1, \ldots, c_k) = \max(c_1, \ldots, c_k),$$
$$f_2(c_1, \ldots, c_k) = (c_1 + \cdots + c_k)/k,$$
$$f_3(c_1, \ldots, c_k) = (c_1 \times \cdots \times c_k)^{1/k},$$

respectively. Note that the functions f_1, f_2, and f_3 are monotone and continuous.

1.1 Previous Work

El-Yaniv, et al. [3] initially investigated the single-objective time series search problem. For the single-objective time series search problem, prices are revealed time by time and the goal of the algorithm is to select one of them as with high price as possible. Assume that $m > 0$ and $M > m$ are the minimum and maximum values of possible prices, respectively, and let $\phi = M/m$ be the *fluctuation ratio* of possible prices. Under the assumption that $M > m > 0$ are known to the algorithms, El-Yaniv, et al. [3] presented a deterministic *reservation price policy* RPP, which is $\sqrt{\phi}$-competitive algorithm and best possible, and a randomized algorithm *exponential threshold* EXPO, which is $O(\log \phi)$-competitive.

In a straightforward manner, Tiedemann, et al. [9] defined the multi-objective time series search problem by generalizing the single-objective time series search problem. For the multi-objective (k-objective) time series search problem, a vector of k (possibly dependent) prices are revealed time by time and the goal of the algorithm is to select one of the price vectors as with low competitive ratio as possible with respect to the monotone continuous function f. Tiedemann, et al. [9] presented best possible online algorithms for the multi-objective time series search problem with respect to the monotone continuous functions f_1, f_2, and f_3, i.e., a best possible online algorithm for the multi-objective time series search problem with respect to the monotone continuous function f_1 [9, Theorems 1 and 2], a best possible online algorithm for the bi-objective time series search problem with respect to the monotone continuous function f_2 [9, Theorems 3 and 4] and a best possible online algorithm for the bi-objective time series search problem with respect to the monotone continuous function f_3 [9, §3.2].

1.2   Our Contribution

We first observe that the definition and framework of competitive analysis given by Tiedemann, et al. [9, Definitions 1, 2, and 3] do not necessarily capture the efficiency of algorithms for multi-objective online problems. Then we introduce modified definition and framework of competitive analysis for multi-objective online problems and verify that each result [9] for the multi-objective time series search problem holds under the modified framework of competitive analysis.

As mentioned in Subsect. 1.1, Tiedemann, et al. [9] showed best possible online algorithms for the multi-objective time series search problem with respect to the monotone continuous functions f_1, f_2 and f_3, however, the optimality for the algorithm with respect to each of the monotone continuous functions f_1, f_2 and f_3 is discussed separately and independently. In this paper, we present a simple online algorithm for the multi-objective time series search problem with respect to any monotone continuous function $f : \mathbf{R}^k \to \mathbf{R}$ and then show that under the modified framework of competitive analysis, the proposed algorithm is *best possible* for any monotone continuous function $f : \mathbf{R}^k \to \mathbf{R}$ (in Theorems 1 and 2). With respect to the monotone continuous functions f_1, f_2, and f_3, we also derive best possible values of the competitive ratio under the modified framework of competitive analysis in Theorems 3, 4, and 5, respectively.

By Theorems 1 and 2, we note that (1) Theorem 3 gives another proof of the result that the algorithm [9, Theorem 1] is best possible for the multi-objective time series search problem with respect to f_1, (2) Theorem 4 disproves the result that the algorithm [9, Theorem 3] is best possible for the bi-objective time series search problem with respect to f_2, and (3) Theorem 5 gives a best possible online algorithm for the multi-objective time series search problem with respect to f_3, which is an extension of the result that the algorithm [9, Theorem 3] is best possible for the bi-objective time series search problem with respect to f_3.

2   Preliminaries

For the subsequent discussions, we present some notations and terminologies. For any pair of integers $a \leq b$, we use $[a, b]$ to denote a set $\{a, \ldots, b\}$ and for any pair of vectors $\boldsymbol{x} = (x_1, \ldots, x_k) \in \mathbf{R}^k$ and $\boldsymbol{y} = (y_1, \ldots, y_k) \in \mathbf{R}^k$, we use $\boldsymbol{x} \preceq \boldsymbol{y}$ to denote a componentwise order, i.e., $x_i \leq y_i$ for each $i \in [1, k]$. It is immediate that $\preceq$ is a partial order on $\mathbf{R}^k$. A function $f : \mathbf{R}^k \to \mathbf{R}$ is said to be *monotone* if $f(\boldsymbol{x}) \leq f(\boldsymbol{y})$ for any pair of vectors $\boldsymbol{x} \in \mathbf{R}^k$ and $\boldsymbol{y} \in \mathbf{R}^k$ such that $\boldsymbol{x} \preceq \boldsymbol{y}$.

2.1   Multi-Objective Online Problems

Tiedemann, et al. [9] formulated a framework of multi-objective online problems by using that of multi-objective optimization problems [2]. In this subsection, we present multi-objective maximization problems (multi-objective minimization problems can be defined analogously).

Let $\mathcal{P}_k = (\mathcal{I}, \mathcal{X}, h)$ be a multi-objective optimization (maximization) problem, where $\mathcal{I}$ is a set of inputs, $\mathcal{X}(I) \subseteq \mathbf{R}^k$ is a set of feasible solutions for each input $I \in \mathcal{I}$, and $h : \mathcal{I} \times \mathcal{X} \to \mathbf{R}^k$ is a function such that $h(I, \boldsymbol{x}) \in \mathbf{R}^k$ represents the objective of each solution $\boldsymbol{x} \in \mathcal{X}(I)$. For an input $I \in \mathcal{I}$, an algorithm ALG_k for $\mathcal{P}_k$ outputs a feasible solution $\mathrm{ALG}_k[I] \in \mathcal{X}(I)$. For an input $I \in \mathcal{I}$ and a feasible solution $\mathrm{ALG}_k[I]$, let $\mathrm{ALG}_k(I) = h(I, \mathrm{ALG}_k[I]) \in \mathbf{R}^k$ be the objective associate with $\mathrm{ALG}_k[I] \in \mathcal{X}(I)$. A feasible solution $\boldsymbol{x}_{max} \in \mathcal{X}(I)$ is said to be *maximal* if there exists no feasible solution $\boldsymbol{x} \in \mathcal{X}(I) \backslash \{\boldsymbol{x}_{max}\}$ such that $h(I, \boldsymbol{x}_{max}) \preceq h(I, \boldsymbol{x})$ and for any input $I \in \mathcal{I}$, let $\mathrm{OPT}_k[I] \subseteq \mathbf{R}^k$ be the set of maximal solutions to $\mathcal{P}_k$.

A multi-objective online problem can be defined in a way similar to a single-objective online problem [1]. We regard a multi-objective online problem as a multi-objective optimization problem in which the input is revealed bit by bit and an output must be produced in an online manner, i.e., after each new part of input is revealed, a decision affecting the output must be made.

2.2   Competitive Analysis for Multi-Objective Online Problems

Tiedemann, et al. [9] defined a notion of competitive analysis for multi-objective online problems[1]. In this subsection, we introduce the notion of competitive analysis for multi-objective online problems with respect to maximization problems (the corresponding minimization problem can be defined analogously).

Definition 1 ([9]). *Let $\mathcal{P}_k = (\mathcal{I}, \mathcal{X}, h)$ be a multi-objective maximization problem. For a vector $\boldsymbol{c} = (c_1, \dots, c_k) \in \mathbf{R}^k$, a multi-objective online algorithm ALG_k for $\mathcal{P}_k$ is $\boldsymbol{c}$-competitive if for every input sequence $I \in \mathcal{I}$ and every maximal solution $\boldsymbol{x} \in \mathrm{OPT}_k[I]$, $h(I, \boldsymbol{x})_i \leq c_i \cdot \mathrm{ALG}_k(I)_i + \alpha_i$ holds for each $i \in [1, k]$, where $\boldsymbol{\alpha} = (\alpha_1, \dots, \alpha_k) \in \mathbf{R}^k$ is a constant vector independent of inputs $I \in \mathcal{I}$.*

Let $f : \mathbf{R}^k \to \mathbf{R}$ be a monotone continuous function. For a multi-objective online algorithm ALG_k for $\mathcal{P}_k$, the *competitive ratio* of the algorithm ALG_k with respect to f is the infimum of $f(\boldsymbol{c})$ over all possible vectors $\boldsymbol{c} = (c_1, \dots, c_k) \in \mathbf{R}^k$ such that ALG_k is $\boldsymbol{c}$-competitive. We use $\mathcal{C}[\mathrm{ALG}_k]$ to denote the set of all possible vectors $\boldsymbol{c} = (c_1, \dots, c_k) \in \mathbf{R}^k$ such that ALG_k is $\boldsymbol{c}$-competitive.

Definition 2 ([9]). *For a monotone continuous function $f : \mathbf{R}^k \to \mathbf{R}$ and an online algorithm ALG_k for a multi-objective maximization problem $\mathcal{P}_k$, the* competitive ratio *of ALG_k with respect to f is $\mathcal{R}^f(\mathrm{ALG}_k) = \inf_{\boldsymbol{c} \in \mathcal{C}[\mathrm{ALG}_k]} f(\boldsymbol{c})$.*

---

[1] Tiedemann, et al. [9] introduced notions of (strong) $\boldsymbol{c}$-competitive and (strong) competitive ratio. In this paper, we do not deal with the notion of $\boldsymbol{c}$-competitive and competitive ratio. Thus for simplicity, we refer to strong $\boldsymbol{c}$-competitive and strong competitive ratio as $\boldsymbol{c}$-competitive and competitive ratio, respectively.

Several examples of a monotone continuous function $f : \mathbf{R}^k \to \mathbf{R}$ are given by Tiedemann, et al. [9], e.g., $f_1(c_1, \ldots, c_k) = \max(c_1, \ldots, c_k), f_2(c_1, \ldots, c_k) = (c_1 + \cdots + c_k)/k, f_3(c_1, \ldots, c_k) = (c_1 \times \cdots \times c_k)^{1/k}$. We refer to the competitive ratio of an algorithm ALG_k with respect to f_1, f_2, and f_3 as *worst component* competitive ratio, *arithmetic mean component* competitive ratio, and *geometric mean component* competitive ratio, respectively.

2.3 Multi-objective Time Series Search Problem

A single-objective time series search problem is initially investigated by El-Yaniv, et al. [3] and it is defined as follows: An online player ALG is searching for the maximum price in a sequence of prices. At the beginning of each time period $t \in [1, T]$, a price p_t is revealed to the online player ALG and it must decide whether to accept or reject the price p_t. If the online player ALG accepts the price p_t, then the game ends and the return for ALG is p_t. We assume that prices are chosen from the *real* interval $I = [m, M]$, where $0 < m \le M$, and m and M are known to the online player ALG[2]. If the online player ALG rejects the price p_t for every $t \in [1, T]$, then the return for ALG is defined to be m. A multi-objective time series search problem [9] can be defined by a natural extension of the single-objective time series search problem. In the multi-objective time series search problem, a vector $\boldsymbol{p}_t = (p_t^1, \ldots, p_t^k) \in \mathbf{R}^k$ is revealed to the online player ALG_k at the beginning of each time period $t \in [1, T]$, and the online player ALG_k must decide whether to accept or reject the price vector $\boldsymbol{p}_t$. If the online player ALG_k accepts the price vector $\boldsymbol{p}_t$, then the game ends and the return for ALG_k is $\boldsymbol{p}_t$. As in the case of a single-objective time series search problem, assume that prices p_t^i are chosen from the *real* interval $I_i = [m_i, M_i]$ with $0 < m_i \le M_i$ for each $i \in [1, k]$, and that the online player ALG_k knows m_i and M_i for each $i \in [1, k]$. If the online player ALG_k rejects the price vector $\boldsymbol{p}_t$ for every $t \in [1, T]$, then the return for ALG_k is defined to be the *minimum* price vector $\boldsymbol{p}_{min} = (m_1, \ldots, m_k)$. Without loss of generality, we assume that $M_1/m_1 \ge \cdots \ge M_k/m_k$.

3   Observations on the Competitive Analysis

For the multi-objective (k-objective) time series search problem, it is natural to regard that m_i and M_i are part of the problem (not part of inputs) for each $i \in [1, k]$. Set $\alpha_i = M_i$ for each $i \in [1, k]$ in Definition 1. Then we can take $c_1 = \cdots = c_k = 0$. This implies that any algorithm ALG for the multi-objective (k-objective) time series search problem is $(0, \ldots, 0)$-competitive, i.e., for any monotone continuous function $f : \mathbf{R}^k \to \mathbf{R}$, the competitive ratio of the algorithm ALG is $f(0, \ldots, 0)$. Thus in Definition 1, we fix $\alpha_i = 0$ for each $i \in [1, k]$.

For simplicity, assume that $k = 2$ and $I_1 = I_2 = [m, M]$, where $0 < m < M$. Consider a simple algorithm ALG_2 that accepts the first price vector.

---

[2] It is possible to show that if only the fluctuation ratio $\phi = M/m$ is known (but not m or M) to ALG, then no better competitive ratio than the trivial one of ϕ is achievable.

Example 1. Let $\mathcal{I}_1 = \{s_1, s_2\}$ be the set of input sequences. In the sequence s_1, $\boldsymbol{p}_1 = (m, M)$, $\boldsymbol{p}_2 = (M, m)$, and $\boldsymbol{p}_3 = (m, m)$ are revealed to the algorithm ALG2 at $t = 1$, $t = 2$, and $t = 3$, respectively, and in the sequence s_2, $\boldsymbol{q}_1 = (M, m)$, $\boldsymbol{q}_2 = (m, m)$, and $\boldsymbol{q}_3 = (m, M)$ are revealed to the algorithm ALG2 at $t = 1$, $t = 2$, and $t = 3$, respectively. Note that for the sequence s_1, the algorithm ALG2 accepts $\boldsymbol{p}_1 = (m, M)$ which is maximal in s_1 and for the sequence s_2, the algorithm ALG2 accepts $\boldsymbol{p}_2 = (M, m)$ which is also maximal in s_2. From Definition 1, we have that the algorithm ALG2 is $(M/m, M/m)$-competitive.

Example 2. Let $\mathcal{I}_2 = \{\sigma\}$ be the set of input sequences. In the sequence σ, price vectors $\boldsymbol{r}_1 = (m, m)$, $\boldsymbol{r}_2 = (m, M)$, and $\boldsymbol{r}_3 = (M, m)$ are revealed at $t = 1$, $t = 2$, and $t = 3$ to the algorithm ALG2, respectively. Note that the algorithm ALG2 accepts $\boldsymbol{r}_1 = (m, m)$ which is not maximal in σ. From Definition 1, we have that the algorithm ALG2 is $(M/m, M/m)$-competitive.

Notice that in Example 1, the algorithm ALG2 accepts price vectors which is maximal in the sequences s_1 and s_2, however, in Example 2, the algorithm ALG2 accepts a price vector which is not maximal in the sequence σ. Thus it follows that for any monotone continuous function $f : \mathbf{R}^2 \to \mathbf{R}$, the competitive ratio of the algorithm ALG2 is $f(M/m, M/m)$ for both Examples 1 and 2, which does not necessarily capture the efficiency of online algorithms. To derive a more realistic framework, we modify the definition of competitive ratio.

Let ALG$_k$ be an online algorithm for a multi-objective maximization problem $\mathcal{P}_k = (\mathcal{I}, \mathcal{X}, h)$ and $\mathcal{CR}^f(\text{ALG}_k; I)$ be the competitive ratio of the algorithm ALG$_k$ for an input sequence $I \in \mathcal{I}$ with respect to a function $f : \mathbf{R}^k \to \mathbf{R}$, i.e.,

$$\mathcal{CR}^f(\text{ALG}_k; I) = \sup_{\boldsymbol{x} \in \text{OPT}_k[I]} f\left(\frac{h(I, \boldsymbol{x})_1}{\text{ALG}_k(I)_1}, \ldots, \frac{h(I, \boldsymbol{x})_k}{\text{ALG}_k(I)_k}\right).$$

Definition 3. *For a monotone continuous function $f : \mathbf{R}^k \to \mathbf{R}$ and an online algorithm* ALG$_k$ *for a multi-objective maximization problem $\mathcal{P}_k$, the* competitive ratio *of* ALG$_k$ *with respect to f is $\mathcal{CR}^f(\text{ALG}_k) = \sup_{I \in \mathcal{I}} \mathcal{CR}^f(\text{ALG}_k; I)$.*

In fact, all of the proofs on the competitive ratio [9] hold under Definition 3. In the rest of the paper, we analyze the algorithms according to Definition 3.

4   Online Algorithm: Balanced Price Policy

In this section, we present a simple online algorithm Balanced Price Policy BPP$_k$ (in Fig. 1) for the multi-objective (k-objective) time series search problem with respect to an arbitrary monotone continuous function $f : \mathbf{R}^k \to \mathbf{R}$.

Let $\mathbf{I}_k = I_1 \times \cdots \times I_k$ and $z_f^k = \sup_{(x_1, \ldots, x_k) \in \mathcal{S}_f^k} f(M_1/x_1, \ldots, M_k/x_k)$, where

$$\mathcal{S}_f^k = \left\{(x_1, \ldots, x_k) \in \mathbf{I}_k : f\left(\frac{M_1}{x_1}, \ldots, \frac{M_k}{x_k}\right) = f\left(\frac{x_1}{m_1}, \ldots, \frac{x_k}{m_k}\right)\right\}.$$

$$
\boxed{
\begin{aligned}
&\textbf{for } t = 1, 2, \ldots, T \textbf{ do} \\
&\quad \left| \; \text{Accept } \boldsymbol{p}_t = (p_t^1, \ldots, p_t^k) \text{ if } f(\tfrac{M_1}{p_t^1}, \ldots, \tfrac{M_k}{p_t^k}) \le f(\tfrac{p_t^1}{m_1}, \ldots, \tfrac{p_t^k}{m_k}). \right. \\
&\textbf{end}
\end{aligned}
}
$$

Fig. 1. Balanced price policy BPP_k

By setting $x_i = \sqrt{m_i M_i} \in I_i = [m_i, M_i]$ for each $i \in [1, k]$, it is immediate that

$$
f\left(\frac{M_1}{x_1}, \ldots, \frac{M_k}{x_k}\right) = f\left(\frac{M_1}{\sqrt{m_1 M_1}}, \ldots, \frac{M_k}{\sqrt{m_k M_k}}\right) = f\left(\sqrt{\frac{M_1}{m_1}}, \ldots, \sqrt{\frac{M_k}{m_k}}\right);
$$

$$
f\left(\frac{x_1}{m_1}, \ldots, \frac{x_k}{m_k}\right) = f\left(\frac{\sqrt{m_1 M_1}}{m_1}, \ldots, \frac{\sqrt{m_k M_k}}{m_k}\right) = f\left(\sqrt{\frac{M_1}{m_1}}, \ldots, \sqrt{\frac{M_k}{m_k}}\right).
$$

For any f, it follows that $(\sqrt{m_1 M_1}, \ldots, \sqrt{m_k M_k}) \in \mathcal{S}_f^k$. So we have that $\mathcal{S}_f^k \ne \emptyset$ and $z_f^k = \sup_{(x_1, \ldots, x_k) \in \mathcal{S}_f^k} f(M_1/x_1, \ldots, M_k/x_k)$ is well-defined.

4.1 The Algorithm BPP_k is Best Possible

In this subsection, we show that the algorithm BPP_k is *best possible* for any integer $k \ge 1$. More precisely, we show that $\mathcal{CR}^f(\mathrm{BPP}_k) \le z_f^k$ (in Theorem 1) and that $\mathcal{CR}^f(\mathrm{ALG}_k) \ge z_f^k$ for any algorithm ALG_k (in Theorem 2).

Theorem 1. $\mathcal{CR}^f(\mathrm{BPP}_k) \le z_f^k$ *for any integer* $k \ge 1$.

Proof: Let $\mathcal{I}$ be the set of input sequences. Define $\mathcal{I}_{\mathrm{acc}}, \mathcal{I}_{\mathrm{rej}} \subseteq \mathcal{I}$ as follows:

$$
\mathcal{I}_{\mathrm{acc}} = \left\{ (\boldsymbol{p}_1, \ldots, \boldsymbol{p}_T) \in \mathcal{I} : \bigvee_{t \in [1,T]} \left[f\left(\frac{M_1}{p_t^1}, \ldots, \frac{M_k}{p_t^k}\right) \le f\left(\frac{p_t^1}{m_1}, \ldots, \frac{p_t^k}{m_k}\right) \right] \right\};
$$

$$
\mathcal{I}_{\mathrm{rej}} = \left\{ (\boldsymbol{p}_1, \ldots, \boldsymbol{p}_T) \in \mathcal{I} : \bigwedge_{t \in [1,T]} \left[f\left(\frac{M_1}{p_t^1}, \ldots, \frac{M_k}{p_t^k}\right) > f\left(\frac{p_t^1}{m_1}, \ldots, \frac{p_t^k}{m_k}\right) \right] \right\}.
$$

For each $I = (\boldsymbol{p}_1, \ldots, \boldsymbol{p}_T) \in \mathcal{I}_{\mathrm{acc}}$, where $\boldsymbol{p}_t = (p_t^1, \ldots, p_t^k) \in \mathbf{R}^k$ for each $t \in [1, T]$, the algorithm BPP_k terminates at the earliest $t \in [1, T]$ to accept $\boldsymbol{p}_t = (p_t^1, \ldots, p_t^k)$ such that $f(M_1/p_t^1, \ldots, M_k/p_t^k) \le f(p_t^1/m_1, \ldots, p_t^k/m_k)$. Thus

$$
\mathcal{CR}^f(\mathrm{BPP}_k; I) = \sup_{\boldsymbol{x} \in \mathrm{OPT}_k[I]} f\left(\frac{h(I, \boldsymbol{x})_1}{p_t^1}, \ldots, \frac{h(I, \boldsymbol{x})_k}{p_t^k}\right) \le f\left(\frac{M_1}{p_t^1}, \ldots, \frac{M_k}{p_t^k}\right).
$$

To show that $f(M_1/p_t^1, \ldots, M_k/p_t^k) \le z_f^k$, consider the following subcases:

(1.1) $f(M_1/p_t^1, \ldots, M_k/p_t^k) = f(p_t^1/m_1, \ldots, p_t^k/m_k)$;

(1.2) $f(M_1/p_t^1, \ldots, M_k/p_t^k) < f(p_t^1/m_1, \ldots, p_t^k/m_k)$.

For the subcase (1.1), it is immediate that $\boldsymbol{p}_t \in \mathcal{S}_f^k$ and $f(M_1/p_t^1, \ldots, M_k/p_t^k) \leq z_f^k$. For the subcase (1.2), let $\mathcal{J} = \{j \in [1, k] : M_j/p_t^j \leq p_t^j/m_j\}$. Notice that $\mathcal{J} \neq \emptyset$[3]. For simplicity, we assume that $\mathcal{J} = \{1, 2, \ldots, u\}$ for $u \geq 1$. By setting $p_t^j = m_j$ for each $j \in \mathcal{J}$, we have that

$$f\left(\frac{M_1}{m_1}, \ldots, \frac{M_u}{m_u}, \frac{M_{u+1}}{p_t^{u+1}}, \ldots, \frac{M_k}{p_t^k}\right) \geq f\left(1, \ldots, 1, \frac{p_t^{u+1}}{m_{u+1}}, \ldots, \frac{p_t^k}{m_k}\right).$$

Since f is monotone and continuous, there exist $q_t^i \in [m_i, p_t^i]$ $(1 \leq i \leq u)$ so that

$$
\begin{aligned}
f\left(\frac{M_1}{p_t^1}, \ldots, \frac{M_k}{p_t^k}\right) &\leq f\left(\frac{M_1}{q_t^1}, \ldots, \frac{M_u}{q_t^u}, \frac{M_{u+1}}{p_t^{u+1}}, \ldots, \frac{M_k}{p_t^k}\right) \\
&= f\left(\frac{q_t^1}{m_1}, \ldots, \frac{q_t^u}{m_u}, \frac{p_t^{u+1}}{m_{u+1}}, \ldots, \frac{p_t^k}{m_k}\right) \leq f\left(\frac{p_t^1}{m_1}, \ldots, \frac{p_t^k}{m_k}\right).
\end{aligned}
$$

Then it turns out that $(q_t^1, \ldots, q_t^u, p_t^{u+1}, \ldots, p_t^k) \in \mathcal{S}_f^k$ and it follows that

$$f\left(\frac{M_1}{p_t^1}, \ldots, \frac{M_k}{p_t^k}\right) \leq f\left(\frac{M_1}{q_t^1}, \ldots, \frac{M_u}{q_t^u}, \frac{M_{u+1}}{p_t^{u+1}}, \ldots, \frac{M_k}{p_t^k}\right) \leq z_f^k.$$

For each $I = (\boldsymbol{p}_1, \ldots, \boldsymbol{p}_T) \in \mathcal{I}_{\mathrm{rej}}$, the algorithm BPP_k rejects a price vector $\boldsymbol{p}_t$ for every $t \in [1, T]$ and eventually settles in the minimum price vector $\boldsymbol{p}_{min} = (m_1, \ldots, m_k)$, while the optimal offline algorithm can accept a price vector $\boldsymbol{p}_\tau = (p_\tau^1, \ldots, p_\tau^k) = \max_{t \in [1, T]} f(p_t^1/m_1, \ldots, p_t^k/m_k)$. Thus

$$\mathcal{CR}^f(\mathrm{BPP}_k; I) = \sup_{\boldsymbol{x} \in \mathrm{OPT}_k[I]} f\left(\frac{h(I, \boldsymbol{x})_1}{m_1}, \ldots, \frac{h(I, \boldsymbol{x})_k}{m_k}\right) \leq f\left(\frac{p_\tau^1}{m_1}, \ldots, \frac{p_\tau^k}{m_k}\right).$$

We show that $f(p_\tau^1/m_1, \ldots, p_\tau^k/m_k) \leq z_f^k$. Since the algorithm BPP_k rejects a price vector $\boldsymbol{p}_t$ for every $t \in [1, T]$, it is immediate that

$$f(M_1/p_\tau^1, \ldots, M_k/p_\tau^k) > f(p_\tau^1/m_1, \ldots, p_\tau^k/m_k).$$

Let $\mathcal{H} = \{h \in [1, k] : M_h/p_\tau^h \geq p_\tau^h/m_h\}$. Note that $\mathcal{H} \neq \emptyset$[4]. We assume that $\mathcal{H} = \{1, 2, \ldots, v\}$ for $v \geq 1$. By setting $p_\tau^h = M_h$ for each $h \in \mathcal{H}$, we have that

$$f\left(1, \ldots, 1, \frac{M_{v+1}}{p_\tau^{v+1}}, \ldots, \frac{M_k}{p_\tau^k}\right) \leq f\left(\frac{M_1}{m_1}, \ldots, \frac{M_v}{m_v}, \frac{p_\tau^{v+1}}{m_{v+1}}, \ldots, \frac{p_\tau^k}{m_k}\right).$$

---

[3] If $\mathcal{J} = \emptyset$, then $M_i/p_t^i > p_t^i/m_i$ for each $i \in [1, k]$. Since $f : \mathbf{R}^k \to \mathbf{R}$ is a monotone function, we have that $f(M_1/p_t^1, \ldots, M_k/p_t^k) \geq f(p_t^1/m_1, \ldots, p_t^k/m_k)$, which contradicts the assumption that $f(M_1/p_t^1, \ldots, M_k/p_t^k) < f(p_t^1/m_1, \ldots, p_t^k/m_k)$.

[4] If $\mathcal{H} = \emptyset$, then $M_i/p_\tau^i < p_\tau^i/m_i$ for each $i \in [1, k]$. Since $f : \mathbf{R}^k \to \mathbf{R}$ is a monotone function, we have that $f(M_1/p_t^1, \ldots, M_k/p_t^k) \leq f(p_t^1/m_1, \ldots, p_t^k/m_k)$, which contradicts the assumption that $f(M_1/p_t^1, \ldots, M_k/p_t^k) > f(p_t^1/m_1, \ldots, p_t^k/m_k)$.

Since f is monotone and continuous, there exist $q_\tau^i \in [p_\tau^i, M_i]$ $(1 \le i \le v)$ so that

$$f\left(\frac{p_\tau^1}{m_1}, \ldots, \frac{p_\tau^k}{m_k}\right) \le f\left(\frac{q_\tau^1}{m_1}, \ldots, \frac{q_\tau^v}{m_v}, \frac{p_\tau^{v+1}}{m_{v+1}}, \ldots, \frac{p_\tau^k}{m_k}\right)$$
$$= f\left(\frac{M_1}{q_\tau^1}, \ldots, \frac{M_v}{q_\tau^v}, \frac{M_{v+1}}{p_\tau^{v+1}}, \ldots, \frac{M_k}{p_\tau^k}\right) \le f\left(\frac{M_1}{p_\tau^1}, \ldots, \frac{M_k}{p_\tau^k}\right).$$

Then it turns out that $(q_\tau^1, \ldots, q_\tau^v, p_\tau^{v+1}, \ldots, p_\tau^k) \in \mathcal{S}_f^k$ and it follows that

$$f\left(\frac{p_\tau^1}{m_1}, \ldots, \frac{p_\tau^k}{m_k}\right) \le f\left(\frac{q_\tau^1}{m_1}, \ldots, \frac{q_\tau^v}{m_v}, \frac{p_\tau^{v+1}}{m_{v+1}}, \ldots, \frac{p_\tau^k}{m_k}\right) \le z_f^k.$$

Since $\mathcal{I}_{\mathrm{acc}} \cap \mathcal{I}_{\mathrm{rej}} = \emptyset$ and $\mathcal{I}_{\mathrm{acc}} \cup \mathcal{I}_{\mathrm{acc}} = \mathcal{I}$, we have that $\mathcal{CR}^f(\mathrm{BPP}_k; I) \le z_f^k$ for any $I \in \mathcal{I}$, and it follows that $\mathcal{CR}^f(\mathrm{BPP}_k) = \sup_{I \in \mathcal{I}} \mathcal{CR}^f(\mathrm{BPP}_k; I) \le z_f^k$. ∎

Theorem 2. $\mathcal{CR}^f(\mathrm{ALG}_k) \ge z_f^k$ for any integer $k \ge 1$ and any algorithm ALG_k.

Proof: Let ALG_k be an arbitrarily online algorithm and $(x_1^*, \ldots, x_k^*) \in \mathcal{S}_f^k$ be a price vector such that $z_f^k = f(M_1/x_1^*, \ldots, M_k/x_k^*)$. The adversary first reveals a price vector $\boldsymbol{p} = (x_1^*, \ldots, x_k^*)$. If the algorithm ALG_k accepts the price vector $\boldsymbol{p}$, then the adversary reveals another price vector $\boldsymbol{p}_{max} = (M_1, \ldots, M_k)$ and accepts $\boldsymbol{p}_{max}$. Let $I = (\boldsymbol{p}, \boldsymbol{p}_{max})$ be an input sequence. Then from the definition of $\boldsymbol{p}$, we have that $\mathcal{CR}^f(\mathrm{ALG}_k; I) = f(M_1/x_1^*, \ldots, M_k/x_k^*) = z_f^k$. If the algorithm ALG_k rejects $\boldsymbol{p}$, then the adversary accepts the price vector $\boldsymbol{p}$ but reveals no further price vectors until the algorithm ALG_k eventually settles in the minimum price vector $\boldsymbol{p}_{min} = (m_1, \ldots, m_k)$. Let $J = (\boldsymbol{p})$ be an input sequence. From the definition of the price vector $\boldsymbol{p}$, we have hat $z_f^k = f(x_1^*/m_1, \ldots, x_k^*/m_k)$. Then it follows that $\mathcal{CR}^f(\mathrm{ALG}_k; J) = f(x_1^*/m_1, \ldots, x_k^*/m_k) = z_f^k$. Thus $\mathcal{CR}^f(\mathrm{ALG}_k) = \sup_{I \in \mathcal{I}} \mathcal{CR}^f(\mathrm{ALG}_k; I) \ge z_f^k$ for any algorithm ALG_k. ∎

From Theorems 1 and 2, we immediately have the following result.

Corollary 1. $\mathcal{CR}^f(\mathrm{BPP}_k) = z_f^k$ for any integer $k \ge 1$.

4.2 Discussions

As mentioned in Subsect. 1.1, El-Yaniv, et al. [3] presented the algorithm RPP for the single-objective time series search problem (see Fig. 2). We refer to p^* as the *reservation price*, where p^* is the solution of $M/p = p/m$.

For the monotone continuous functions f_1, f_2, and f_3, we have that $f_1(x) = f_2(x) = f_3(x) = x$ if $k = 1$. This implies that the algorithm BPP_1 coincides with the algorithm RPP with respect to f_1, f_2, and f_3, however, this is not necessarily the case for arbitrary nondecreasing continuous functions $f : \mathbf{R} \to \mathbf{R}$. Consider the following nondecreasing continuous function $g : \mathbf{R} \to \mathbf{R}$ (see Fig. 3).

> **for** $t = 1, 2, \ldots, T$ **do**
> | Accept p_t if $p_t \geq p^* = \sqrt{Mm}$.
> **end**

Fig. 2. Reservation price policy RPP

From the assumption that $0 < m < M$, it follows that $M/m > 1$ and we can take any constant c such that $1 < c < \sqrt{M/m}$. Then we have that

$$g(M/p) \begin{cases} > g(p/m) & \text{for } m \leq p < \sqrt{Mm}/c; \\ = g(p/m) & \text{for } \sqrt{Mm}/c \leq p \leq c\sqrt{Mm}; \\ > g(p/m) & \text{for } c\sqrt{Mm} < p \leq M. \end{cases}$$

Thus the algorithm BPP$_1$ does not coincide with the algorithm RPP [3] with respect to the function $g : \mathbf{R} \rightarrow \mathbf{R}$ in Fig. 3.

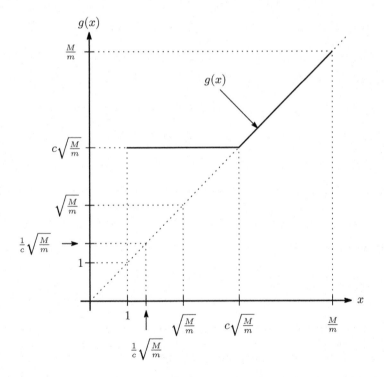

Fig. 3. Counterexample for nondecreasing continuous function $g : \mathbf{R} \rightarrow \mathbf{R}$

5 Analysis for Competitive Ratio

In this section, we derive exact values of the competitive ratio for the multi-objective time series search problem with respect to the worst component competitive analysis, the arithmetic mean component competitive analysis, and the geometric mean component competitive analysis.

All the proofs of Theorems 3, 4, and 5 are omitted due to the page limitation, but those can be found in [5].

5.1 Worst Component Competitive Ratio

In this section, we show that $\mathcal{CR}^{f_1}(\text{BPP}_k) = z_{f_1}^k = \max\{\sqrt{M_1/m_1}, M_2/m_2\}$. This implies that the algorithm RPP-HIGH [9, Algorithm 1] can be regarded as a special case of the algorithm BPP_k with respect to the function $f_1(c_1, \ldots, c_k) = \max(c_1, \ldots, c_k)$. For the function f_1, let

$$\mathcal{S}_{f_1}^k = \left\{ (x_1, \ldots, x_k) \in \mathbf{I}_k : \max\left(\frac{M_1}{x_1}, \ldots, \frac{M_k}{x_k} \right) = \max\left(\frac{x_1}{m_1}, \ldots, \frac{x_k}{m_k} \right) \right\};$$

$$z_{f_1}^k = \sup_{(x_1, \ldots, x_k) \in \mathcal{S}_{f_1}^k} \left[\max\left(\frac{M_1}{x_1}, \ldots, \frac{M_k}{x_k} \right) \right].$$

Theorem 3. $z_{f_1}^k = \max\{\sqrt{M_1/m_1}, M_2/m_2\}$ *for any integer* $k \geq 2$.

With respect to the function f_1, Tiedemann, et al. [9] presented the algorithm RPP-HIGH and showed that $\mathcal{CR}^{f_1}(\text{RPP-HIGH}) = \max\{\sqrt{M_1/m_1}, M_2/m_2\}$ [9, Theorems 1 and 2]. By combining Corollary 1 and Theorem 3, we have that $\mathcal{CR}^{f_1}(\text{BPP}_k) = z_{f_1}^k = \max\{\sqrt{M_1/m_2}, M_2/m_2\}$, and this is another proof for the optimality on the worst component competitive ratio.

5.2 Arithmetic Mean Component Competitive Ratio

For $c_1, \ldots, c_k \in \mathbf{R}$, let $f_2(c_1, \ldots, c_k) = (c_1 + \cdots + c_k)/k$. For the function f_2, let

$$\mathcal{S}_{f_2}^k = \left\{ (x_1, \ldots, x_k) \in \mathbf{I}_k : \frac{1}{k}\left(\frac{M_1}{x_1} + \cdots + \frac{M_k}{x_k} \right) = \frac{1}{k}\left(\frac{x_1}{m_1} + \cdots + \frac{x_k}{m_k} \right) \right\};$$

$$z_{f_2}^k = \sup_{(x_1, \ldots, x_k) \in \mathcal{S}_{f_2}^k} \frac{1}{k}\left(\frac{M_1}{x_1} + \cdots + \frac{M_k}{x_k} \right) = \frac{1}{k} \sup_{(x_1, \ldots, x_k) \in \mathcal{S}_{f_2}^k} \left(\frac{M_1}{x_1} + \cdots + \frac{M_k}{x_k} \right).$$

From Corollary 1, we have that $\mathcal{R}_s^{f_2}(\text{BPP}_k) = z_{f_2}^k$ with respect to f_2. It would be difficult to explicitly represent $z_{f_2}^k$ by $m_1, \ldots, m_k, M_1, \ldots, M_k$. So we consider the case that $k = 2$ and we give an explicit form of $z_{f_2}^2$ by m_1, m_2, M_1, M_2.

Theorem 4. *With respect to the function* f_2 *for* $k = 2$, *the following holds:*

$$z_{f_2}^2 = \frac{1}{2}\left[\sqrt{\left\{ \frac{1}{2}\left(\frac{M_2}{m_2} - 1 \right) \right\}^2 + \frac{M_1}{m_1}} + \frac{1}{2}\left(\frac{M_2}{m_2} + 1 \right) \right].$$

With respect to the function f_2 for $k = 2$, Tiedemann, et al. [9] presented the algorithm RPP-MULT and showed that $\mathcal{CR}^{f_2}(\text{RPP-MULT}) \leq \sqrt[4]{(M_1 M_2)/(m_1 m_2)}$ [9, Theorem 3] (this holds under Definition 2, but can be shown by Definition 3). Note that $\sqrt[4]{(M_1 M_2)/(m_1 m_2)} < z_{f_2}^2$. Then from Theorems 2 and 4, we have that $\mathcal{CR}^{f_2}(\text{ALG}_2) \geq z_{f_2}^2$ for any algorithm ALG$_2$, which disproves the result [9, Theorem 3]. This is because in the proof of the result [9, Theorem 3], the maximum in Equation (9) cannot be achieved at $\sqrt{M_1 z^*/M_2}$, where $z^* = \sqrt{m_1 M_2 m_2 M_1}$.

5.3   Geometric Mean Component Competitive Ratio

For $c_1, \ldots, c_k \in \mathbf{R}$, let $f_3(c_1, \ldots, c_k) = (\prod_{i=1}^{k} c_i)^{1/k}$. For the function f_3, let

$$
\mathcal{S}_{f_3}^k = \left\{ (x_1, \ldots, x_k) \in I_1 \times \cdots \times I_k : \left(\prod_{i=1}^{k} \frac{M_i}{x_i} \right)^{1/k} = \left(\prod_{i=1}^{k} \frac{x_i}{m_i} \right)^{1/k} \right\};
$$

$$
z_{f_3}^k = \sup_{(x_1, \ldots, x_k) \in \mathcal{S}_{f_3}^k} \left(\prod_{i=1}^{k} \frac{M_i}{x_i} \right)^{1/k}.
$$

With respect to the function f_3 for $k = 2$, it is immediate to see that the algorithm RPP-MULT [9] is identical to the algorithm BPP$_2$. In fact, Tiedemann, et al. [9] showed that $\mathcal{CR}^{f_3}(\text{RPP-MULT}) = \sqrt[4]{(M_1 M_2)/(m_1 m_2)}$ with respect to f_3 for $k = 2$, and this can be generalized to the result that $\mathcal{CR}_s^{f_3}(\text{BPP}_k) = z_{f_3}^k$ for any integer $k \geq 2$ (see Corollary 1 with respect to f_3).

Theorem 5. $z_{f_3}^k = \left(\prod_{i=1}^{k} M_i/m_i \right)^{1/2k}$ for any integer $k \geq 2$.

References

1. Borodin, A., El-Yaniv, R.: Online Computation and Competitive Analysis. Cambridge University Press, Cambridge (1998)
2. Ehrgott, M.: Multicriteria Optimization. Springer, Heidelberg (2005)
3. El-Yaniv, R., Fiat, A., Karp, R.M., Turpin, G.: Optimal search and one-way trading online algorithms. Algorithmica **30**(1), 101–139 (2001)
4. Goldwasser, M.H.: A survey of buffer management policies for packet switches. ACM SIGACT New **41**(1), 100–128 (2010)
5. Hasegawa, S., Itoh, T.: Optimal Online Algorithms for the Multi-Objective Time Series Search Problem. CoRR abs/1506.04474 (2015)
6. Koutsoupias, E.: The k-server conjecture. Comput. Sci. Rev. **3**(2), 105–118 (2009)
7. Mohr, E., Ahmad, I., Schmidt, G.: Online algorithms for conversion problems: a survey. Surv. Oper. Res. Manag. Sci. **19**(2), 87–104 (2014)
8. Sleator, D.D., Tarjan, R.: Amortized efficiency of list update and paging rules. Commun. ACM **28**(2), 202–208 (1085)
9. Tiedemann, M., Ide, J., Schöbel, A.: Competitive analysis for multi-objective online algorithms. In: Rahman, M.S., Tomita, E. (eds.) WALCOM 2015. LNCS, vol. 8973, pp. 210–221. Springer, Heidelberg (2015)
10. Young, N.E.: Online paging and caching. In: Kao, M.-Y. (ed.) Encyclopedia of Algorithms, pp. 601–604. Springer, Heidelberg (2008)

Fully Dynamically Maintaining Minimal Integral Separator for Threshold and Difference Graphs

Tiziana Calamoneri[✉], Angelo Monti, and Rossella Petreschi

Computer Science Department, "Sapienza" University of Rome, Rome, Italy
{calamo,monti,petreschi}@di.uniroma1.it

Abstract. This paper deals with the well known classes of threshold and difference graphs, both characterized by separators, i.e. node weight functions and thresholds. We show how to maintain minimum the value of the separator when the input (threshold or difference) graph is fully dynamic, i.e. edges/nodes are inserted/removed. Moreover, exploiting the data structure used for maintaining the minimality of the separator, we handle the operations of disjoint union and join of two threshold graphs.

Keywords: Fully dynamic graphs · Threshold graphs · Difference graphs · Chain graphs · Threshold signed graphs · Graph operations

1 Introduction

In many applications of graph algorithms, *graphs* are *fully dynamic*, i.e. both edges and nodes may be inserted or eliminated.

Typically, one would like to answer to a precise query on the fully dynamic graph, so the goal is to update the data structure after dynamic changes, rather than having to recompute it from scratch each time.

Threshold graphs constitute a very important and well studied graph class, since they find applications in several fields, such as psychology, parallel processing, scheduling, and graph theory. For this reason, threshold graphs have been defined many times in the literature (see, e.g. [2,5]), and have been widely studied. *Difference graphs* (also known as *chain graphs*)–that are strictly related to threshold graphs, though incomparable–had similar destiny and have been independently introduced (see, e.g. [3,9]). For a comprehensive survey on threshold graphs, difference graphs and related topics, see [6].

Among the numerous equivalent definitions of threshold and difference graphs, many of them exploit a node weight function and a threshold. This pair is called a *separator* and of course it is not unique. It is of interest to determine a *minimum separator*, i.e. a separator with minimum value of the threshold. Orlin [7] presented an algorithm for minimizing the threshold w.r.t. one of these definitions in linear time in the number of nodes.

Partially supported by the Italian Ministry of Education and University, PRIN project "AMANDA: Algorithmics for MAssive and Networked DAta".

M. Kaykobad and R. Petreschi (Eds.): WALCOM 2016, LNCS 9627, pp. 313–324, 2016.
DOI: 10.1007/978-3-319-30139-6_25

314   T. Calamoneri et al.

In this paper we consider a different (equivalent) definition, also based on a threshold and a node weight function. After a pre-computation, we have always available the minimum separator after fully dynamically changing the graph. Both the pre-computation and each linear time operation of addition/deletion of either an edge or a node with all its incident edges are performed in linear time w.r.t. the number of different degrees of the current graph.

So, this is a contribution to the problem of the dynamic maintenance of threshold and difference graphs. To the best of our knowledge, few works deal with this topic. Namely, in [8] the problem of dynamically recognizing some classes of graphs (and among them threshold graphs) is handled. In [4] the authors consider the problem of adding/deleting edges with the aim of transforming a given graph into a threshold graph with the minimum number of changes.

We conclude this paper with a section that, exploiting the data structure used for maintaining the minimality of the separator, handles the operations of disjoint union and join of two threshold graphs.

2   Preliminaries

In this section, we list some definitions and properties, all from [6]. For the sake of clarity, we reorganized them in order to optimize the presentation.

Definition 1. *A graph $G = (V, E)$ is a* threshold *graph if there is a mapping $a : V \to \mathbb{R}^+$ and a positive real number S such that*

$$a(v) < S \text{ for all } v \in V \tag{1}$$

$$\{v, w\} \in E \text{ if and only if } a(v) + a(w) \geq S \tag{2}$$

The pair (a, S) will be called separator *for graph G.*

In Fig. 1a a threshold graph with one of its separators is depicted.

Definition 2. *A graph $G = (V, E)$ is a* difference *graph if there is mapping $a : V \to \mathbb{R}$ and a positive real number T such that*

$$|a(v)| < T \text{ for all } v \in V \tag{3}$$

$$\{v, w\} \in E \text{ if and only if } |a(v) - a(w)| \geq T \tag{4}$$

The pair (a, T) will be called separator *for graph G.*

The node set of a difference graph G can be partitioned as $V = U \cup W$, where $U = \{v \in V : a(v) \geq 0\}$ and $W = \{v \in V : a(v) < 0\}$; both U and W induce a stable set and hence G is bipartite with bipartition (U, W). A difference graph with one of its separators is shown in Fig. 1b.

Although Definitions 1 and 2 are very similar, the two defined classes are incomparable. Nevertheless they are strictly related, as shown by the following theorem:

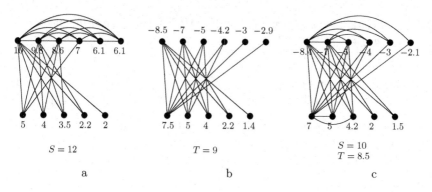

Fig. 1. a. A threshold graph; b. A difference graph; c. A threshold signed graph.

Theorem 1. *A bipartite graph $G = (U \cup W, E)$ is a difference graph if and only if adding to G all possible edges with both ends in the same side of the bipartition (either U or W) yields a threshold graph.*

Given a graph $G = (V, E)$, we denote by $\deg(v)$ the *degree* of node v.

Definition 3. *Let $G = (V, E)$ be a graph whose distinct node-degrees are $\delta_1 < \ldots < \delta_m$, and let $\delta_0 = 0$ (even if no node of degree 0 exists). Let $D_i = \{v \in V \text{ s.t. } \deg(v) = \delta_i\}$ for $i = 0, \ldots, m$; D_i is called i-th box; the sequence $D_0, \ldots, D_m$ is called the* degree partition *of G.*

Definition 4. *Let $G = (U \cup W, E)$ be a bipartite graph and let X be either U or W. The distinct node-degrees for the nodes in X are $\delta_1^X < \ldots < \delta_{t_X}^X$ and let $\delta_0^X = 0$ (even if no node of degree 0 exists in the partition X). Let $D_i^X = \{v \in X \text{ s.t. } \deg(v) = \delta_i^X\}$ for $0 \leq i \leq t_X$; the sequence $D_0^U, \ldots, D_{t_U}^U, D_0^W, \ldots, D_{t_W}^W$ is called the* bipartite degree partition *of G.*

Lemma 1. *Let $G = (V, E)$ be a threshold graph with degree partition $D_0, \ldots, D_m$ and whose node set is partitioned into a clique K and a stable set I; let $x \in D_i$ and $y \in D_j$ be two distinct nodes.*

1. *$D_0 \cup \ldots \cup D_{\lfloor m/2 \rfloor} = I$ and $D_{\lfloor m/2 \rfloor+1} \cup \ldots \cup D_m = K$;*
2. *If $e = \{x, y\} \notin E$, the graph $G' = (V, E \cup \{e\})$ is a threshold graph if and only if $i + j = m$;*
3. *If $e = \{x, y\} \in E$, the graph $G' = (V, E \setminus \{e\})$ is a threshold graph if and only if $i + j = m + 1$;*
4. *$e = \{x, y\} \in E$ if and only if $i + j \geq m + 1$.*

Lemma 2. *Let $G = (U \cup W, E)$ be a difference graph with degree partition $D_0^U, \ldots, D_{t_U}^U, D_0^W, \ldots, D_{t_W}^W$, and let $x \in D_i^U$ and $y \in D_j^W$ be two distinct nodes.*

1. *$t_U = t_W = t$;*
2. *If $e = \{x, y\} \notin E$, the graph $G' = (U \cup W, E \cup \{e\})$ is a difference graph if and only if $i + j = t$;*

3. If $e = \{x, y\} \in E$, the graph $G' = (U \cup W, E \setminus \{e\})$ is a difference graph if and only if $i + j = t + 1$;

4. $e = \{x, y\} \in E$ if and only if $i + j \geq t + 1$.

Given a graph G, we recall that the *neighborhood* $N(v)$ of a node v is the set of all neighbors of v, and its *closed neighborhood* is $N[v] = N(v) \cup \{v\}$.

Lemma 3. *Given a graph* $G = (V, E)$, *the* vicinal preorder *is a binary relation on the nodes of* V *such that* $u \succeq v \Leftrightarrow N[u] \supseteq N(v)$. *The vicinal preorder is total on* V *if* G *is a threshold graph* G *and on* U *and* V *if* G *is a difference graph.*

We now introduce a superclass of both threshold and difference graphs.

Definition 5. *[1] A graph* $G = (V, E)$ *is a* threshold signed graph *if there is a mapping* $a : V \rightarrow \mathbb{R}$ *and two positive real numbers* S *and* T *such that*

$$|a(v)| < min\{S, T\} \tag{5}$$

$$\{v, w\} \in E \text{ iff either } |a(v) + a(w)| \geq S \text{ or } |a(v) - a(w)| \geq T. \tag{6}$$

The triple (a, S, T) *will be called* separator *for graph* G.

Consider $X = \{x \in V \text{ s.t. } a(x) < 0\}$ and $Y = \{x \in V \text{ s.t. } a(x) \geq 0\}$.

As highlighted in Fig. 1c, we can see a threshold signed graph as constituted by two threshold graphs, G^- and G^+ respectively induced by X and Y, that are connected by a difference graph D. Notice that for X we consider the opposite of the a's values.

3   A Data Structure for Computing Minimal Integral Separator for Threshold or Difference Graphs

Although in Definition 1 a threshold graph $G = (V, E)$ is a graph having a separator with non-negative *real* values, it is common to equivalently require the separator to have non-negative *integral* values (i.e. an integral separator) [6]. We say that an integral separator (a, S) for G is *minimum* if for any other integral separator (a', S') for G we have $S \leq S'$. In the following theorem we show that the value of S of a minimum integral separator (a, S) of G is given by the cardinality of the degree partition of G plus 1.

Theorem 2. *Let* $G = (V, E)$ *be a threshold graph with degree partition* $D_0, \ldots, D_m$. *The pair* (a, S), *where* $S = m + 1$ *and for each node* $v \in V$, $a(v) = i$ *if* $v \in D_i$, *is a minimal integral separator of* G.

Proof. First of all, we prove that (a, S) is a separator, i.e. that it satisfies the two inequalities of Definition 1. Note that for each $v \in V$ it holds $0 \leq a(v) \leq m < S$ thus the pair (a, S) satisfies Inequality 1. Moreover, Inequality 2 follows from Item 4 in Lemma 1. Trivially, (a, S) is integral.

Let us now prove that (a, S) is minimal. By contradiction, let (a, S) be not minimal, and let (a', S') be an integral separator for G such that $S' < S$. Observe that only isolated nodes can have weight equal to zero (indeed, if $\{u, v\} \in E$ and $a'(u) = 0$ then $a'(u) + a'(v) = a'(v) < S'$ from Inequality 1, but this contradicts Inequality 2). Moreover, notice that two nodes u and v having the same weight necessarily behave in the same way (i.e. for any other node $w \in V$, it holds that $\{u, w\} \in E$ if and only if $\{v, w\} \in E$), so nodes having different degree cannot have the same weight. All this implies that the function a' on the non isolated nodes assumes at least m different strictly positive weights. Thus Inequality 1 implies that $S' \geq m+1$. The chain $m+1 = S > S' \geq m+1$ proves the minimality of (a, S).   □

From the same reasonings, we deduce the following theorem for a difference graph G.

Theorem 3. *Let $G = (U \cup W, E)$ be a difference graph with bipartite degree partition $D_1^U, \ldots, D_t^U, D_1^W, \ldots, D_t^W$. The pair (a, S), where $S = 2t + 1$ and, for each node $v \in U$, $a(v) = -i$ if $v \in D_i^U$ and $a(v) = i$ if $v \in D_i^W$, is a minimal integral separator of G.*

Notice that Orlin [7] shows how to minimize the threshold for threshold graphs considering an equivalent definition requiring that the sum of the weights of the nodes of any independent set has to be smaller than the threshold. The value of this threshold is larger than the value of S computed in this paper.

Now we present two data structures for representing threshold and difference graphs allowing us to compute in a natural way minimal integral separators for these graph classes, according to Theorems 2 and 3. Since a threshold graph $G = (V, E)$ is univocally determined by its degree partition $D_0, \ldots, D_m$, we may store G using two arrays $\delta[0..m]$ and $\mu[0..m]$, where $\delta[i]$ represents the degree δ_i of the nodes in the i-th box D_i, and $\mu[i]$ represents its cardinality, $|D_i|$. Similarly, let $G = (U \cup W, E)$ be a difference graph with bipartite degree partition $D_1^U, \ldots, D_t^U, D_1^W, \ldots, D_t^W$. This partition univocally determines G, and G may be represented using the arrays $\delta_X[0..t]$ and $\mu_X[0..t]$, where $\delta_X[i]$ represents the degree of the nodes in the i-th box D_i^X, and $\mu_X[i]$ represents its cardinality $|D_i^X|$, $X \in \{U, W\}$. If $G = (V, E)$ is a threshold graph, i is the weight associated to all the $\mu[i]$ nodes belonging to the box D_i of degree δ_i. Similarly, if $G = (U \cup W, E)$ is a difference graph, i is the weight associated to all the $\mu_X[i]$ nodes belonging to box D_i^X of degree $\delta_X[i]$, with $X \in \{U, W\}$.

Exploiting these two data structures, the following theorem holds:

Theorem 4. *Given a threshold (difference) graph G by means of its (bipartite) degree partition, its minimal integral separator can be found in time linear w.r.t. the number of different degrees in G.*

In Sects. 4 and 5, each time we speak about a graph G (either threshold or difference), G is represented by means of the arrays δ and μ.

4 Adding/deleting an Edge to Threshold/difference Graphs

In this section we study how to get a new graph, obtained by adding/deleting an edge from a graph that is either a threshold or a difference graph, and to keep immediately available the knowledge of the minimum separator for the new graph. In order to make easier the exposition, preliminarily we consider two functions, operating on the data structures introduced in Sect. 3.

By $\mathbf{IncreaseDeg}(\delta, \mu, i, dim)$ we denote the operation of updating arrays $\delta[0..dim]$ and $\mu[0..dim]$ when the degree of a node in box D_i, $0 \leq i \leq dim$, is increased by one. $\mathbf{IncreaseDeg}$ can have as consequence the appearance of a new box (if $i = dim$ or if the degree of the nodes in D_{i+1} is different from the degree of nodes in D_i plus one). On the other hand, this increment can also have as consequence the disappearance of box D_i (if $i \neq 0$, $|D_i| = 1$ and the degree of the nodes in D_{i+1} is equal to the degree of the nodes in D_i plus one). Symmetrically, we may consider the operation $\mathbf{DecreaseDeg}(\delta, \mu, i, dim)$ of updating arrays $\delta[0..dim]$ and $\mu[0..dim]$ when the degree of a node in box D_i, $0 < i \leq dim$, is decreased by one.

Let us now consider a threshold graph G. Let $\{x, y\}$, $x \in D_i$ and $y \in D_j$, the edge to add/delete to/from G. Items 2 and 3 of Lemma 1 give a characterization of the indices i and j to ensure that the modified graph is still a threshold graph; namely, $i + j = m$ in case of insertion, and $i + j = m + 1$ in case of deletion.

We present two operations, $\mathbf{InsEdge}(\delta, \mu, i, m)$ and $\mathbf{DelEdge}(\delta, \mu, i, m)$, that update the data structure when an edge is added between a node in box D_i and a node in box D_{m-i} and when an edge is deleted between a node in box D_i and a node in box D_{m+1-i}, respectively. Observe that with the insertion of an edge, the degrees of its endpoints are increased by one. Thus we can call twice subroutine $\mathbf{IncreaseDeg}$, once on a node in box D_i and once on a node in box D_{m-i}. We have just to take into account that the increment of the degree of node in box D_i can change the index of the box of the other endpoint. Analogous considerations hold for the deletion of an edge. These observations give rise to the following simple algorithms:

$\mathbf{InsEdge}(\delta, \mu, i, m)$	$\mathbf{DelEdge}(\delta, \mu, i, m)$
$\quad j \leftarrow m - i;$	$\quad j \leftarrow m + 1 - i;$
$\quad a \leftarrow m;$	$\quad a \leftarrow m;$
$\quad \mathbf{IncreaseDeg}(\delta, \mu, i, m);$	$\quad \mathbf{DecreaseDeg}(\delta, \mu, i, m);$
$\quad CASE(m - a)$	$\quad CASE(m - a)$
$\quad\quad -1 : \mathbf{IncreaseDeg}(\delta, \mu, j - 1, m);$	$\quad\quad -1 : \mathbf{DecreaseDeg}(\delta, \mu, j - 1, m);$
$\quad\quad\ \ 0 : \mathbf{IncreaseDeg}(\delta, \mu, j, m);$	$\quad\quad\ \ 0 : \mathbf{DecreaseDeg}(\delta, \mu, j, m);$
$\quad\quad +1 : \mathbf{IncreaseDeg}(\delta, \mu, j + 1, m);$	$\quad\quad +1 : \mathbf{DecreaseDeg}(\delta, \mu, j + 1, m);$

Since after the execution of $\mathbf{IncreaseDeg}$ ($\mathbf{DecreaseDeg}$), m may potentially vary from m to $m \pm 1$, with the execution of $\mathbf{InsEdge}$ ($\mathbf{DelEdge}$) the number of boxes can potentially vary from m to $m \pm 2$. In Fig. 2 we show that all the five possibilities may actually occur.

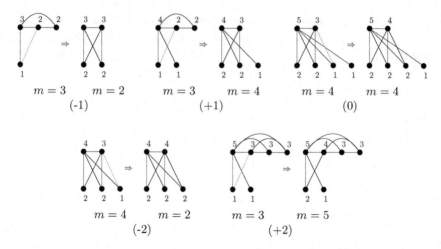

Fig. 2. Examples proving that all 5 cases in algorithm **InsEdge** are possible. Grey edges represent the edges that are going to be added. (In order to consider **DelEdge**, figures must be read from right to left.)

Assume now that G is a difference graph. The algorithms for adding/ eliminating an edge in G are based on the same idea presented for the algorithms on threshold graphs, but they are even simpler because the data structure used for representing these graphs keeps separated the bipartition (and so, adding a new box after the first call of **IncreaseDeg** does not affect the index of the other endpoint). Notice that Item 1 of Lemma 2 ensures that the number of boxes in the two classes is the same t. So, for difference graphs, t can either remain unaltered or to change to $t \pm 1$ and all the three possibilities may actually occur. The two algorithms for inserting and deleting an edge in a difference graph follow:

D – InsEdge($\delta_U, \mu_U, \delta_W, \mu_W, i, t$)	**D – DelEdge**($\delta_U, \mu_U, \delta_W, \mu_W, i, t$)
$j \leftarrow t - i;$	$j \leftarrow t + 1 - i;$
IncreaseDeg(δ_U, μ_U, i, t);	**DecreaseDeg**(δ_U, μ_U, i, t);
IncreaseDeg($\delta_W, \mu_W, j, t,$);	**DecreaseDeg**(δ_W, μ_W, j, t);

Note. *All the algorithms described in this section are correct and maintain the minimality of the integral separators.*

5 Adding/deleting a Node to Threshold/difference Graphs

In this section we will work with nodes in an analogous way as we did in Sect. 4 with edges. Also in this case, we keep immediately available the knowledge of the minimum separator for the new graph. We start defining four functions, operating on the data structures introduced in Sect. 3.

By **+Node**(δ, μ, d, dim) we denote the operation of giving space to a new node of degree d either in a threshold graph or in a partition of a difference graph, without caring about the update of its neighbors (that will be done with another subroutine). This subroutine looks for the box where the new node must be inserted: if there exists a box D_i with degree d, μ_i is simply increased by one; otherwise a new box for the new node is created.

By $\mathbf{IncreaseDegOfSetNode}(\delta, \mu, d, dim)$ we denote the operation of augmenting by one the degree of the d nodes of highest degree either in a threshold graph or in a partition of a difference graph. This subroutine increases by one the degree of all the boxes D_i s.t. $d - \sum_{s=j+1}^{m} |D_s| \geq 0$, while nodes of boxes $D_1, \ldots D_{j-1}$ remain unchanged. For what concerns D_j, it is in general split into two boxes (precisely $d - \sum_{s=j+1}^{m} |D_s|$ nodes leave D_j to form a new box with degree augmented by one). We can define even the symmetric functions: by -$\mathbf{Node}(\delta, \mu, i, dim)$ we denote the operation eliminating from the data structure storing either a threshold or a difference graph a node in box D_i, $0 \leq i \leq dim$, regardless of its neighbors (whose degree will be updated with another subroutine). By $\mathbf{DecreaseDegOfSetNode}(\delta, \mu, d, dim)$ we denote the operation of decreasing by one the degree of the d nodes of highest degree either in a threshold graph or in a partition of a difference graph.

Let now $G = (V, E)$ be a threshold graph. Adding a new node of degree d to G yields a threshold graph if and only if the d neighbors of the new node are the d nodes with highest degrees (this can be easily deduced from Item 4 of Lemma 1). So, we can call $\mathbf{IncreaseDegOfSetNode}$ and observe that, after its execution, m could be increased by one. Then, we have to update the data structure by inserting the new node by means of $+\mathbf{Node}$, and even in this case m could be increased by one. So, the number of the different degrees can potentially vary from m to $m + 2$. Figure 3 shows that all three possibilities can occur.

By $\mathbf{InsNode}(\delta, \mu, d, m)$ we denote the operation of updating the data structure storing threshold graph G when a node of degree d is added to the graph. The previous reasonings can be repeated when G is a difference graph (assuming, w.l.o.g., that the new node is inserted in partition U), so giving rise to $\mathbf{D\text{-}InsNode}(\delta_U, \mu_U, \delta_W, \mu_W, i, t)$, that is the operation of updating the data structure when a node of degree d is added to the difference graph.

$\mathbf{InsNode}(\delta, \mu, d, m)$	$\mathbf{D - InsNode}(\delta_U, \mu_U, \delta_W, \mu_W, d, t)$
$+\mathbf{Node}(\delta, \mu, d, m)$	$+\mathbf{Node}(\delta_U, \mu_U, d, t)$
$\mathbf{IncreaseDegOfSetNode}(\delta, \mu, d, m)$	$\mathbf{IncreaseDegOfSetNode}(\delta_W, \mu_W, d, t)$

Now we consider the problem of deleting nodes to a threshold graph. Any node-induced subgraph G' of G is a threshold graph (indeed for the graph G' use the mapping a restricted to the nodes of G' and the same value S). Thus the class of threshold graphs is closed under the deletion of an arbitrary node. Given a threshold graph, by $\mathbf{DelNode}(\delta, \mu, i, m)$ we denote the operation of updating the data structure when a node is deleted from box D_i, $0 \leq i \leq m$. This deletion is performed by -$\mathbf{Node}$ that can have as consequence the disappearance of box D_i (if $i \neq 0$ and $|D_i| = 1$). Thus m can decrease by one. Moreover, the $\delta[i]$ nodes with highest degree must have their degrees decreased by one. These nodes belong to boxes $D_m, \ldots, D_{m+1-i}$. It can occur that the degree of nodes

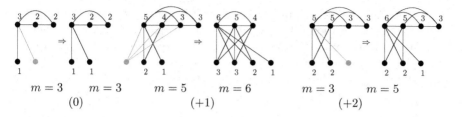

$$m = 3 \qquad m = 3 \qquad m = 5 \qquad m = 6 \qquad m = 3 \qquad m = 5$$
$$(0) \qquad\qquad (+1) \qquad\qquad\qquad (+2)$$

Fig. 3. Examples proving that all 3 cases in algorithm **InsNode** are possible. Grey nodes and edges represent the objects that are going to be added (in order to consider **DelNode** figure must be read from right to left).

in box $m + 1 - i$ becomes equal to the degree of the nodes in box $m - i$ and, in this case, the two boxes merge and the number of boxes further decrease by one. Hence after deleting a node, the number of boxes in the degree partition can potentially vary from m to $m - 2$ and all cases can occur, as shown in Fig. 3.

Analogous reasonings can be done when G is a difference graph, and define **D-DelNode**$(\delta_U, \mu_U, \delta_W, \mu_W, i, t)$ as the operation of updating the data structure when a node is deleted (assuming w.l.o.g. that the new node is deleted from partition U).

DelNode(δ, μ, i, m)
   $d \leftarrow \delta[i]$
   $-$**Node**(δ, μ, i, m)
   DecreaseDegOfSetNode(δ, μ, d, m)

D $-$ **DelNode**$(\delta_U, \mu_U, \delta_W, \mu_W, i, t)$
   $d \leftarrow \delta_U[i]$
   $-$**Node**(δ_U, μ_U, i, t)
   DecreaseDegOfSetNode(δ_W, μ_W, d, t)

Note. *All the algorithms described in this section are correct and maintain the minimality of the integral separators.*

6   Disjoint Union and Join of Two Threshold Graphs

Given two graphs with disjoint node sets $G_1 = (V_1, E_1)$ and $G_2 = (V_2, E_2)$, their *disjoint union* is the graph $G_1 \cup G_2 = (V_1 \cup V_2, E_1 \cup E_2)$; their *join* is the graph $G_1 + G_2$ obtained adding to their disjoint union all the edges that connect the nodes of the first graph with the nodes of the second graph. We observe that, if G_1 and G_2 are both threshold graphs, $G_1 \cup G_2$ and $G_1 + G_2$ are threshold signed graphs, where the difference graph connecting the two threshold graphs is either the null graph (in $G_1 \cup G_2$) or the complete bipartite graph (in $G_1 + G_2$).

In this section, exploiting the considerations done for the data structures presented in Sect. 3 for threshold and difference graphs, we introduce a data structure for representing a threshold signed graph. Thanks to this data structure, we handle the dynamic operations of disjoint union and join of two threshold graphs.

Given a threshold signed graph $G = (X \cup Y, E)$, two nodes u and w are *false twins* if they have the same neighborhood, i.e. $N(u) = N(w)$; they are *true twins* if they have the same closed neighborhood, i.e. $N[u] = N[w]$. We say that u and w are simply *twins* if they are either true or false twins and they belong to the same set, X or Y. Let us consider the partition of the node set into equivalence classes, $B_1, \ldots, B_\Delta$, induced by the relation of being twins. Even though there is not a tie between the degree partition of a threshold signed graph and its structure, as in the case of threshold and difference graphs, it is possible to extend the reasonings done in the proof of Theorem 2 to this class of graphs. Indeed, if v is an isolated node it is not restrictive to assume $a(v) = 0$ and, obviously, if $a(v) = 0$ then v is an isolated node. Moreover, it is easy to see that two nodes having the same value of a are necessarily twins. From the other hand, if there are two twins u and w having $a(u) \neq a(w)$ (w.l.o.g. let $a(u) < a(w)$), we can easily modify function a in order to assign them the same value (that is $a(w)$ if u and w are connected and $a(u)$ otherwise). So, from now on, we consider only node weight functions assigning value 0 to each isolated node and the same value to each set of twins.

As consequence of all these reasonings, we may store a threshold signed graphs by means of two arrays $\alpha[0..\Delta]$ and $\mu[0..\Delta]$: in $\alpha[i]$ there is the value of the weight assigned to the $\mu[i]$ nodes of B_i, $0 \leq i \leq \Delta$; if there are no isolated nodes $\alpha[0]$ and $\mu[0]$ are set to 0. Variables S and T store the two thresholds.

Let us now go back to consider the operations of disjoint union and join of two threshold graphs G_1 and G_2 stored as $\delta_1[0..m_1]$, $\mu_1[0..m_1]$ and $\delta_2[0..m_2]$, $\mu_2[0..m_2]$, respectively. Assume first that G_1 and G_2 have the same threshold S (i.e. $m_1 = m_2 = m$). Informally, the array $\alpha[1..\Delta]$ of both $G_1 \cup G_2$ and $G_1 + G_2$ is obtained by opportunely transcribing the values of the node weight function of the single threshold graphs (deduced through Theorem 2), the array $\mu[1..\Delta]$ is obtained by copying the values of μ_1 and μ_2, while threshold S is kept unaltered. For what concerns threshold T, in the case of $G_1 \cup G_2$ it is set to a sufficiently large value in order to guarantee that no edges are in the difference subgraph, in the case of $G_1 + G_2$ it is set to a sufficiently small value in order to guarantee that the difference subgraph is a complete bipartite graph. In this latter case, T assumes a too small value, contradicting Property 5 of Definition 5, so we need to modify the values of the node weight function and of the two thresholds in order to restore the property.

The following lemmas formalize the operations of disjoint union and join of two threshold graphs:

Lemma 4. *Let be given two threshold graphs G_1 and G_2 by means of $\delta_1[0..m]$, $\mu_1[0..m]$ and $\delta_2[0..m]$, $\mu_2[0..m]$ and let $S = m + 1$ be their common threshold. The following function determines the threshold signed graph $G_1 \cup G_2$:*

DisjointUnion$(\delta_1, \mu_1, \delta_2, \mu_2, m)$
$\mu[0] \leftarrow \mu_1[0] + \mu_2[0]; \; \alpha[0] \leftarrow 0;$
$FOR \; i = 1 \; TO \; m \; DO$
    $\alpha[i] \leftarrow -i; \; \mu[i] \leftarrow \mu_1[i];$
$FOR \; i = 1 \; TO \; m \; DO$
    $\alpha[m + i] \leftarrow i; \; \mu[m + i] \leftarrow \mu_2[i];$
$\Delta \leftarrow 2m;$
$S \leftarrow m + 1;$
$T \leftarrow 2m + 1;$
$RETURN \; (\alpha, \mu, S, T, \Delta).$

Proof. Both S and T determined by function **DisjointUnion** are greater than the modulo of each $\alpha[i]$, $i = 0, \ldots, \Delta$ as far as they are defined.

Moreover, the two threshold subgraphs G^- and G^+ of $G_1 \cup G_2$ are exactly the same as G_1 and G_2, respectively. Finally, no edge can satisfy the condition $\alpha[u] + \alpha[v] \geq T$ in view of the definition of T, so the difference subgraph D is empty. It follows that α, μ, S and T correctly define $G_1 \cup G_2$.   □

Lemma 5. *Let be given two threshold graphs G_1 and G_2 by means of $\delta_1[0..m]$, $\mu_1[0..m]$ and $\delta_2[0..m]$, $\mu_2[0..m]$ and let $S = m + 1$ be their common threshold. The following function determines the threshold signed graph $G_1 + G_2$:*

```
Join(δ₁, μ₁, δ₂, μ₂, m)
    α[0] ← 0; μ[0] ← 0;
    IF μ₁[0] ≠ 0
        THEN flag1 ← 0
        ELSE flag1 ← 1
    FOR i = flag1 TO m DO
        α[i + 1 − flag1] ← −i; μ[i + 1 − flag1] ← μ₁[i];
    IF μ₂[0] ≠ 0
        THEN flag2 ← 0
        ELSE flag2 ← 1
    FOR i = flag2 TO m DO
        α[m + i + 2 − flag1-flag2] ← i; μ[m + i + 2 − flag1-flag2] ← μ₂[i];
    Δ ← 2m + 2−flag1-flag2;
    S ← m + 1;
    T ← min₁≤i≤Δ{|α[i]|};
    k ← m − T + 1;
    FOR i = 1 TO m + 1 − flag1 DO
        α[i] ← α[i] − k;
    FOR i = m + 2−flag1 TO 2m + 2−flag1-flag2 DO
        α[i] ← α[i] + k;
    S ← S + 2k;
    T ← T + 2k;
    RETURN (α, μ, S, T, Δ).
```

Proof. Preliminarily, observe that $G_1 + G_2$ cannot have isolated nodes, so we set $\mu[0]$ to 0; moreover, if either G_1 or G_2 contain isolated nodes, a box needs to be added: we do this exploiting the two boolean variables flag1 and flag2.

So, even in this case, the two threshold graphs G^- and G^+ of $G_1 + G_2$ are exactly the same as G_1 and G_2, respectively. T is set to the modulo of the smallest node weight in order to guarantee that the difference graph D is a complete bipartite graph. In this way, T results in a value that contradicts Property 5 of Definition 5. By incrementing the modulo of each $\alpha[i]$ of an opportune value k and S and T by $2k$, we are able to restore the inequality.   □

It remains to handle the case in which G_1 and G_2 have different thresholds. In this case, we prepose to the functions described in the proofs of Lemmas 4 and 5 a preprocessing phase that equalize their thresholds as detailed in the following lemma, where with the notation $a' = xa + y$ (where x and y are integer values and a is a node weight function) we compactly mean that, for each node v, $a'(v) = xa(v) + y$. We want to underline that now on we represent a threshold graph in terms of its separator, instead of in terms of our data structure, because the description of the equalization appears more comprehensive.

Lemma 6. *Let be given two thresholds graphs G_1 and G_2 and let (a_1, S_1) and (a_2, S_2) be their integral separators with $S_1 < S_2$. Then $(a'_1, S'_1) = (2a_1 + S_2 - S_1, 2S_2)$ is an integral separator for G_1 and $(a'_2, a'_2) = (2a_2, 2S_2)$ is an integral separator for G_2.*

Proof. Let $\{v, w\}$ be an edge in G_1, i.e. $a_1(v) + a_1(w) \geq S_1$; then $a'_1(v) + a'_1(w) = 2a_1(v) + S_2 - S_1 + 2a_1(w) + S_2 - S_1 \geq 2S_2 = S'_1$. In the same way, let v and w be not connected in G_1, i.e. $a_1(v) + a_1(w) < S_1$; then $a'_1(v) + a'_1(w) = 2a_1 + S_2 - S_1 < 2S_2 = S'_1$. Finally, the pair (a'_1, S'_1) is a feasible integral separator since, for any node v, $a_1(v) < S_1$ implies $a'_1(v) < S'_1$. Analogous reasonings lead to prove that (a'_2, S'_2) is an integral separator for G_2.   □

Notice that the values of the node weight function and of the two thresholds S and T of the resulting threshold signed graphs will be integral but not necessarily minimal.

References

1. Benzaken, C., Hammer, P.L., de Werra, D.: Threshold characterization of graphs with Dilworth number two. J. Graph Theory **9**, 245–267 (1985)
2. Chvátal, V., Hammer, P.L.: Aggregation of inequalities in integer programming. Ann. Discrete Math. **1**, 145–162 (1977)
3. Cogis, O.: Ferrers digraphs and threshold graphs. Discrete Math. **38**, 33–46 (1982)
4. Heggernes, P., Papadopoulos, C.: Single-edge monitor sequences of graphs and linear-time algorithms for minimal completions and deletions. Theor. Comput. Sci. **410**, 1–15 (2009)
5. Henderson, P.B., Zalcstein, Y.: A graph-theoretic characterization of the PV_{chunk} class of synchronizing primitives. SIAM J. Comput. **6**, 88–108 (1977)
6. Mahadev, N.V.R., Peled, U.N.: Threshold Graphs and Related Topics, Ann. Discrete Math., vol. 56, North-Holland (1995)
7. Orlin, J.: The minimal integral separator of a threshold graph. Ann. Discrete Math. **1**, 415–419 (1977)
8. Shamir, R., Sharan, R.: A fully dynamic algorithm for modular decomposition and recognition of cographs. Discrete Appl. Math. **136**, 329–340 (2004)
9. Yannakakis, M.: The complexity of the partial order dimension problem. SIAM J. Algebraic Discrete Methods **3**, 351–358 (1982)

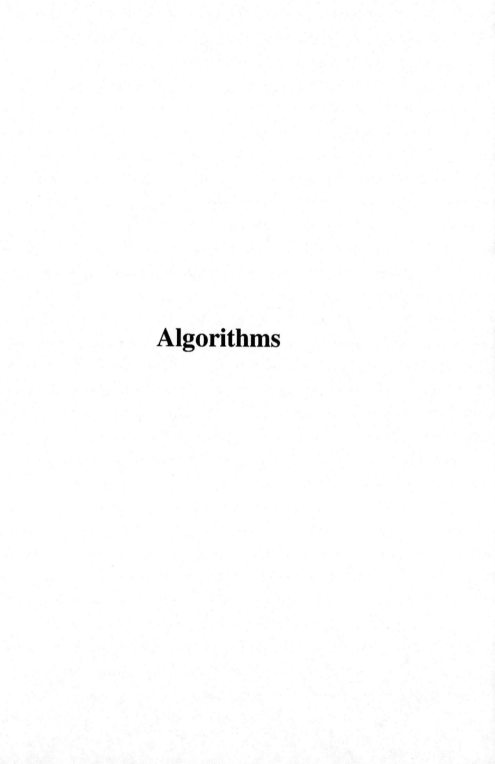

Algorithms

A Lagrangian Relaxation-Based Heuristic to Solve Large Extended Graph Partitioning Problems

Oliver G. Czibula[1,2](✉), Hanyu Gu[2], and Yakov Zinder[2]

[1] Ausgrid, 570 George Street, Sydney 2000, NSW, Australia
oliver.g.czibula@student.uts.edu.au
[2] School of Mathematical and Physical Sciences, University of Technology Sydney,
15 Broadway, Ultimo, NSW 2007, Australia

Abstract. The paper is concerned with the planning of training sessions in large organisations requiring periodic retraining of their staff. The allocation of students must take into account student preferences as well as the desired composition of study groups. The paper presents a bicriteria Quadratic Multiple Knapsack formulation of the considered practical problem, and a novel solution procedure based on Lagrangian relaxation. The paper presents the results of computational experiments aimed at testing the optimisation procedure on real world data originating from Australia's largest electricity distributor. Results are compared and validated against a Genetic Algorithm based matheuristic.

1 Introduction

This paper is concerned with the optimisation of class formation at large organisations, typically with thousands of workers of different types, that require periodic retraining of their staff. Finding good solutions to this problem is important as it allows more effective training sessions to be provided.

This research is motivated by the problem of providing training to workers at Ausgrid, Australia's largest electricity distributor. Due to the multitude of hazards that exist when working with high voltages at heights or in confined spaces, Ausgrid is required by Australian law to deliver regular safety, technical, and professional training to all its employees who work on or near the electricity network. Ausgrid provides regular training to thousands of employees, contractors, and third parties. Many of these people have very different learning outcomes from courses, different learning styles, different levels of education or English proficiency, and different levels of technical proficiency for certain tasks.

Consider an example where field workers and upper management are undertaking a particular course: while the core material would remain the same for both groups, if they are taught in separate classes the trainers can take the opportunity to better tailor the delivery to the specific needs of their group, allowing for more productive training sessions. It is often not possible to run segregated classes due to the scarcity of training resources and the associated

M. Kaykobad and R. Petreschi (Eds.): WALCOM 2016, LNCS 9627, pp. 327–338, 2016.
DOI: 10.1007/978-3-319-30139-6_26

cost of delivering additional classes, therefore some blending of different student types is often necessary. Ideally, differing student types should be combined into a single class only when they have a high compatibility with one another.

In similar problems where students are preferentially assigned to classes, it is customary to give each student-class pair a weight (or cost). Most of the training provided by Ausgrid has a limited period in which it is valid, and therefore courses should be periodically retrained. Workers are only permitted to work in roles for which they have up-to-date training. The date at which a worker's training expires is considered their due date for that course. If there are a number of scheduled classes for a given course that a worker requires, we can designate a cost of assigning the worker to any one of those classes: classes that run on or before their due date have low or zero cost, while classes that run after the due date have a high cost that increases the later the class is scheduled.

The considered problem has a bi-criteria objective: we wish to find an assignment of students to classes that minimises incompatibility between student types within classes, and that minimises the assignment cost of individual students to classes. Each class has a minimum and maximum number of students it can hold, and of student types it can be assigned.

Johnson et al. [10] discuss the problem of partitioning the vertices of a graph $G(V, E)$ with nonnegative weights w_v, $v \in V$ and costs c_e, $e \in E$, into K disjoint clusters partitioning V, such that the sum of the weights in each cluster is bounded between $w_{\min}$ and $w_{\max}$ and the sum of the costs within each cluster is minimised. This problem is known as the graph partitioning problem, and is known to be NP-hard [4,7]. Johnson et al. proposed a column generation approach, and tested the approach on graphs with between 30 and 61 nodes, and between 47 and 187 edges. For the 12 test cases the authors considered, their proposed approach provided integer solutions for all but two, and for those, solutions obtained by a branch-and-bound scheme were very close to the fractional solutions provided by column generation.

The problem considered in this paper can be modelled as a generalised graph partitioning problem. Each student j is represented by a vertex $v_j \in V$; the preference between students i and j is represented by an undirected edge $e = (v_i, v_j) \in E$ with edge cost c_e. The problem is to find a partition $\Gamma = \{W_1, W_2, \ldots, W_N\}$ of V that solves

$$\text{Minimise:} \quad \alpha \sum_{i=1}^{N} \sum_{e \in E(W_i)} c_e + \beta \sum_{i=1}^{N} \sum_{v_k \in W_i} w_{v_k}^i \qquad (1)$$

$$\text{Subject To:} \quad w_{min} \leq |W_i| \leq w_{max} \quad i = 1, \ldots, k \qquad (2)$$

where N is the number of classes, W_i is the set of students assigned to class i, $E(W_i) = \{(v_k, v_l) | v_k \in W_i, v_l \in W_i\}$, $w_{v_k}^i$ is cost of assigning student k to class i, α and β are weights for the assigning cost and preference cost respectively. A special case of the problem considered in this paper, in which $w_{v_k}^1 = w_{v_k}^2 = \ldots =$

$w_{v_k}^N$, is equivalent to the graph partitioning problem. Therefore our problem is also NP-hard.

Chopra and Rao [3] discuss several forms of the graph partitioning problem as well as IP models for each. The authors do not assume a complete graph, allowing them to take advantage of the graph structure when clustering. They also discuss several valid inequalities and facet-defining inequalities for the GPP.

The considered problem can also be modelled as an extended Quadratic Multiple Knapsack Problem (QMKP) with additional constraints. The QMKP is a generalization and combination of the well-known multiple knapsack problem and the quadratic knapsack problem. The QMKP received little attention in the literature until recently, and most solution approaches are based on metaheuristics [2,6,9]. The main contributions of this paper include: *(i)* formulation of the considered practical problem as an extended bicriteria QMKP. *(ii)* design of a Lagrangian relaxation (LR) based, fast heuristic capable of solving large, real-world instances.

Caprara et al. [1] discuss an LR-based approach to solving the QMKP exactly, whereby a tighter upper bound (for their maximisation objective) is computed using the subgradient method in linear expected time. The presented approach is able to solve instances with up to 400 binary variables exactly. The authors note that the presented approach can also be used to solve Max Clique problems almost as fast as with the cutting plane approaches available at the time.

Julstrom [11] discusses greedy, genetic, and greedy genetic algorithms for the QMKP. The author presents two greedy heuristics that build solutions by choosing objects according to their value densities, and two genetic algorithm (GA) heuristics. One GA is a standard implementation, whereas the other is extended with greedy techniques that probabilistically favour objects of high value density. The four algorithms are tested on 20 problem instances, and the extended GA is reported to perform best on all but one test case.

The remainder of this paper is organised as follows: Sect. 2 introduces a quadratic programming formulation and its linearisation for the considered problem; Sect. 3 introduces a Lagrangian relaxation formulation for the quadratic programming model; Sect. 4 discusses the proposed LR-based heuristic technique; Sect. 5 presents a GA matheuristic for the problem; Sect. 6 presents the computational results of the proposed heuristic on a number of industry-inspired test cases; and Sect. 7 discusses our conclusions and outlines possibilities for future research.

2   Quadratic Programming Formulation

Let $\mathcal{N} = \{1, \cdots, N\}$, $\mathcal{M} = \{1, \cdots, M\}$, and $\mathcal{K} = \{1, \cdots, K\}$ be the set of classes available, the set of students to be assigned, and the set of student types respectively. Denote the cost of assigning students $j \in \mathcal{M}$ to class $i \in \mathcal{N}$ by $c_{i,j}$, the cost of pairing student types $k \in \mathcal{K}$ and $l \in \mathcal{K}$ together in the same class by $b_{k,l}$. Each student has exactly one type. The set of students who are of type k is represented by T_k, $k \in \mathcal{K}$. Each student must be assigned to exactly one class, but not all classes need to be run. Each class $i \in \mathcal{N}$ that is run must contain at least a_i and at most b_i students, and at least p_i and at most q_i student types. Students or student types cannot be assigned to any class that is not run. In this problem, we assume there are more scheduled classes than needed, i.e. a feasible solution always exists. The following Quadratic Program describes the problem:

$$\text{(QP) Minimise:} \quad \alpha \sum_{i=1}^{N} \sum_{k=1}^{K} \sum_{l=1}^{K} b_{k,l} Y_{i,k} Y_{i,l} + \beta \sum_{i=1}^{N} \sum_{j=1}^{M} c_{i,j} X_{i,j} \tag{3}$$

$$\text{Subject To:} \quad \sum_{i=1}^{N} X_{i,j} = 1 \quad j = 1, \dots, M \tag{4}$$

$$a_i Z_i \leq \sum_{j=1}^{M} X_{i,j} \leq b_i Z_i \quad i = 1, \dots, N \tag{5}$$

$$p_i Z_i \leq \sum_{k=1}^{K} Y_{i,k} \leq q_i Z_i \quad i = 1, \dots, N \tag{6}$$

$$X_{i,j} \leq Y_{i,k} \quad i = 1, \dots, N; k = 1, \dots, K; j \in T_k \tag{7}$$

$$X_{i,j} \in \{0, 1\} \quad i = 1, \dots, N; j = 1, \dots, M \tag{8}$$

$$Y_{i,k} \in \{0, 1\} \quad i = 1, \dots, N; k = 1, \dots, K \tag{9}$$

$$Z_i \in \{0, 1\} \quad i = 1, \dots, N \tag{10}$$

where the binary variable $X_{i,j}$ is defined to be 1 if student j is assigned to class i, or 0 otherwise; the binary variable $Y_{i,k}$ is defined to be 1 if student type k is assigned to class i, or 0 otherwise; The binary variable Z_i is defined to be 1 if class i is run, or 0 otherwise.

In the objective function (3), the quadratic term represents the cost of pairing student types together, and the linear term represents the cost of assigning students to classes, weighted by coefficients α and β, respectively.

The constraints (4) express the requirement that each student be assigned to exactly one class. The constraints (5) and (6) express the requirement that each running class has between a_i and b_i students and between p_i and q_i student types, respectively, if the class is run, or zero otherwise. The constraints (7) express the requirement that a student may only be assigned to a class if that student's type has also been assigned to that class.

It is possible to linearise the quadratic term in (3) by introducing $\hat{Y}_{i,k,l} = Y_{i,k} Y_{i,l}$ together with constraints:

$$\hat{Y}_{i,k,l} \leq Y_{i,k} \quad i = 1, \dots, N; k = 1, \dots, K; l = 1, \dots, K \tag{11}$$

$$\hat{Y}_{i,k,l} \leq Y_{i,l} \quad i = 1, \dots, N; k = 1, \dots, K; l = 1, \dots, K \tag{12}$$

$$\hat{Y}_{i,k,l} \geq Y_{i,k} + Y_{i,l} - 1 \quad i = 1, \dots, N; k = 1, \dots, K; l = 1, \dots, K \tag{13}$$

to give the linearised model:

$$\text{(LQP) Minimise: } \alpha \sum_{i=1}^{N} \sum_{k=1}^{K} \sum_{l=1}^{K} b_{k,l} \hat{Y}_{i,k,l} + \beta \sum_{i=1}^{N} \sum_{j=1}^{M} c_{i,j} X_{i,j} \qquad (14)$$

$$\text{Subject To: } (4) - (13) \qquad (15)$$

$$\hat{Y}_{i,k,l} \in \{0,1\} \quad i = 1, \ldots, N; k = 1, \ldots, K; l = 1, \ldots, K \qquad (16)$$

The (QP) model has $N(K+M+1)$ variables and $M+2N+MNK$ constraints, whereas the (LQP) model has $N(K + M^2 + M + 1)$ variables and $M + 2N + MNK + 3NK^2$ constraints.

The time required to find an optimal solution to the (QP) and (LQP) models grows rapidly, where even some small test cases with just a few dozen students can take hours to solve. As the problem instances we hope to solve are significantly larger than this, we propose to use a heuristic approach.

3 Lagrangian Relaxation

To improve the convergence of the subgradient algorithm, the equalities (4) are first converted into inequalities:

$$\sum_{i=1}^{N} X_{i,j} \geq 1 \quad j = 1, \ldots, M \qquad (17)$$

By moving the constraints (4) into the objective function, we obtain the Lagrangian relaxation model:

$$\text{(QR) Minimise: } \alpha \sum_{i=1}^{N} \sum_{k=1}^{K} \sum_{l=1}^{K} b_{k,l} Y_{i,k} Y_{i,l} + \beta \sum_{i=1}^{N} \sum_{j=1}^{M} c_{i,j} X_{i,j} + \sum_{j=1}^{M} \lambda_j \left(1 - \sum_{i=1}^{N} X_{i,j}\right)$$
$$\qquad (18)$$

$$\text{Subject To: } (5) - (10) \qquad (19)$$

where $\lambda_j \geq 0$, $j = 1, \ldots, M$, is the Lagrangian multiplier corresponding to the cost of not assigning student j to any class. The quadratic term in (18) can be linearised in the same way as with (3).

In (QP), only constraints (4) couple together the N sub-problems of assigning students to a *particular* class i. In (QR), these constraints are moved to the objective function, therefore it is possible to express (QR) as N smaller, class-specific sub-problems (QR$_i$):

$$\text{(QR}_i\text{) Minimise: } \alpha \sum_{k=1}^{K} \sum_{l=1}^{K} b_{k,l} Y_{i,k} Y_{i,l} + \beta \sum_{j=1}^{M} c_{i,j} X_{i,j} - \sum_{j=1}^{M} \lambda_j X_{i,j} \qquad (20)$$

$$\text{Subject To:} \quad a_i Z_i \leq \sum_{j=1}^{M} X_{i,j} \leq b_i Z_i \tag{21}$$

$$p_i Z_i \leq \sum_{k=1}^{K} Y_{i,k} \leq q_i Z_i \tag{22}$$

$$X_{i,j} \leq Y_{i,k} \quad k = 1, \ldots, K; j \in T_k \tag{23}$$

$$X_{i,j} \in \{0, 1\} \quad j = 1, \ldots, M \tag{24}$$

$$Y_{i,k} \in \{0, 1\} \quad k = 1, \ldots, K \tag{25}$$

$$Z_i \in \{0, 1\} \tag{26}$$

where the N combined solutions to (QR$_i$) for $1 \leq i \leq N$ forms the solution to (QR).

Solving the Lagrangian relaxation (QR) provides a lower bound to the optimal objective value of (QP). We wish to find the tightest possible lower bound for (QP) by solving the Lagrangian dual problem, the solution of which are the optimal Lagrangian multipliers λ^* that give the largest objective value of (QR) [5,8]. A typical way to numerically approximate λ^* is by using the iterative subgradient algorithm.

In the subgradient algorithm, λ is iteratively updated with the relationship

$$\lambda^{k+1} = \lambda^k + \frac{s^k \cdot \epsilon_k (\eta^* - \eta^k)}{||s^k||^2} \tag{27}$$

where λ^k is the value of the Lagrangian multipliers λ at the kth iteration; η^* is an upper bound of the optimal objective value of the Lagrangian dual problem, and η^k is the objective value of (QR) at the kth iteration; and ϵ_k is a positive scaling factor, typically with the initial value of $\epsilon_0 = 2$ and halved whenever the objective value hasn't been improved in certain number of iterations; s^k is a subgradient of the Lagrangian dual at iteration k given by

$$s_j^k = 1 - \sum_{i=1}^{N} X_{i,j}^k \quad j = 1, \ldots, M \tag{28}$$

where X^k is from the solution of (QR) at the kth iteration.

4   Lagrangian Heuristic

In our proposed heuristic, we solve the Lagrangian dual using the subgradient algorithm. In doing so, we generate many solutions in this iterative process, and we record the following characteristics about them:

– The number of times class i is running (ξ_i), and the number of times it is not ($\bar{\xi}_i$).
– The number of times type k is assigned to class i ($\phi_{i,k}$), and the number of times it is not ($\bar{\phi}_{i,k}$).

- The number of times students j_1 and j_2 are assigned to the same class (ψ_{j_1,j_2}), and the number of times they are not ($\bar{\psi}_{j_1,j_2}$).

Based on these values, one or more assumptions are made about the final solution with some probability. The higher the value, the greater the probability the assumption will be made. Each value has a corresponding counter-value, such as the number of times class i is running (ξ_i), and the number of times class i is not running ($\bar{\xi}_i$). An assumption can be made according to the value/counter-value that has greater magnitude, for example if $\xi_i = 7649$ and $\bar{\xi}_i = 1008$ then, since $\xi_i > \bar{\xi}_i$, we would be making the assumption that class i will be run, with some probability. The probability with which we make an assumption according to some value v (or counter-value $\bar{v}$) is $\frac{v}{v+\bar{v}}^2$ (or $\frac{\bar{v}}{v+\bar{v}}^2$), so if $\xi_i = 7649$ and $\bar{\xi}_i = 1008$ then with probability ($\frac{7649}{7649+1008}$)$^2 \approx 0.7807$ we will make the assumption that class i will run in the final solution.

Each kind of imposed assumption reduces the number of feasible solutions of the model being solved. In the case of the first kind of assumption, the Z_i variable is fixed to 1, if class i is assumed running, or if class i is assumed not running then the Z_i and the related $X_{i,j}$ and $Y_{i,k}$ variables, $1 \leq j \leq M$ and $1 \leq k \leq K$, can be all fixed to zero. For the assumption that type k is or is not assigned to class i, the constraints (7) can be omitted for $j \in T_k$, and all the related $X_{i,j}$ variables are also fixed to zero if type k is assumed to be not assigned to class i. For the assumption that students j_1 and j_2 should be assigned together, we introduce the additional constraints $X_{i,j_1} = X_{i,j_2}$ for $1 \leq i \leq N$, and for the assumption that they should be assigned separately, we introduce the constraints $X_{i,j_1} + X_{i,j_2} \leq 1$ for $1 \leq i \leq N$.

Once an assumption is made, the model is updated and the Lagrangian dual is once again solved using the subgradient algorithm. With each new assumption the size of the model becomes smaller or the number of feasible solutions is reduced. Once the model is sufficiently small, the optimal solution, subject to the assumptions, can be found using a commercial solver.

The initial value of $\lambda = \lambda^0$ is likely to be quite distant from λ^*. It is expected that there would be an initial period of convergence from λ^0 towards the neighbourhood of λ^*, followed by a period of convergence within this neighbourhood. Since these early values of λ^k are likely to produce fairly poor solutions, we treat this early period of convergence towards the neighbourhood of λ^* as a "burn-in stage", and do not record these solutions.

To calculate an upper bound for the Lagrangian dual in (27), we relax constraints (7) from the (QP) model. The resulting capacitated assignment problem can be solved quickly, and a repair heuristic is used to make the solution feasible with respect to the original problem. The objective value of this feasible solution is used as η^* in the updating rule (27) of the subgradient algorithm.

Our LR-based heuristic is described as follows:

LR-based Heuristic:

Step 1. Initialise an empty set of assumptions A.

Step 2. Construct the (QP) model, together with assumptions A.

i. If the size of the model is sufficiently small to solve in practically acceptable time, attempt to solve it:

   (a) If a solution with sufficiently small relative IP gap can found within a short time limit, terminate the algorithm returning the optimal solution to the (QP) subject to assumptions A.

   (b) If no solution with sufficiently small relative IP gap can be found within a short time limit, continue to Step 3.

   (c) If no feasible solution exists, determine which assumptions are causing the infeasibility, remove them, and continue to Step 3.

ii. If the size of the model is expected to be too large to solve in practically acceptable time, continue to Step 3.

Step 3. Run LD-Subroutine and retrieve all the values of ξ, $\bar{\xi}$, ϕ, $\bar{\phi}$, ψ, and $\bar{\psi}$.

Step 4. For each of the value and counter-value pairs $\{v, \bar{v}\}$ obtained in Step 3, add an assumption to A with probability $\frac{v}{v+\bar{v}}^2$ (or $\frac{\bar{v}}{v+\bar{v}}^2$) and then return to Step 2.

LD-Subroutine:

Step 1. Initialise the λ vector to its starting value λ^0, $\epsilon_0 := 2$, determine an estimate for η^*, initialise the iteration counter $k := 0$, and initialise all ξ, $\bar{\xi}$, ϕ, $\bar{\phi}$, ψ, and $\bar{\psi}$ values to zero.

Step 2. Construct the N (QR$_i$) models according to the input data and the set of assumptions A, for $1 \leq i \leq N$.

Step 3. Solve the N (QR$_i$) models, and update λ according to (27).

Step 4. If $\epsilon_k \leq \frac{1}{2}$, update the values of ξ, $\bar{\xi}$, ϕ, $\bar{\phi}$, ψ, and $\bar{\psi}$ according to observations about solution X^k.

Step 4. If the η^k has not been improved in the last $10 \times N$ iterations, then $\epsilon_{k+1} := \frac{1}{2}\epsilon_k$.

Step 5. If $\epsilon_k < 0.1$ then terminate LD-Subroutine and return the values of ξ, $\bar{\xi}$, ϕ, $\bar{\phi}$, ψ, and $\bar{\psi}$; otherwise set $k := k + 1$ and return to Step 2.

5   Genetic Algorithm Based Matheuristic

Although Genetic Algorithms (GA) are often used in solving the QKP and QMKP [9,13,14], these publications on GA are not directly applicable to our problem due to the existence of lower bounds on class sizes. Moreover, the existence of lower and upper bounds on class sizes renders classical operations of crossover and mutation inefficient, i.e. both operations produce too many infeasible solutions. The computational experiments indicated that the common approach of introducing penalty for infeasibility does not improve the performance of the classical version of GA. These observations lead to the development of a matheuristic, presented in this paper, which is an amalgamation of GA and IP. This matheuristic was compared with the LR-based approach described above.

    In the developed version of GA, feasible solutions in the initial population are be generated using a two-step procedure. First, we decide which classes will be run, and how many students will be in each class by solving a straightforward IP. For all test cases, including all cases where CPLEX failed to solve the original QP problem, this IP was solved in under a second. Next, we randomly assign students to classes according to the numbers obtained in the previous step.

The crossover operator, which addresses the challenge imposed by the existence of lower and upper bounds on class sizes, is defined as follows. As with conventional crossover operators, the designed crossover operates on two solutions, referred to as parents. The results of the designed crossover is a single solution, referred to as a child. As with the procedure that generates solutions for the initial population, the crossover operator involves two stages. First, an IP is solved that determines which classes should be run in the child, and the number of students in each class. In this IP, those classes that are run in both parents must run in the child, and those classes that are not run in either parent will not run in the child.

Next, students are assigned to classes in numbers specified by the IP. In this stage, first, students are assigned starting with students who are assigned to the same class in both parents. Each of these students is assigned to the that class in the child as well. The remaining students are assigned one at a time. If at least one class chosen for the student in the parents is available for this student in the child, then the student is assigned to one of these classes. If there are two such classes, the actual class is chosen at random. Students who were not allocated are then randomly allocated to the classes in the child solution according numbers specified by the IP.

6   Computational Results

Using Ausgrid's training data as a template, we generated a series of random test cases[1]. Each test case had between 100 and 500 students, with between 4 and 12 student types, and between 56 and 98 classes.

We used IBM ILOG CPLEX 12.5.0.0 64-bit on an Intel i7-4790K quad-core 4.00 Ghz system with 16 GB of RAM, running Windows 7 Professional. Our code was written in C# 4.0, and interacted with CPLEX using the IBM ILOG Concert API. We used default CPLEX settings, except we increased the maximum allowed memory usage to the total amount of free physical memory. Since the proposed heuristic is probabilistic, we applied it to each of the test cases 10 times. The pseudorandom number generator we used was the MT19937 Mersenne Twister [12]. For the weighted objective function, we used $\alpha = \beta = 1$. We also applied the GA matheuristic to each of the test cases 10 times for the same amount of time that the LR-based heuristic used on average.

Table 1 shows the results of the computational experiments. The tables shows the test case (Case), the number of students (Std), the number of classes (Cls), the number of student types (Typ), the number of variables in the QP (Vars), the minimum (tMin), average (tAvg), and maximum (tMax) solution time for

---

[1] All test cases and solution files are available on request.

Table 1. The results of the computational experiments for many of the test cases.

Case	Std	Cls	Typ	Vars	tMin	tAvg	tMax	lrMin	lrMax	gaMin	gaMax
1	300	56	3	62776	12.1	125.4	327.1	1700.5	14101.5	7979	13904.5
2	300	66	3	73986	10.3	13.4	18.4	1483.5	3009	20452	26405.5
4	300	88	3	98648	47.5	110.1	383.8	1526.5	6277	12868.5	20827.5
5	300	98	3	109858	39.7	94.4	261.1	1328.5	3012	19947.5	21808.5
7	291	66	4	73392	20.8	29	48.4	1883.5	2405	13872	20788
8	291	78	4	86736	13.1	24.7	62.5	2160	2739.5	16303.5	22527.5
10	291	98	4	108976	19.2	54.4	147.6	1904	6210	14350.5	22959.5
11	206	56	4	57512	6.5	43.4	241.5	1437	17346.5	5930.5	10101
13	206	78	4	80106	9.8	14.6	20.5	1443.5	3482	12167	16141
14	206	88	4	90376	10	15	23	1236	12347	13018.5	16949
16	300	56	6	62776	8.6	14.7	19.4	1597	7193.5	21518.5	31016
17	300	66	6	73986	13.5	44.3	113.4	1574.5	3314	11135.5	21437
19	300	88	6	98648	35.4	140.5	303.9	1341	3251	10727	18855
20	300	98	6	109858	22.3	35.6	73.2	1136.5	3634.5	20977	25647
22	400	66	5	80586	39.9	99	290.6	2181	5061	13182.5	28905
23	400	78	5	95238	69.8	194.1	349.9	2078	2665.5	11723.5	25204
25	400	98	5	119658	41.8	304.8	642.1	1665	7581	11361.5	26601
26	500	56	6	73976	14.4	205.8	638.5	3004.5	16429	16148	38892.5
28	500	78	6	103038	37.6	129.6	377.8	3117	4130.5	28708	43548.5
29	500	88	6	116248	46.1	175.2	361.3	2389.5	7289	23430	45908.5
31	132	56	7	53368	51.2	159.1	411	1454.5	1654.5	4324.5	7100
32	132	66	7	62898	19.7	61.1	138.6	1383	2549.5	5418	6608
34	132	88	7	83864	30.7	90.7	186.4	1518.5	2106	4081	6520
35	132	98	7	93394	60	317.4	985.1	1292	1890	4297.5	6680
37	400	66	5	80586	52.5	145.4	285.9	1889	3319.5	11636.5	24165
38	400	78	5	95238	29	194.7	547.5	2262	3835.5	23962	26256.5
40	400	98	5	119658	57.9	279.1	546.7	1639	5139.5	11082.5	25742
41	500	56	8	73976	64.9	224.9	587	3786.5	11623	14888.5	37871.5
43	500	78	8	103038	480.3	774.3	954.3	3095.5	7006.5	13194	29130.5
44	500	88	8	116248	58.3	594.7	991.5	2574	3868.5	10993.5	32387.5
46	300	56	6	62776	25.1	45.1	69.6	2274.5	18798	9859	21706.5
47	300	66	6	73986	33.6	57.2	85.3	2183.5	8768	10534	19975.5
49	300	88	6	98648	45.7	130.2	466	2308	7697.5	9481.5	19685
50	300	98	6	109858	58.9	193.2	397.4	1778.5	3082	10332	18988
52	168	66	6	65274	19	44.7	74.1	1637	2614	6321.5	8415
53	168	78	6	77142	41.2	172.9	532.7	1605.5	12674	6406	8189
55	168	98	6	96922	46.1	95.1	224.9	1404.5	4050.5	7505.5	9456
56	500	56	5	73976	44.7	168	1084.7	2825	13742	15216.5	40438
58	500	78	5	103038	21.7	72.7	199.1	2550.5	3835	32732.5	44762.5
59	500	88	5	116248	45.3	122.9	276.3	2465.5	13736.5	36664.5	47615.5
61	300	56	7	62776	149.9	640.6	904.9	2286	3305.5	7636.5	12557
62	300	66	7	73986	90.6	678.1	978	2312.5	2786.5	10760.5	14360
64	300	88	7	98648	789	1071.9	1556.5	2215	2666.5	9325	12500.5
65	300	98	7	109858	123.3	719.3	1256.1	1999.5	4367.5	9902.5	16164
67	400	66	8	80586	72.6	188.6	519.3	2105.5	3174.5	16810.5	24442.5
68	400	78	8	95238	60	161.5	416.2	1992.5	3147.5	14011	26611.5
70	400	98	8	119658	33.3	239.1	597.3	1584.5	2056.5	11480.5	27258
71	132	56	7	53368	26.8	56.4	120.8	1237	9558	5303	7409.5
73	132	78	7	74334	13.4	25	65	1230.5	1910.5	6667.5	7684.5
74	132	88	7	83864	32.6	48.2	63.6	1296	1665	5611.5	6721
76	300	56	9	62776	84.4	544.1	1542.1	2189	13773	6696	11942.5
77	300	66	9	73986	44.1	200.3	438.4	1982.5	2979	10571.5	16390
79	300	88	9	98648	109.9	542.2	938	2254	3810.5	7113	15953.5
80	300	98	9	109858	95	534.7	802.9	1958.5	3169.5	8544	12726
82	400	66	11	80586	79.5	663.1	1082.2	2577	6110	10912	16338
83	400	78	11	95238	52.5	540.4	1021.6	2681.5	3984	10771.5	23656.5
84	400	88	11	107448	267.5	1078.4	1563.9	2512.5	5764.5	10662	14386.5
85	400	98	11	119658	83.4	818.2	1316.3	2404.5	5749	10067.5	23096
86	500	56	11	73976	181.8	543.3	946.7	3735	17272.5	16703	17796
87	500	66	11	87186	185.1	961.3	1643.4	3624	8932.5	14413.5	20936
88	500	78	11	103038	339.4	788	1465	3709.5	7593	13059.5	30584.5
89	500	88	11	116248	1130.2	1501	2035.3	3185	16049.5	10998	21648.5
90	500	98	11	129458	821.2	1347.1	2401	2724.5	9361	10298.5	29047

the LR-based heuristic, the minimum (lrMin) and maximum (lrMax) objective value obtain by applying the LR-based heuristic, and the minimum (gaMin) and maximum (gaMax) objective value obtain by applying the GA matheuristic. Since the two approaches tested are quite different, time is reported in CPU time. It is clear from the results that the LR-based heuristic outperformed the GA matheuristic.

In Fig. 1, the horizontal axis depicts the 90 test cases. The vertical axis gives the time taken, in seconds. The graph shows the range of solution times across the 10 runs of the LR-based heuristic. The solid line shows the average time across the 10 runs of the heuristic. The largest value was about 40 min, however the overall average was only about 272 s. In contrast, the attempts to obtain exact solutions by solving the corresponding quadratic programming problem using CPLEX failed in most cases given a six hour limit.

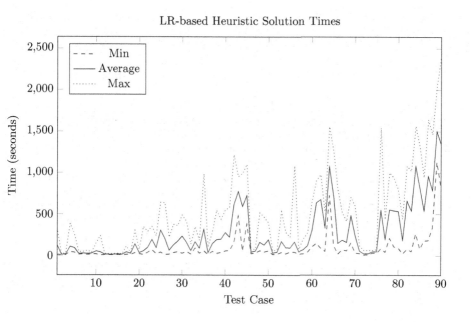

Fig. 1. The time taken for the LR-based heuristic to produce a solution.

7   Conclusions

In this paper we presented a heuristic solution approach, based on the Lagrangian relaxation, for the problem of assigning students to classes. This problem arises in large organisations that require training and retraining of staff. The objective function reflects the preference of assigning certain groups of students to the same class, which often occurs in practice. The proposed heuristic was tested by computational experiments on a number of randomly generated test cases, based on data supplied by Ausgrid. The proposed heuristic was able to provide solutions to all test cases in a practically acceptable time. In contrast, the straight forward quadratic programming based approach failed in most cases with a time limit of six hours. In both cases, CPLEX was used as the solver. The Lagrangian relaxation-based heuristic was also compared with a specifically designed matheuristic based on Genetic Algorithms. This comparison indicated the superiority of the Lagrangian relaxation-based heuristic. The integer programming components of the matheuristic were also solved with CPLEX.

The proposed Lagrangian relaxation-based heuristic includes a number of parameters, and future research can be focussed on an investigation of the influence of these parameters on the performance of the entire procedure.

References

1. Caprara, A., Pisinger, D., Toth, P.: Exact solution of the quadratic knapsack problem. INFORMS J. Comput. **11**(2), 125–137 (1999)
2. Chen, Y., Hao, J.K.: Iterated responsive threshold search for the quadratic multiple knapsack problem. Ann. Oper. Res. **226**, 101–131 (2015)
3. Chopra, S., Rao, M.R.: The partition problem. Math. Program. **59**(1–3), 87–115 (1993)
4. Dahlhaus, E., Johnson, D.S., Papadimitriou, C.H., Seymour, P.D., Yannakakis, M.: The complexity of multiway cuts. In: Proceedings of the twenty-fourth annual ACM Symposium on Theory of computing, pp. 241–251. ACM (1992)
5. Fisher, M.L.: The lagrangian relaxation method for solving integer programming problems. Manage. Sci. **50**(12-supplement), 1861–1871 (2004)
6. García-Martínez, C., Rodriguez, F., Lozano, M.: Tabu-enhanced iterated greedy algorithm: a case study in the quadratic multiple knapsack problem. Eur. J. Oper. Res. **232**, 454–463 (2014)
7. Garey, M.R., Johnson, D.S.: Computers and Intractability: A Guide to NP-completeness. W. H. Freeman and co., New York (1979)
8. Guignard, M.: Lagrangean relaxation. Top **11**(2), 151–200 (2003)
9. Hiley, A., Julstrom, B.A.: The quadratic multiple knapsack problem and three heuristic approaches to it. In: Proceedings of the 8th Annual conference on Genetic and Evolutionary Computation, pp. 547–552. ACM (2006)
10. Johnson, E.L., Mehrotra, A., Nemhauser, G.L.: Min-cut clustering. Math. Program. **62**(1–3), 133–151 (1993)
11. Julstrom, B.A.: Greedy, genetic, and greedy genetic algorithms for the quadratic knapsack problem. In: Proceedings of the 7th Annual Conference on Genetic and Evolutionary Computation, pp. 607–614. ACM (2005)
12. Matsumoto, M., Nishimura, T.: Mersenne twister: a 623-dimensionally equidistributed uniform pseudo-random number generator. ACM Trans. Model. Comput. Simul. (TOMACS) **8**(1), 3–30 (1998)
13. Saraç, T., Sipahioglu, A.: A genetic algorithm for the quadratic multiple knapsack problem. In: Mele, F., Ramella, G., Santillo, S., Ventriglia, F. (eds.) BVAI 2007. LNCS, vol. 4729, pp. 490–498. Springer, Heidelberg (2007)
14. Singh, A., Baghel, A.S.: A new grouping genetic algorithm for the quadratic multiple knapsack problem. In: Cotta, C., van Hemert, J. (eds.) EvoCOP 2007. LNCS, vol. 4446, pp. 210–218. Springer, Heidelberg (2007)

Semimetric Properties of Sørensen-Dice and Tversky Indexes

Alonso Gragera[1] and Vorapong Suppakitpaisarn[1,2][(✉)]

[1] Department of Computer Science, The University of Tokyo, Tokyo, Japan
{alonso,vorapong}@is.s.u-tokyo.ac.jp
[2] JST ERATO Kawarabayashi Large Graph Project, Tokyo, Japan

Abstract. In this work we prove a semimetric property for distances used for finding dissimilarities between two finite sets such as the Sørensen-Dice and the Tversky indexes. The Jaccard-Tanimoto index is known to be one of the most common distances for the task. Because the distance is a metric, when used, several algorithms can be applied to retrieve information from the data. Although the Sørensen-Dice index is known to be more robust than the Jaccard-Tanimoto when some information is missing from datasets, the distance is not a metric as it does not satisfy the triangle inequality. Recently, there are several machine learning algorithms proposed which use non-metric distances. Hence, instead of the triangle inequality, it is required that the distance satisfies the approximate triangle inequality with some small value of ρ. This motivates us to find the value of ρ for the Sørensen-Dice index. In this paper, we prove that this value is 1.5. Besides, we can find the value for some of the Tversky index.

Keywords: Distance · Metric · Approximate triangle inequality · Sørensen-Dice index · Tversky index

1 Introduction

In this work, we consider distances used for finding a dissimilarity between two finite sets. The most common dissimilarity between sets is called Jaccard-Tanimoto index (JT index) [1,2]. The Jaccard-Tanimoto distance between a finite set A and a finite set B, $d_{JT}(A, B)$, can be defined as follows:

$$d_{JT}(A, B) := 1 - \frac{|A \cap B|}{|A \cup B|}.$$

Because the JT distance is known to be a metric [3], we can use algorithms proposed for any metric space to obtain information from the data. Examples of those algorithms include the approximation algorithms for facility location problem [4], nearest neighbor problem [5], and the Steiner tree problem [6].

Besides the JT index, there are other dissimilarity indexes considered in literature. Among them, the most well-known indexes include the Sørensen-Dice

© Springer International Publishing Switzerland 2016
M. Kaykobad and R. Petreschi (Eds.): WALCOM 2016, LNCS 9627, pp. 339–350, 2016.
DOI: 10.1007/978-3-319-30139-6_27

index [7,8] (SD index), in which the distance between a finite set A and a finite set B, $d_{SD}(A, B)$, can be defined as follows:

$$d_{SD} := 1 - \frac{2|A \cap B|}{|A| + |B|} = 1 - \frac{2|A \cap B|}{|A \cup B| + |A \cap B|}.$$

The SD dissimilarity is known to be robust in datasets of which some data points are missing [9]. Because of that, the distance is widely used in the ecological community data [10]. In our previous work [11], for some Go board position G, we know the set of positions P_G which is the set of positions that professional Go players choose to play when the board position is G. Because there is a lot of information missing in the database, we need to use a robust dissimilarity such as the SD distance.

Consider the case when $A = B = \{1, 2, 3\}$, suppose not all elements can be observed so A is observed as $A' = \{1, 2\}$ and B as $B' = \{2, 3\}$. For such case, the JT dissimilarity d_{JT} is $1 - 1/3 \approx 0.67$, while the SD dissimilarity is $1 - (2 \cdot 1)/(2 + 2) \approx 0.5$. We can see from this example that with only one element missing the JT dissimilarity can increase to $2/3$, while the ST dissimilarity is still as small as $1/2$.

Although the ST index is robust, the dissimilarity is not a metric. That is because the dissimilarity does not satisfy the triangle inequality [12]. For example, when $A = \{1\}$, $B = \{1, 2\}$ and $C = \{3\}$, we have $d_{ST}(A, C) = 1$, $d_{ST}(A, B) = 1/3$, and $d_{ST}(B, C) = 1/3$. By that, we have $d_{ST}(A, C) > d_{ST}(A, B) + d_{ST}(B, C)$. Because the distance is not a metric, we cannot use the algorithms devised for metric spaces.

Recently, there are several algorithms proposed for semi-metric distances that satisfy the ρ-approximate triangle inequality [13–15] for some $\rho > 1$. A distance D satisfies the inequality, if for any finite sets A, B, C,

$$d(A, C) \leq \rho \left(d(A, B) + d(B, C) \right).$$

Those algorithms are more efficient, when the value of ρ is smaller. Knowing the value of ρ for a specific dissimilarity can help in analyzing the efficiency of the algorithms, when it is applied to the distance.

1.1   Our Contribution

By the previous example, we know that the value of ρ for the SD index must be at least $3/2$. In this paper, we will show that the lower bound is tight, meaning that the SD index satisfies the $3/2$-approximate triangle inequality.

As discussed previously, the SD index is more robust than the JT index. We extend that idea to propose a dissimilarity called **robust Jaccard index (RJ index)**. We define the RJ dissimilarity between a finite set A and a finite set B, $d_{RJ,\alpha}(A, B)$, as follows:

$$d_{RJ,\alpha}(A, B) := 1 - \frac{\alpha |A \cap B|}{|A \cup B| + (\alpha - 1)|A \cap B|}.$$

Clearly, when $\alpha = 1$, the dissimilarity $d_{RJ,1}$ is equal to the JT index. When $\alpha = 2$, the dissimilarity $d_{RJ,2}$ is equal to the SD index. When $\alpha \to \infty$, $d_{RJ,\infty}(A, B)$ is always 0 when $A \cap B \neq \emptyset$ and the distance is 1 when $A \cap B = \emptyset$. The dissimilarity is very robust because we still get the distance equals 0 even when a lot of information are missing and $|A \cap B| = 1$. When $A' = \{1, 2\}$, $B' = \{2, 3\}$, $d_{RJ,\alpha} = 2/(2 + \alpha)$. The distance will get smaller and more robust when α is larger.

The robust Jaccard index is known to be a subclass of the Tversky index [16]. The Tversky index is proposed for differentiating the importance between two sets obtained as inputs [17]. They do not consider the robustness as one of the applications of the distance, but we believe that their index can also be used for this purpose.

The RJ index is a semi-metric. We can find an example to show that the value of ρ of $d_{RJ,\alpha}$ is at least $(\alpha + 1)/2$. We also show that the lower bound is tight. The distance $d_{RJ,\alpha}$ satisfies the $(\alpha + 1)/2$-approximate triangle inequality.

The remainder of this paper is divided into five sections. Section 2 describes the robust Jaccard index and its interpretation and discusses the relation of our proposed coefficient with other similarity indexes. Section 3 consist of the proof of its semimetric properties. Then, the final section summarizes this paper and present the future direction of our research.

2 Robust Jaccard Index

In order to be able to properly study how the concept of robustness of a set similarity index, and how it affects its metric properties, we propose a new coefficient that captures its fundamental idea.

Definition 1 (Robust Jaccard index). *For sets X and Y the robust Jaccard index is a number between 0 and 1 given by:*

$$S_{RJ,\alpha}(X, Y) := \frac{\alpha|X \cap Y|}{|X \cup Y| + (\alpha - 1)|X \cap Y|}.$$

By using this index, we not only expect to make the following proofs clearer to the reader; but also provide a powerful yet easy to use tool that can be directly applied when a more customized approach for dealing with uncertainty in experimental data is required.

2.1 Relation with Other Indexes

In this section, we proceed to investigate the relation between our similarity index and the three best-known indexes on sets; the Jaccard-Tanimoto, Sørensen-Dice and Tversky.

Definition (Jaccard-Tanimoto Index). *For sets X and Y the Jaccard-Tanimoto index is a number between 0 and 1 given by:*

$$S_{JT}(X, Y) := \frac{|X \cap Y|}{|X \cup Y|}.$$

The robust Jaccard index is equivalent to the Jaccard-Tanimoto index when $\alpha = 1$.

Definition (Sørensen-Dice Index). *For sets X and Y the Sørensen-Dice index is a number between 0 and 1 given by:*

$$S_{SD}(X,Y) := \frac{2|X \cap Y|}{|X| + |Y|}.$$

The robust Jaccard index is equivalent to the Sørensen-Dice index when $\alpha = 2$.

Definition (Tversky Index). *For sets X and Y the Tversky index is a number between 0 and 1 given by:*

$$S_{T,\beta,\gamma}(X,Y) := \frac{|X \cap Y|}{|X \cap Y| + \beta|X - Y| + \gamma|Y - X|}.$$

Proposition 1. *The Tversky index is equivalent to the robust Jaccard index when $\beta = \gamma = \frac{1}{\alpha}$, i.e. $S_{T,1/\alpha,1/\alpha} = S_{RJ,\alpha}$.*

Proof. By the definitions of Tversky and robust Jaccard index, we have

$$
\begin{aligned}
S_{T,1/\alpha,1/\alpha}(X,Y) &= \frac{|X \cap Y|}{|X \cap Y| + \frac{1}{\alpha}|X - Y| + \frac{1}{\alpha}|Y - X|} \\
&= \frac{|X \cap Y|}{|X \cap Y| + \frac{1}{\alpha}(|X \cup X| - |X \cap Y|)} \\
&= \frac{\alpha|X \cap Y|}{|X \cup X| - (\alpha - 1)|X \cap Y|} = S_{RJ,\alpha}
\end{aligned}
$$

$\square$

3 Metric Properties

Once that we have defined the robust Jaccard similarity index, it is only natural to define a distance (or dissimilarity) coefficient as well.

Definition 2 (Robust Jaccard Distance). *For sets X and Y and the robust Jaccard similarity index, the expression as a distance is given by:*

$$d_{RJ}(X,Y) = 1 - S_{RJ}(X,Y)$$

Then one could start questioning about the metric properties of this distance. In our case, since a counter example that show that it is not a metric can be easily found, we are more interested on the possibility of it being a semimetric.

Definition (Semimetric). *A semimetric on X is a function $d : X \times X \to \mathbb{R}$ that satisfies the first three axioms of a metric, but not necessarily the triangle inequality:*

(1) $d(x, y) \geq 0$
(2) $d(x, y) = 0 \iff x = y$
(3) $d(x, y) = d(y, x)$

Among all possible semimetrics, some of the most interesting ones are the so called "near-metrics", that are useful to guarantee a good performance in several approximation algorithms [13–15].

Definition (ρ-relaxed Semimetric). *ρ-relaxed semimetric is a semimetric that also satisfies a ρ-relaxed triangle inequality:*

(4) $d(x, z) \leq \rho(d(x, y) + d(y, z))$.

In order to make the following proofs simpler, let us define $X_{A,B}, Y_{B,C}$, and $Z_{A,C}$ as follows:

$$X_{A,B} = \frac{S_{RJ,\alpha}(A, B)}{\alpha} = \frac{|A \cap B|}{|A \cup B| + (\alpha - 1)|A \cap B|},$$

$$Y_{B,C} = \frac{S_{RJ,\alpha}(B, C)}{\alpha} = \frac{|B \cap C|}{|B \cup C| + (\alpha - 1)|B \cap C|},$$

$$Z_{A,C} = \frac{S_{RJ,\alpha}(A, C)}{\alpha} = \frac{|A \cap C|}{|A \cup C| + (\alpha - 1)|A \cap C|}.$$

Also, $f(A, B, C) := \frac{\frac{1}{\alpha} - Z_{A,C}}{\frac{2}{\alpha} - X_{A,B} - Y_{B,C}}$, and let

$$A^*, B^*, C^* := \arg \max_{A,B,C} f(A, B, C).$$

In Lemma 3, we show that

$$f(A^*, B^*, C^*) = \frac{\alpha + 1}{2}.$$

Using that result, we can obtain our main result stated in the following theorem.

Theorem 1. *For $\alpha \in \mathbb{Z}_+$, $d_{RJ,\alpha}$ is a ρ-relaxed semimetric, with $\rho = (\alpha + 1)/2$, i.e., for any finite sets A, B, C,*

$$d_{RJ,\alpha}(A, C) \leq \frac{\alpha + 1}{2} \left(d_{RJ,\alpha}(A, B) + d_{RJ,\alpha}(B, C) \right).$$

Furthermore, there are finite sets A, B, C such that

$$d_{RJ,\alpha}(A, C) = \frac{\alpha + 1}{2} \left(d_{RJ,\alpha}(A, B) + d_{RJ,\alpha}(B, C) \right).$$

Proof. Equation (4) can be rewritten as

$$1 - \alpha Z_{A,C} \le \rho(1 - \alpha X_{A,B} + 1 - \alpha Y_{B,C})$$

$$\rho \ge \frac{1 - \alpha Z_{A,C}}{2 - \alpha X_{A,B} - \alpha Y_{B,C}}$$

$$= \frac{\frac{1}{\alpha} - Z_{A,C}}{\frac{2}{\alpha} - X_{A,B} - Y_{B,C}}.$$

By the above inequality and Lemma 3, we can prove this theorem.   □

By substituting the value of α in the previous theorem by 2, we have the following corollary.

Corollary 1. d_{SD} *is a 3/2-relaxed semimetric, i.e., for any finite sets* A, B, C,

$$d_{SD}(A, C) \le \frac{3}{2}\left(d_{SD}(A, B) + d_{SD}(B, C)\right).$$

Furthermore, there are finite sets A, B, C *such that*

$$d_{SD}(A, C) = \frac{3}{2}\left(d_{SD}(A, B) + d_{SD}(B, C)\right).$$

In order to prove Lemma 3, we show several properties of A^*, B^*, C^* though Propositions 2–5, Lemmas 2 and 3. As shown in Fig. 1(a), we show that $B^*\backslash(A^* \cup B^*)$, $(A^* \cup C^*)\backslash B^*$, $A^*\backslash(B^* \cup C^*)$, and $C^*\backslash(A^* \cup B^*)$ must be empty sets in Propositions 2, 3, 4, 5 respectively. By those propositions, we can consider A^*, B^*, C^* in the form shown in Fig. 1(b). Then, we will prove that all sets A^*, B^*, C^* in that form must have ρ less than or equal to $(\alpha + 1)/2$ in Lemmas 1 and 2.

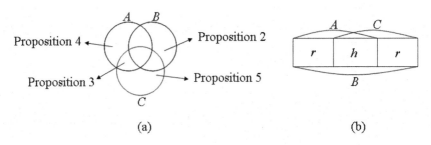

Fig. 1. An outline of our proof of Theorem 1

Proposition 2. *If* $B' \setminus (A' \cup C') \ne \emptyset$ *then* $f(A', B', C') < \max\limits_{A,B,C} f(A, B, C).$

Proof. Assume $e \in B' \setminus (A' \cup C')$. Let $A'' = A'$, $B'' = B' \setminus \{e\}$ and $C'' = C'$. Since $|A' \cap B'| = |A'' \cap B''|$ and $|A' \cup B'| - 1 = |A'' \cup B''|$, we have

$$X_{A'',B''} = \frac{|A'' \cap B''|}{|A'' \cup B''| + (\alpha - 1)|A'' \cap B''|}$$

$$= \frac{|A' \cap B'|}{|A' \cup B'| + (\alpha - 1)|A' \cap B'| - 1}$$

$$> \frac{|A' \cap B'|}{|A' \cup B'| + (\alpha - 1)|A' \cap B'|} = X_{A',B'}.$$

Similarly, $|B' \cap C'| = |B'' \cap C''|$ and $|B' \cup C'| - 1 = |B'' \cup C''|$, we have

$$Y_{B'',C''} = \frac{|B'' \cap C''|}{|B'' \cup C''| + (\alpha - 1)|B'' \cap C''|}$$

$$= \frac{|B' \cap C'|}{|B' \cup C'| + (\alpha - 1)|B' \cap C'| - 1}$$

$$> \frac{|B' \cap C'|}{|B' \cup C'| + (\alpha - 1)|B' \cap C'|} = X_{B',C'}.$$

Since $A'' = A'$ and $C'' = C'$, we have $Z_{A'',C''} = Z_{A',C'}$. Then,

$$f(A'', B'', C'') = \frac{1/\alpha - Z_{A'',C''}}{2/\alpha - X_{A'',C''} - Y_{A'',C''}}$$

$$> \frac{1/\alpha - Z_{A',C'}}{2/\alpha - X_{A',C'} - Y_{A',C'}} = f(A', B', C')$$

$$\square$$

To show the next proposition, we need the following lemma.

Lemma 1. *Let* N, M *be positive integers such that* $M \geq hN$ *and* c *be a natural number smaller or equal to* h. *We have*

$$\frac{N+1}{M+c} \geq \frac{N}{M}.$$

Furthermore, when c *is strictly smaller than* h. *We have*

$$\frac{N+1}{M+c} > \frac{N}{M}.$$

Proof. From the first part of this lemma statement, we have

$$(N+1)M \geq N(M+c)$$
$$NM + M \geq NM + Nc$$
$$M \geq Nc.$$

By the same argument, we can also prove the second part of this lemma. $\square$

Proposition 3. *If* $(A' \cap C') \setminus B' \neq \emptyset$ *then* $f(A', B', C') < \max\limits_{A,B,C} f(A, B, C)$.

Proof. Assume $e \in (A' \cap C') \setminus B'$. Let $A'' = A'$, $B'' = B' \cup \{e\}$ and $C'' = C'$. Then, $|A' \cap B'| + 1 = |A'' \cap B''|$ and $|A' \cup B'| = |A'' \cup B''|$. By Lemma 1, we have

$$
\begin{aligned}
X_{A'',B''} &= \frac{|A'' \cap B''|}{|A'' \cup B''| + (\alpha - 1)|A'' \cap B''|} \\
&= \frac{|A' \cap B'| + 1}{|A' \cup B'| + (\alpha - 1)|A' \cap B'| + \alpha - 1} \\
&> \frac{|A' \cap B'|}{|A' \cup B'| + (\alpha - 1)|A' \cap B'|} = X_{A',B'}.
\end{aligned}
$$

Similarly, $|B' \cap C'| + 1 = |B'' \cap C''|$ and $|B' \cup C'| = |B'' \cup C''|$. By Lemma 1,

$$
\begin{aligned}
Y_{B'',C''} &= \frac{|B'' \cap C''|}{|B'' \cup C''| + (\alpha - 1)|B'' \cap C''|} \\
&= \frac{|B' \cap C'| + 1}{|B' \cup C'| + (\alpha - 1)|B' \cap C'| + \alpha - 1} \\
&> \frac{|B' \cap C'|}{|B' \cup C'| + (\alpha - 1)|B' \cap C'|} = Y_{B',C'}.
\end{aligned}
$$

Since $A'' = A'$ and $C'' = C'$, we have $Z_{A'',C''} = Z_{A',C'}$. Then,

$$
\begin{aligned}
f(A'', B'', C'') &= \frac{1/\alpha - Z_{A'',C''}}{2/\alpha - X_{A'',C''} - Y_{A'',C''}} \\
&> \frac{1/\alpha - Z_{A',C'}}{2/\alpha - X_{A',C'} - Y_{A',C'}} = f(A', B', C')
\end{aligned}
$$

$\square$

Up until now, we know that $B^* - (A^* \cup C^*)$ and $(A^* \cup C^*) - B^*$ must be an empty set. In the next proposition, we will show that there must be sets A^*, B^*, C^* that maximize function f and $A^* - (B^* \cup C^*)$ is also an empty set.

Proposition 4. *If* $A' \setminus (B' \cap C') \neq \emptyset$ *then there exists* A, B, C *such that* $B - (A \cup C) = \emptyset, (A \cup C) - B = \emptyset, A - (B \cup C) = \emptyset$ *and* $f(A, B, C) \geq f(A', B', C')$.

Proof. Assume $e \in A' \setminus (B' \cap C')$. Let $A'' = A' \setminus \{e\}$, $B'' = B' \cup \{e\}$ and $C'' = C' \cup \{e\}$. Since $|A'' \cap B''| = |A' \cap B'|$ and $|A'' \cup B''| = |A' \cup B'|$, we have $X_{A'',B''} = X_{A',B'}$.

By $|B'' \cap C''| = |B' \cap C'| + 1$, $|B'' \cup C''| = |B' \cup C''| + 1$, and Lemma 1, we have

$$
\begin{aligned}
Y_{B'',C''} &= \frac{|B'' \cap C''|}{|B'' \cup C''| + (\alpha - 1)|B'' \cap C''|} \\
&= \frac{|B' \cap C'| + 1}{|B' \cup C'| + (\alpha - 1)|B' \cap C'| + \alpha} \\
&\geq \frac{|B' \cap C'|}{|B' \cup C'| + (\alpha - 1)|B' \cap C'|} = Y_{B',C'}.
\end{aligned}
$$

Since $|A'' \cap C''| = |A' \cap C'|$ and $|A'' \cup C''| = |A' \cup C'|$, we have $Z_{A'',C''} = Z_{A',C'}$. Then,

$$
\begin{aligned}
f(A'', B'', C'') &= \frac{1/\alpha - Z_{A'',C''}}{2/\alpha - X_{A'',C''} - Y_{A'',C''}} \\
&\geq \frac{1/\alpha - Z_{A',C'}}{2/\alpha - X_{A',C'} - Y_{A',C'}} = f(A', B', C').
\end{aligned}
$$

If $A'' - (B'' \cup C'')$ is an empty set, then we can prove this proposition. If not, we can take another element out of $A'' - (B'' \cup C'')$ and add that element to $(B'' \cap C'') - A''$. By the same argument, we know that the value of function f is not decreased. We can do the same action until the set $A'' - (B'' \cup C'')$ becomes an empty set. After the loop, we will have A, B, C that satisfy the lemma statement.

By the previous proposition, we know there exists at least one set of A^*, B^*, C^* such that $B^* - (A^* \cup C^*) = \emptyset, (A^* \cup C^*) - B^* = \emptyset, A^* - (B^* \cup C^*) = \emptyset$ and $A^*, B^*, C^* = \arg\max A, B, C f(A, B, C)$. The next theorem will show that $C^* - (A^* \cup B^*)$ can also be an empty set.

Proposition 5. *Suppose that $B' - (A' \cup C') = \emptyset, (A' \cup C') - B' = \emptyset, A' - (B' \cup C') = \emptyset$. Then, if $C' - (A' \cup B') \neq \emptyset$, there are finite sets A, B, C such that $B - (A \cup C) = \emptyset, (A \cup C) - B = \emptyset, A - (B \cup C) = \emptyset$, and $f(A, B, C) \geq f(A', B', C')$.*

Proof. Assume $e \in C' \setminus (A' \cup B')$. Let $A'' = A' \cup \{e\}$, $B'' = B' \cup \{e\}$ and $C'' = C' \setminus \{e\}$. By $|A'' \cap B''| = |A' \cap B'| + 1$, $|A'' \cup B''| = |A'' \cup B''| + 1$, and Lemma 1, we have

$$
\begin{aligned}
X_{A'',B''} &= \frac{|A'' \cap B''|}{|A'' \cup B''| + (\alpha - 1)|A'' \cap B''|} \\
&= \frac{|A' \cap B'| + 1}{|A' \cup B'| + (\alpha - 1)|A' \cap B'| + \alpha} \\
&\geq \frac{|A' \cap B'|}{|A' \cup B'| + (\alpha - 1)|A' \cap B'|} = X_{A',B'}.
\end{aligned}
$$

Since $|B'' \cap C''| = |B' \cap C'|$ and $|B'' \cup C''| = |B' \cup C'|$, we have $Y_{B'',C''} = Y_{B',C'}$. Since $|A'' \cap C''| = |A' \cap C'|$ and $|A'' \cup C''| = |A' \cup C'|$, we have $Z_{A'',C''} = Z_{A',C'}$. Then,

$$
\begin{aligned}
f(A'', B'', C'') &= \frac{1/\alpha - Z_{A'',C''}}{2/\alpha - X_{A'',C''} - Y_{A'',C''}} \\
&\geq \frac{1/\alpha - Z_{A',C'}}{2/\alpha - X_{A',C'} - Y_{A',C'}} = f(A', B', C').
\end{aligned}
$$

If $A'' - (B'' \cup C'')$ is an empty set, then we can prove this proposition. If not, we can take another element out of $A'' - (B'' \cup C'')$ and add that element to $(B'' \cap C'') - A''$. By the same argument, we know that the value of function

f is not decreased. We can do the same action until the set $A'' - (B'' \cup C'')$ becomes an empty set. After the loop, we will have A, B, C that satisfy the lemma statement.

By Propositions 2–5, we know that if $e \in (A^* \cup B^* \cup C^*)$, either $e \in (A^* \cap B^*) \setminus C^*$, $e \in (A^* \cap B^* \cup C^*)$, or $e \in (B^* \cap C^*) \setminus A^*$. Meaning that

$$B^* = A^* \cup C^*.$$

With this result, we can state the following lemmas:

Lemma 2. *When* $\alpha > 1$, $|A^*| = |C^*|$

Proof. Let assume that $|A^* \cap C^*| = h$ and $|A^*| + |C^*| = m$. With this, Z_{A^*,C^*} is always equal to $h/(m + (\alpha - 2)h)$, and $A^*, B^*, C^* = \arg \max_{A,B,C}(X_{A,B} + Y_{B,C})$.

Let $0 \le k \le 1$. We can denote now $|A^*| = km$, and $|C^*| = (1-k)m$. Thus,

$$X_{A,B} + Y_{B,C} = \frac{km}{(\alpha - 1)km + m - h} + \frac{(1-k)m}{(\alpha - 1)(1-k)m + m - h}.$$

Because

$$\frac{d(X_{A,B} + Y_{B,C})}{dk} = \frac{(\alpha - 1)(2k-1)m^2(h-m)((\alpha+1)m - 2h)}{(m(\alpha(k-1) - k) + h)^2(m(-\alpha k + h - 1) + h)^2},$$

and for $\alpha > 1$ it can only be equal to 0 when $k = \frac{1}{2}$.

Therefore $|A^*| = \frac{1}{2}m = |C^*|$.   □

Lemma 3. *When* $\alpha > 1$, $f(A^*, B^*, C^*) = (\alpha + 1)/2$.

Proof. Recall that $|A_* \cap C_*| = h$ and $|A^*| = |C^*|$.

In that case $|A^* \setminus C^*| = |A^*| - |A^* \cap C^*| = |C^*| - |A^* \cup C^*| = |C^* \setminus A^*|$. Thus, we can denote $|A^* \setminus C^*| = |C^* \setminus A^*| = r$.

Also recall that $\rho = \frac{\frac{1}{\alpha} - Z_{A^*,C^*}}{\frac{2}{\alpha} - X_{A^*,B^*} - Y_{B^*,C^*}}$, so

$$X_{A^*,B^*} = \frac{r + h}{(\alpha + 1)r + \alpha h},$$

$$Y_{B^*,C^*} = \frac{r + h}{(\alpha + 1)r + \alpha h},$$

$$Z_{A^*,C^*} = \frac{h}{2r + \alpha h}.$$

Then

$$\rho = \frac{\frac{1}{\alpha} - \frac{h}{2r + \alpha h}}{\frac{2}{\alpha} - \frac{2r + 2h}{(\alpha + 1)r + \alpha h}} = \frac{(\alpha + 1)r + \alpha h}{2r + \alpha h}.$$

Therefore ρ is maximized when $|A^* \cap C^*| = h = 0$.   □

The proof of Lemma 2 works only for the case when $\alpha > 1$, but we know from [3] that the RJ distance is a metric ($\rho = 1$) when $\alpha = 1$. That makes our theorem hold for any positive integer α.

We can easily construct an example to show that the bound obtained in Theorem 1 is tight. That is when $A = \{1\}$, $B = \{1, 2\}$, $C = \{2\}$. We have

$$d_{RJ,\alpha}(A, B) = d_{RJ,\alpha}(B, C) = d_{RJ,\alpha}(B, C) = 1 - \frac{\alpha}{1 + \alpha} = \frac{1}{\alpha},$$

and $d_{RJ,\alpha}(B, C) = 1$. Then,

$$1 \leq \rho \left(\frac{1}{\alpha} + \frac{1}{\alpha} \right)$$

$$\rho \geq \frac{\alpha}{2}.$$

4 Conclusions and Future Work

In this paper, we have proposed a family of similarity indexes, named as the robust Jaccard index, for datasets with missing information. The only parameter of the index is α. When α gets larger, the dissimilarity becomes more robust. However, we have shown in this paper that the value of ρ in the approximate triangle inequality also becomes larger, when the value of α increases. Because, when the value of ρ gets larger, the algorithms proposed for this semi-metric spaces will be less efficient, we have to trade between the robustness and the efficiency of algorithms.

Because of that, we plan to perform experiments to see what is the optimal value of α in each dataset. Besides, We are aiming to find the value of ρ for the general Trevsky distance in our future work. We want to find the relationship between the symmetricity, robustness, and efficiency with those results.

Acknowledgement. The authors would like to thank Mr. Naoto Osaka and Prof. Hiroshi Imai for several useful comments during the course of this research.

References

1. Jaccard, P.: Lois de distribution florale dans la zone alpine. Corbaz (1902)
2. Tanimoto, T.: An elementary mathematical theory of classification and prediction. Technical report, IBM Report (1958)
3. Lipkus, A.H.: A proof of the triangle inequality for the Tanimoto distance. J. Math. Chem. **26**(1–3), 263–265 (1999)
4. Jain, K., Vazirani, V.V.: Primal-dual approximation algorithms for metric facility location and k-median problems. In: FOCS 1999, pp. 2–13 (1999)
5. Ruiz, E.V.: An algorithm for finding nearest neighbours in (approximately) constant average time. Pattern Recogn. Lett. **4**(3), 145–157 (1986)
6. Sankoff, D., Rousseau, P.: Locating the vertices of a steiner tree in an arbitrary metric space. Math. Program. **9**(1), 240–246 (1975)

7. Sørensen, T.: A method of establishing groups of equal amplitude in plant sociology based on similarity of species and its application to analyses of the vegetation on danish commons. Biol. Skr. **5**, 1–34 (1948)

8. Dice, L.R.: Measures of the amount of ecologic association between species. Ecology **26**(3), 297–302 (1945)

9. McCune, B., Grace, J.B., Urban, D.L.: Analysis of Ecological Communities, vol. 28. MjM software design, Gleneden Beach (2002)

10. Looman, J., Campbell, J.: Adaptation of Sorensen's K (1948) for estimating unit affinities in prairie vegetation. Ecology, 409–416 (1960)

11. Gragera, A.: Approximate matching for Go board positions. In: GPW 2015 (2015)

12. Schubert, A., Telcs, A.: A note on the Jaccardized Czekanowski similarity index. Scientometrics **98**(2), 1397–1399 (2014)

13. Braverman, V., Meyerson, A., Ostrovsky, R., Roytman, A., Shindler, M., Tagiku, B.: Streaming k-means on well-clusterable data. In: SODA 2011, pp. 26–40 (2011)

14. Mettu, R.R., Plaxton, C.G.: The online median problem. SIAM J. Comput. **32**(3), 816–832 (2003)

15. Jaiswal, R., Kumar, M., Yadav, P.: Improved analysis of D2-sampling based PTAS for k-means and other clustering problems. Inf. Process. Lett. **115**(2), 100–103 (2015)

16. Tversky, A., Gati, I.: Similarity, separability, and the triangle inequality. Psychol. Rev. **89**(2), 123 (1982)

17. Jimenez, S., Becerra, C., Gelbukh, A., Bátiz, A.J.D., Mendizábal, A.: Softcardinality-core: Improving text overlap with distributional measures for semantic textual similarity. In: SEM 2013, pp. 194–201 (2013)

Finding Mode Using Equality Comparisons

Varunkumar Jayapaul[1(✉)], Venkatesh Raman[2], and Srinivasa Rao Satti[3]

[1] Chennai Mathematical Institute,
H1, SIPCOT IT Park, Siruseri, Chennai 603 103, India
varunkumarj@cmi.ac.in
[2] The Institute of Mathematical Sciences, CIT Campus,
Taramani, Chennai 600 113, India
vraman@imsc.res.in
[3] Seoul National University, 1 Gwanak-ro, Gwanak-gu, Seoul 151-744, Korea
ssrao@cse.snu.ac.kr

Abstract. We consider the problem of finding the mode (an element that appears the maximum number of times) in a list of elements that are not necessarily from a totally ordered set. Here, the relation between elements is determined by 'equality' comparisons whose outcome is $=$ when the two elements being compared are equal and $\neq$ otherwise. In sharp contrast to the $\Theta(\frac{n \lg n}{m})$ bound known in the classical three way comparison model where elements are from a totally ordered set, a recent paper gave an $O(\frac{n^2}{m})$ upper bound and $\Omega(\frac{n^2}{m})$ lower bound for the number of comparisons required to find the mode, where m is the frequency of the mode. While the number of comparisons made by the algorithm is roughly $\frac{n^2}{m}$, it is not clear how the necessary bookkeeping required can be done to make the rest of the operations take $\Theta(\frac{n^2}{m})$ time.

In this paper, we give two mode finding algorithms, one taking at most $\frac{2n^2}{m}$ comparisons and another taking at most $\frac{3n^2}{2m} + O(\frac{n^2}{m^2})$ comparisons. The bookkeeping required for both the algorithms are simple enough to be implemented in $O(\frac{n^2}{m})$ time. The second algorithm generalizes a classical majority finding algorithm due to Fischer and Salzberg.

1   Introduction

Selection problems (finding min, max, median, mode) constitute one of the fundamental data processing tasks that have been well studied in the classical and modern models of computation [4,9]. While there has been a lot of work on algorithms for finding the median, finding the mode (an element that appears the maximum number of times) is an equally interesting and an important problem. In the standard comparison model, given a set of n elements from a total order, a classical paper by Dobkin and Munro [6] gives a $\Theta(\frac{n \log n}{m})$ algorithm and a matching lower bound for finding the mode whose frequency is m. In this paper, we consider the problem under the constraint that elements in the given list do not form a total order. Here, the only way the relation between a pair of elements is determined is by making an equality comparison. While this is a natural variant that occurs when dealing with heterogenous sets of elements, to the

© Springer International Publishing Switzerland 2016
M. Kaykobad and R. Petreschi (Eds.): WALCOM 2016, LNCS 9627, pp. 351–360, 2016.
DOI: 10.1007/978-3-319-30139-6_28

best of our knowledge the only other problem studied extensively in this model is the problem of determining the majority element (an element that appears more than $\lfloor \frac{n}{2} \rfloor$ times) if exists, and there is a classical linear time algorithm for this [5]. Exact comparison complexity and average case complexity of this problem have been studied [1–3,7,13].

In general, the lack of transitivity of the 'not equal' operation throws interesting challenges in this model. An earlier paper [8] studied the mode problem in this model, and showed that $\Omega(\frac{n^2}{m})$ (equality) comparisons are necessary and this many comparisons are sufficient to find an element that appears at least m times. However, it appears difficult to make the bookkeeping required to implement the remaining operations in $O(\frac{n^2}{m})$ time.

In the next section, we summarize the earlier algorithm and formulate a data structure problem on a family of sets that is required to implement the algorithm in detail. In Sect. 3.1 we first give a simple algorithm that makes at most $\frac{2n^2}{m}$ comparisons but it can be run in $O(\frac{n^2}{m})$ time where m is the frequency of the mode. Then, in Sect. 3.2 we give another algorithm which improves the comparison bound to at most $\frac{3n^2}{2m} + O(\frac{n^2}{m^2})$. This is a generalization of the classical Fischer-Salzberg [7] algorithm to find the majority of a given list of elements. Both our algorithms require the knowledge of m, the frequency of the mode. In Sect. 4, we explain how this assumption can be worked around to give a general algorithm even without the knowledge of m with only twice the number of comparisons. Section 5 concludes with remarks and open problems.

1.1   Related Work

As referred earlier, we know of only the majority problem [5] studied with $=, \neq$ comparisons. In one of the earliest papers studying optimal algorithms on sets, Reingold [12] proved lower bounds for determining the intersection/union of two sets if only $=, \neq$ comparisons are allowed. Munro and Spira [10] considered optimal algorithms and lower bounds to find the mode and the spectrum (the frequencies of all elements), albeit in the three way comparison model. Misra and Gries [11] gave algorithms to determine an element that appears at least $\frac{n}{k}$ times for various values of k, in the three way comparison model.

2   Finding Mode Using $O(\frac{n^2}{m})$ Comparisons

The following theorem summarizes some of the main results given in [8].

Theorem 1. *Given a list of n elements and an integer k, all elements (if any) with frequency at least k can be found using at most $O(\frac{n^2}{k})$ comparisons. In particular, the mode of a given list of elements can be found in $O(\frac{n^2}{m})$ comparisons, where m is the frequency of the mode, even if the algorithm does not know m. Furthermore $\Omega(\frac{n^2}{m})$ comparisons are required to find the mode even if the algorithm knows m.*

The algorithm that achieves the optimum bound is described in pseudocode below. Basically, it maintains the given input in a circular list, and for an element at position i, it maintains the list of elements $EQ(i)$ that are known to be equal to it, and the list $NEQ(i)$ of elements that are known not to be equal to it. Then it finds the first element (wrapping around if necessary) after i, not in $EQ(i) \cup NEQ(i)$ and compares the element with it. Based on the outcome of the comparison, the sets EQ and NEQ, of not just i, but other appropriate elements as well, are updated. Then the algorithm proceeds to the next element at position $i + 1$ and continues in a circular fashion. The algorithm stops when it finds an element with frequency m if m is known, or when it finds an element with frequency at least roughly $\frac{n}{r}$ after r 'rounds' of comparisons. We refer to [8] for details.

Initialize $r = 0$; for $i = 1$ to n $eq(a_i) = \{i\}$; $neq(a_i) = \emptyset$;
Repeat
    $r = r + 1$
    for $i = 1$ to n
        find the next j if any, starting from $i + 1$, wrapped around after n
        if necessary, such that $j \notin eq(a_i) \cup neq(a_i)$.
        if such a j is found then
          if $a_i = a_j$ then
            for all $x \in eq(a_i) \cup eq(a_j)$,
            $eq(a_x) \leftarrow eq(a_i) \cup eq(a_j)$ and
            $neq(a_x) \leftarrow neq(a_i) \cup neq(a_j)$
          else if $a_i \neq a_j$ then
            for all $x \in eq(a_i), neq(a_x) \leftarrow neq(a_x) \cup eq(a_j)$ and
            for all $y \in eq(a_j), neq(a_y) \leftarrow neq(a_y) \cup eq(a_i)$
        endfor
until there exists an element i such that $|eq(a_i)| \geq \frac{(n-1)}{r}$

It is shown in [8] that the above algorithm finds the mode in at most $n\lceil\frac{n}{m}\rceil$ comparisons. To actually implement the above algorithm to take $O(\frac{n^2}{m})$ time, we need a data structure to maintain the sets $EQ(i)$ and $NEQ(i)$ so that the updation of the two sets after each comparison, and the query of the smallest element larger than i (wrapping around if necessary) not in $EQ(i) \cup NEQ(i)$ can be supported in constant time. We can maintain all elements that are known to be equal to each other as a single set referred to by elements in the set (as their $EQ()$ sets). Thus if two positions i and j refer to different $EQ()$ sets, then they are mutually disjoint. Similarly, the $NEQ(i)$ sets themselves are disjoint union of some $EQ(j)$ sets. Now after every comparison, several EQ sets may need to be updated. We leave it as an interesting open problem to implement the above algorithm to take overall $O(\frac{n^2}{m})$ time.

3   Finding Mode in $O(\frac{n^2}{m})$ Time

Here we provide two algorithms that not only use $O(\frac{n^2}{m})$ comparisons, but also spend only $O(\frac{n^2}{m})$ time for the rest of the operations. The first one takes at most $\frac{2n^2}{m}$ comparisons, and the number of comparisons made by the second one is $\frac{3n^2}{2m} + O(\frac{n^2}{m^2})$. The first one is relatively simple to argue correctness, and the second algorithm generalizes a classical majority finding algorithm.

For now, we assume that m is known, and in Sect. 4, we explain how this assumption can be removed.

3.1   A Simple Mode Finding Algorithm

Let k be the smallest integer such that $\lfloor \frac{n}{k} \rfloor \leq m - 1$. I.e. $\lfloor \frac{n}{k} \rfloor \leq m - 1 < \lfloor \frac{n}{k-1} \rfloor$. Let $a_1, a_2, \ldots a_n$ be the given list of n elements. We give an algorithm that finds all elements with frequency more than $\lfloor \frac{n}{k} \rfloor$ from an input of size n. The pseudocode description is given below. Here B is a set of distinct elements with some frequencies associated with each element.

Initialize $B = \{a_1\}$; $i = 1$;
While $i \leq n - 1$
$\quad i = i + 1$
$\quad$ if a_i already appears in B,
$\qquad$ then increment the frequency of the value that equals a_i in B
$\qquad$ else add a_i to B with frequency 1.
$\quad$ If the number of distinct elements in B is k
$\qquad$ then decrement the frequency of each element in B;
$\qquad$ delete elements with (the new) frequency 0.
Endwhile
Find frequency in the entire list, of all elements (if any) of B,
and output those whose frequency over entire input is at least m.

We show that the algorithm finds all elements with frequency more than $\lfloor \frac{n}{k} \rfloor$ using at most $2n(k-1)$ comparisons. As $k \leq \frac{n}{m} + 1$, the total running time of the algorithm is then at most $\frac{2n^2}{m}$.

Suppose after every decrement, one copy of each of the elements is placed in a (separate) set, then as each set has k elements, the total number of such sets is at most $\lfloor \frac{n}{k} \rfloor < m$. Also each of the sets has distinct elements. So if an element has frequency more than $\lfloor \frac{n}{k} \rfloor$, it will have a copy in the final set B. Thus all elements with frequency at least m have a copy in B.

Every new element (after the first $k - 1$ distinct elements) is compared with at most $k - 1$ distinct elements of B for a total of $(n - k + 1)(k - 1) + \frac{(k-1)k}{2}$ comparisons. Also finally B has at most $k - 1$ distinct elements which are compared with the remaining elements for a total of at most $n(k - 1)$ comparisons for the confirmation phase. This results in an overall at most $2n(k-1)$ comparisons. As $m \leq \frac{n}{k-1}$, we have $k - 1 \leq \frac{n}{m}$. Thus we have the following theorem.

Theorem 2. *Given a multiset of n elements and a frequency m, we can find all elements with frequency at least m using at most $\frac{2n^2}{m}$ comparisons and $O(\frac{n^2}{m})$ time.*

3.2 An Improved Algorithm – Generalization of the Fischer-Salzberg Majority Algorithm

Fischer and Salzberg [7] developed an algorithm to find a majority element (if exists) in a list of n elements using at most $\frac{3n}{2} - 2$ comparisons. (Recall that a majority element is an element that appears more than $\lfloor \frac{n}{2} \rfloor$ times.). We generalize this to find the mode to improve the coefficient of $\frac{n^2}{m}$ in Theorem 2 to $\frac{3}{2}$, resulting in at most $\frac{3n}{2}(\lfloor \frac{n}{m} \rfloor)$ comparisons.

As before, let k be the integer such that $\lfloor \frac{n}{k} \rfloor \le m - 1 < \lfloor \frac{n}{k-1} \rfloor$. We will give an algorithm to find an element with frequency at least $m \ge \lfloor \frac{n}{k} \rfloor + 1$ using at most $\frac{3n(k-1)}{2}$ comparisons and other operations. When $k = 2$, the problem degenerates to the majority problem and the bound becomes at most $\frac{3n}{2}$ as in the case of Fischer and Salzberg's algorithm.

Let $a_1, a_2, \ldots a_n$ be the given sequence of n elements. The algorithm maintains a list L, and an array B with the following invariants:

- For any index i, $L[i]$ is not equal to any element in the set $S = \{L[j], |j - i| < k\}$. I.e. in L, any set of consecutive k elements are distinct.
- B, if non-empty, contains up to $k - 1$ distinct elements, all of which appear in the last $k - 1$ elements of L. Each cell in B contains the value of an element x, its last location in L and a frequency f which has the following property: the frequency of x in the input sequence (up to the point we have processed) is the frequency of x in L plus f. We maintain elements of B in a queue by increasing order of their (last occurrence) locations in L.

We can interpret that the input sequence is partitioned into L and B; i.e. every element of the input sequence is in L or in B. Initially B is empty, and L contains the first element in the input sequence. Then it processes each element $a_i (i > 1)$ in the sequence as follows.

- If a_i equals an element in the last $k - 1$ positions of L, then if a_i appears in B, find its occurrence and increment its frequency. If a_i does not appear in B, then create a new entry for in B, with a frequency of 1, and set its last occurrence location to its last occurrence in L.
- If a_i does not equal any element in the last $k - 1$ positions of L, then add a_i to L, and repeat the following step until not possible:
  - Take the element x in the first location of B. This is an element that has the least (last) location in L among those in B. If the last location in L of that element is at least k positions away from the current last position, then add x to the end of L after decrementing its frequency in B. Remove it from B if its frequency becomes 0, and update its last location to the current location in L otherwise. Move that element to the last location of B.

At the end, we claim that only the last $k - 1$ elements of L are possible candidates for the mode, and so we check the frequency of each of those elements with all elements of L and output all those with frequency m. We call the step of finding the frequency of the last $k - 1$ elements of L as the 'confirmation phase'. The following pseudocode gives the details of the algorithm.

```
Initialize L = a₁; i = 1; B = ∅;
While i ≤ n − 1
    i = i + 1
    if aᵢ equals an element in the last k − 1 elements of L
        then if aᵢ ∈ B
            find and increment frequency of aᵢ in B by one
        else if aᵢ ∉ B
            Create a new entry in B with aᵢ as value, frequency as 1, and its position as
            last known position of aᵢ in L.
            Sort B in increasing order of the 'position' field of its elements.
    if aᵢ does not equal any of the last k − 1 elements of L
        Add aᵢ to the end of L
        While |B(first).location − L(last)| > k
            decrease frequency of B(first) by 1
            add B(first).value to the end of list L
            update B(first).location = last + 1
            if B(first)'s frequency is nonzero then
                move B(first) to end of B,
            else remove it from B
        Endwhile
Endwhile
Comment: Confirmation Phase
For all elements in the last k − 1 elements of L
    find their actual frequency in the input
    and output those whose total frequency in the entire input is equal to m
```

In the initial phase, the only comparisons made for each element are to test whether it equals an element in the last $k - 1$ elements of L. We can use the last location entry to find its existence (if at all) in B which involves no comparisons with the element. So at most $n(k-1)$ comparisons and $n(k-1)$ other operations are made in the first processing phase of the algorithm to construct L and B. The confirmation phase takes at most $(k-1)n$ comparisons for a total of $2n(k-1)$ comparisons.

In what follows, we tighten the analysis to show a bound of at most $\frac{n(k-1)}{2}$ for the confirmation phase resulting in an overall comparison bound of $\frac{3n(k-1)}{2}$, and prove the correctness of the algorithm.

First it is clear that the algorithm maintains the two invariants (on L and B) mentioned above after every step. Invariant on L is maintained as and when we add new elements to L (either from the input sequence or from B). B always contains at most $k - 1$ elements. We add an element to B only when a new element does not appear in the last $k - 1$ elements of L, and when that happens, B had less than $k - 1$ elements, as the elements of B appear in the last $k - 1$ elements of L, and so the addition to B does not make the number of elements of B go above $k - 1$. Also when we add new elements to L, if B is non-empty, we add elements from B to L ensuring that elements of B are in the last $k - 1$ elements of L, for if an element of B is not in the last $k - 1$ elements of L, then that element would have been added to the last element of L.

Suppose that the size of L is divisible by k. Then every element in L can appear at most $\frac{n}{k} \leq m - 1$ times (as every consecutive k elements are distinct). And hence B has to be non-empty (for the mode to appear m times), and the only candidates for elements appearing more than $\lfloor \frac{n}{k} \rfloor$ times are those in B which are anyway in the last $k - 1$ elements of L by the invariant on B.

Suppose the size of L is not divisible by k. Then the only possible input elements that appear more than $\lfloor \frac{n}{k} \rfloor \leq m - 1$ times are those in B and those in the last $n - k(\lfloor \frac{n}{k} \rfloor)$ locations of L, and the last $k - 1$ elements of L cover these.

This completes the correctness of the algorithm. Now we give a slight modification of the confirmation phase and give a careful calculation of the number of comparisons made in the confirmation phase and show it to be at most $\frac{n(k-1)}{2}$.

The Confirmation Phase. For any element in the last $k - 1$ locations of L that is not in B, the confirmation phase starts by comparing it with an element that is k locations apart to the left of it (as we know that the intermediate $k - 1$ elements are distinct from it). For an element in B, the confirmation phase starts with its copy in the last $k - 1$ locations and continues as above. And during the confirmation phase, if we find an element which is not equal to the element being compared, then we move left by a position and continue the comparison. And if we find an element which is equal to the element being compared, then we skip $k - 1$ positions to the left and continue our comparison.

Let $\ell \leq k - 1$ be the number of distinct elements in B and let $f_1, f_2, \ldots f_\ell$ be their respective frequencies in B at the end of the algorithm. Let $f = \sum_{i=1}^{\ell} f_i$.

Lemma 1. *Let $\ell \leq k - 1$ be the number of distinct elements in B and let $f_1, f_2, \ldots, f_\ell$ be their respective frequencies in B at the end of the algorithm. Let $f = \sum_{i=1}^{\ell} f_i$. The number of comparisons done by candidates from B in the confirmation phase is at most $f(k - 1) + \ell(\frac{n}{k} - f) + (k - 1)^2$.*

Proof. Let b be an element of B with frequency f_1. Then for b to qualify as a mode, it should have at least $\lfloor \frac{n}{k} \rfloor - f_1$ copies in L, not counting its copy in the last $k - 1$ locations of L. View L as a contiguous sequence of k-sized blocks. There are $\lceil \frac{n-f}{k} \rceil$ such blocks. Hence b may not be present in at most $\lceil \frac{n-f}{k} \rceil - 1 - (\lfloor \frac{n}{k} \rfloor - f_1)$ such blocks and may get not equal outcomes in comparisons with elements in these blocks for a total of at most $k(\lceil \frac{n-f}{k} \rceil - 1 - \lfloor \frac{n}{k} \rfloor + f_1)$ comparisons with not equal outcomes. It is easy to see that it may make at most $k - 1$ more comparisons with not equal outcomes, and has to get at least $\lfloor \frac{n}{k} \rfloor - f_1$ equality comparisons. If the number of comparisons with not equal outcomes or equal outcomes exceeds these quantities, we can stop the confirmation phase of that element.

Now summing these comparisons for every element of B, the number of comparisons made by elements of B in the confirmation phase is at most

$$\sum_{i=1}^{\ell}(\lfloor\frac{n}{k}\rfloor - f_i + (k-1) + k(\lceil\frac{n-f}{k}\rceil - 1 - \lfloor\frac{n}{k}\rfloor + f_i))$$

$$\leq \ell(\lfloor\frac{n}{k}\rfloor(1-k) + (n-f) + k - 1) + f(k-1)$$

$$\leq f(k-1) + \ell(\frac{n}{k} - f + k - 1)$$

$$\leq f(k-1) + \ell(\frac{n}{k} - f) + (k-1)^2 \qquad\qquad \square$$

Now we continue with the analysis of the total number of comparisons in the confirmation phase.

Case 1: $f \geq \frac{n}{2}$.
This implies that $|L| \leq \frac{n}{2}$. The confirmation phase finds the frequency of each of the last $k-1$ elements of L with the other elements of L which, in this case, will take at most $\frac{n(k-1)}{2}$ comparisons.

Case 2: $n \bmod k < f < \frac{n}{2}$.
In this case, $|L| \leq k\lfloor n/k\rfloor$, and hence any element of L that is not in B, appears at most $\lfloor\frac{n}{k}\rfloor$ times and hence they don't qualify to become a mode.

From Lemma 1, the number of comparisons made by elements of B during the confirmation phase is at most $f(k-1) + \ell(\frac{n}{k} - f) + (k-1)^2$.

If $f \geq \frac{n}{k}$, then

$$f(k-1) + \ell(\frac{n}{k} - f) + (k-1)^2 \leq f(k-1) + (k-1)^2$$

$$< \frac{n(k-1)}{2} + (k-1)^2.$$

If $f < \frac{n}{k}$, then

$$f(k-1) + \ell(\frac{n}{k} - f) + (k-1)^2 < \frac{n(k-1-\ell)}{k} + \frac{\ell n}{k} + (k-1)^2$$

$$= \frac{n(k-1)}{k} + (k-1)^2$$

$$\leq \frac{n(k-1)}{2} + (k-1)^2$$

Case 3: $f \leq (n \bmod k)$.
Let S be the set of all the elements which occur in last $k-1$ positions of L, but which do not occur in B. Let $|S| = p$. These p elements need at least $\lfloor\frac{n}{k}\rfloor$ additional copies for each of them to become a candidate for mode. I.e. each

of them should appear in every block of L. Hence, the verification process for each element in S ends in at most $\lfloor \frac{n}{k} \rfloor$ equality comparisons and possibly at most $k-1$ comparisons with not equal outcomes, totally making at most $\frac{pn}{k} + p(k-1)$ comparisons.

The remaining $\ell = (k-1-p)$ elements in the last $k-1$ positions of L have a copy in B. From Lemma 1, the number of comparisons made by these ℓ elements for verification is at most $f(k-1) + \ell(\frac{n}{k} - f) + (k-1)^2$.

$$f(k-1) + \ell(\frac{n}{k} - f) + (k-1)^2 = f(p+\ell) + \ell(\frac{n}{k} - f) + (k-1)^2$$

$$= fp + \frac{\ell n}{k} + (k-1)^2$$

Thus the total number of comparisons made by all the last $(k-1)$ elements of L is

$$\frac{pn}{k} + p(k-1) + fp + \frac{\ell n}{k} + (k-1)^2 = \frac{(p+\ell)n}{k} + 3(k-1)^2$$

$$\leq \frac{n(k-1)}{2} + 3(k-1)^2$$

Thus, the total number of comparisons is at most $\frac{n(k-1)}{2} + 3(k-1)^2$.

Thus in all three cases, the confirmation phase takes at most $\frac{n(k-1)}{2} + O(k^2)$ comparisons. So the total number of comparisons made by the algorithm in Theorem 2 is $\frac{3n(k-1)}{2} + O(k^2)$, which is at most $\frac{3n}{2}\lfloor \frac{n}{m} \rfloor + O(\frac{n^2}{m^2})$ as $k-1 \leq \lfloor \frac{n}{m} \rfloor$.

Theorem 3. *Given a multiset of n elements and a frequency m, all elements with frequency more than m can be found using $\frac{3n}{2}\lfloor \frac{n}{m} \rfloor + 3(\frac{n^2}{m^2})$ comparisons and $O(\frac{n^2}{m})$ time.*

4   Finding the Mode When m is Not Known

The algorithms presented in Theorems 2 and 3 can be made to find the mode, even if the value of m is not known. This is achieved by guessing the values of m, in turn guessing the values of k.

We keep running the algorithm for $k = 1, 2, 4...$ and so on till we find a power of 2 (say x) at which the algorithm returns a non-empty set of all elements with frequency greater than $\frac{n}{2^x}$ thereby finding all elements which have the frequency of the mode.

The total number of comparisons performed by the algorithm in Theorem 2, when m is not known would be $S = \sum_{i=1}^{x}(2n)(2^i)$ which is $\frac{4n^2}{m}$ (as $(k-1) \leq \frac{n}{m}$) which is about twice the amount of time taken by algorithm in Theorem 2 when m is known. Similarly, we can show that the algorithm in Theorem 3 takes $\frac{3n^2}{m}$ comparisons to find the mode (with frequency m) when m is not known.

5   Conclusions

We have given two mode finding algorithms that take $O(\frac{n^2}{m})$ time to find an element with frequency m, when the relation between elements can be determined only by equality comparisons. Our best algorithm takes at most $\frac{n^2}{m}$ comparisons and the lower bound proved in [8] is $\frac{n^2}{2m}$. Tightening this gap is an interesting open problem. Fischer and Salzberg [7] also present a lower bound to find the majority which matches the upper bound for finding majority (i.e. $\frac{3n}{2} - 2$ comparisons). Can this lower bound be generalized to get a better (than $\frac{n^2}{2m}$) lower bound for mode?

Another interesting open problem is to implement the algorithm of [8] summarized in Sect. 2, that takes at most $\frac{n^2}{m}$ comparisons, in $O(\frac{n^2}{m})$ time.

References

1. Alonso, L., Reingold, E.M., Schott, R.: The average-case complexity of determining the majority. SIAM J. Comput. **26**, 1–14 (1997)
2. Alonso, L., Reingold, E.M., Schott, R.: Determining the majority. Inf. Process. Lett. **47**, 253–255 (1993)
3. Alonso, L., Reingold, E.M., Schott, R.: Analysis of Boyer and Moore's MJRTY. Inf. Process. Lett. **113**, 495–497 (2013)
4. Blum, M., Floyd, R.W., Pratt, V.R., Rivest, R.L., Tarjan, R.E.: Time bounds for selection. J. Comput. Syst. Sci. **7**(4), 448–461 (1973)
5. Boyer, R.S., Moore, J.S.: MJRTY—a fast majority vote algorithm. In: Boyer, R.S. (ed.) Essays in Honor of Woody Bledsoe. Automated Reasoning Series, vol. 1, pp. 105–117. Springer, Netherlands (1991)
6. Dobkin, D.P., Munro, J.I.: Determining the mode. Theor. comput. sci. **12**, 255–263 (1980)
7. Fischer, M.J., Salzberg, S.L.: Solution to problem 81-5. J. Algorithms **3**, 376–379 (1982)
8. Jayapaul, V., Munro, J.I., Raman, V., Satti, S.R.: Sorting and selection with equality comparisons. In: Dehne, F., Sack, J.-R., Stege, U. (eds.) WADS 2015. LNCS, vol. 9214, pp. 434–445. Springer, Heidelberg (2015)
9. Knuth, D.E.: The Art of Computer Programming. Sorting and Searching, vol. III. Addison-Wesley, Reading (1973)
10. Munro, J.I., Spira, P.M.: Sorting and searching in multisets. SIAM J. Comput. **5**(1), 1–8 (1976)
11. Misra, J., Gries, D.: Finding repeated elements. Sci. Comput. Program. **2**(2), 143–152 (1982)
12. Reingold, E.M.: On the optimality of some set algorithms. J. ACM **19**(4), 649–659 (1972)
13. Saks, M.E., Werman, M.: On computing majority by comparisons. Combinatorica **11**(4), 383–387 (1991)

Author Index

Printed in the United States
By Bookmasters